IN THE LIGHT OF
SCIENCE

IN THE LIGHT OF
SCIENCE

Our Ancient Quest for Knowledge
and the
Measure of Modern Physics

DEMETRIS NICOLAIDES

Prometheus Books

59 John Glenn Drive
Amherst, New York 14228

Published 2014 by Prometheus Books

Cover design by Grace M. Conti-Zilsberger
Background cover image © Bettman/Corbis
Left and right cover images © Bigstock
Center cover image © Ian Cumming/Ikon Images/Corbis

Inquiries should be addressed to
Prometheus Books
59 John Glenn Drive
Amherst, New York 14228
VOICE: 716–691–0133
FAX: 716–691–0137
WWW.PROMETHEUSBOOKS.COM

18 17 16 15 14   5 4 3 2 1

Library of Congress Cataloging-in-Publication Data Pending

Printed in the United States of America

Dedicated to
my wonderful daughter, Maria-Christina, who is curious about the world,
my loving wife, Anna,
and the memory of my beloved father.

CONTENTS

8 CONTENTS

ACKNOWLEDGMENTS

I am always grateful to Professor Alexander A. Lisyansky, my PhD thesis mentor, who taught me physics and the art of fundamental research. Many thanks to Professors Dennis Organ and Ivana Djuric for their willingness to review the manuscript at its early stages and for their helpful comments and suggestions.

I would like to thank my agent, Nancy Rosenfeld, who believed in my project and has stood by me with useful advice. I have benefited immensely from her experienced recommendations.

I am grateful to Bloomfield College for giving me the opportunity to research this topic and also to my students whose thirst for knowledge keeps me improving.

I thank my parents for their unconditional love and guidance throughout the years. I am grateful to my wife, Anna, for encouraging discussions over coffee early, early in the morning during all the stages of this book project. Her insightful comments led to several improvements.

I am thankful to the staff of Prometheus Books for all their help, including Jade Zora Scibilia, Julia DeGraf, Mark Hall, and Melissa Raé Shofner. I am deeply indebted to my gifted editor, Steven L. Mitchell, for his thorough recommendations, perceptive questioning, and generous editorial assistance, which, without a doubt, have contributed considerably to the enhancement of the manuscript.

But most of all, I am thankful to a nine-year-old, my daughter, my pride and joy, my Maria-Christina, whose curious mind and love for knowledge kept me focused. Anna and our daughter, Maria-Christina, mean the world to me. Thank you!

PROLOGUE

THEME

In an attempt to discover the roots of science and understand the development of our scientific knowledge about nature as a series of logical progressions, this book narrates a concise history of humans from the "beginning" (from when *Homo sapiens* evolved two hundred thousand years ago), by investigating, in the light of science, subtle interconnections among the three most significant cultural landmarks of humanity: (1) the culturally explosive urbanization, ten thousand years ago, and the unavoidable birth of civilization gradually thereafter—phenomenal events that triggered a wealth of new pursuits including religion; (2) the intellectually revolutionary birth of science, 2,600 years ago, which broke the bonds of superstition; and (3) the scientifically extraordinary modern era of quantum physics, relativity, string theory, and "the God Particle"[1]—of mind-boggling science that contributed to a better understanding of the universe and skyrocketed progress in technology, although it also challenged society.

I trace the intellectual continuity in our efforts to know nature by describing human life when it was primitive, outlining the pivotal transition in lifestyle from nomadic hunting-gathering to settled urban communities (i.e., to the birth of civilization), discussing the life-changing birth of religion and its influence on the eye-opening birth of science, as well as the intellectual evolution from mythology to science and its causes, thus offering us the first scientific theories, conceived during the sixth and fifth centuries BCE, analyzed from the context of modern science.

It is a book about the history of science and about science itself both ancient (particularly pre-Socratic theories) and modern (particularly physics) and in light of each other. This comparative approach to science demonstrates that the measure of modern physics is both an expansion and a reflection of various scientific ideas conceived by prominent pre-Socratic philosophers. And also that scientists of the twenty-first century are still grappling with the fundamental problems they raised some twenty-five hundred years ago.

In the Light of Science begins with its more historic part I, and proceeds to its more scientific part II, by blending history, religion, philosophy, science, and overall human culture. The theme of this book is discussed more analytically in its two parts.

PART I: FROM CHAOS TO ORDER

An epoch of sheer struggle for survival began with the evolution of *Homo sapiens* about two hundred thousand years ago. This was followed by the first form of clear self-expression through cave painting around thirty thousand years ago. About twenty thousand years later came the domestication of animals and plants, which led to herding, agriculture, the cultural phenomenon of urbanization, and the consequent birth of civilization. With civilization came the remarkable birth of religion and the development of mythological worldviews, events all of which were starting to occur roughly ten thousand years ago. Soon there would be complex, busy city-states, some four thousand years later, followed by the momentous invention of writing and the beginning of written history, a few centuries later. The enlightening birth of science began about 2,600 years ago.

Overall, this part presents a brief history of humans with the goal of understanding the main events leading to the most critical transition in the evolution of human thought, the shift from mythology to science—from the mythological and apparent worldview of deceptive senses to the rational yet intangible worldview of inventive intellect.

Contributing to this transition was the rise of the Greek civilization. About 2,600 years ago the ancient Greeks had a magnificent intellectual awakening. "Suddenly" the age-old popular mythological worldviews were questioned, rethought, and eventually changed. Nature was no longer seen as a chaos of random, unpredictable, and incomprehensible phenomena attributed to mysterious supernatural forces through myths, superstition, and the chancy decisions of capricious, anthropomorphic gods. On the contrary, nature was viewed as a cosmos: a well-structured, organized, ordered, harmonious, self-contained, self-consistent, and beautiful whole in which the phenomena were *natural* components that obeyed intrinsic causal laws that could be discovered and understood by the practice of rational analysis of nature and without invoking the supernatural. A profound transition in human thought took place that was a consequence of the realization by these Greeks that nature *is* comprehensible. A simple question emerged: what is the nature of nature? The prolific answers Greek thinkers offered ascribed purely naturalistic causes to all phenomena in nature and gave birth to science.

This part examines the birth of science and discusses why Greece may be considered its birthplace. I argue that while the generally accepted conditions (those dealing with geography, economics, religion, and political structure) might have been necessary for the rise of Greek civilization and particularly for the birth of science, they were hardly sufficient. Thus a hypothesis will be proposed: that these together with three other conditions—(1) the influence of the evolving ancient Greek language, (2) the force of intellectual habits, and (3) the intriguing ancient Greek idiosyncrasy—all had a critical role to play.

The factor of language has not been sufficiently appreciated, but neither has the unusual factor of intellectual habits, through which I hope to explore how the phenomenon of natural selection from biological evolution has contributed to the birth and constant development of the scientific outlook at the expense of the mythological one. Moreover, the factor of Greek idiosyncrasy is often overlooked, but I believe that any effort to understand the rise of a civilization is probably incomplete without also an attempt to understand the idiosyncrasy of the people who caused such a rise.

The earliest scientific theories were formulated by the pre-Socratics, the Greek natural philosophers from the sixth and fifth centuries BCE who were the first (at least that we know of) to explain the phenomena of nature solely in terms of naturalistic causes and consequently to contribute significantly in the early genesis of science. That said, there is absolutely no reason to believe that, before or after the Greeks and independently of them, others in the world would not have conceived the scientific interpretation of nature. In fact, the proof of my claim is that all kinds of people from all over the world do, or can learn to do, science. Nonetheless if something profound like practicing science—that is, seeking exclusively *naturalistic* (scientific and rationalistic rather than supernatural) explanations for *all* the phenomena of nature, including the origin and evolution of humankind and of life in general—had not happened completely, systematically, freely, enthusiastically, and most importantly, truly *consciously* (not accidentally or merely instinctively) and *habitually* (not occasionally), as in the case of the pre-Socratics, such good skill would not have evolved into a cultural tradition, would not have spread, and almost certainly would have quickly vanished without demonstrating a significant effect on society. But evolution occurs when a phenomenon leaves a mark on the environment. The first scientific mark, which served as a basis for science thereafter, was left by the ancient Greeks. In this perspective science may be regarded as having been born out of the Greek civilization. Nevertheless, from a grander point of view the story of science—seeking the nature of nature—begins with the evolution of the human species two hundred thousand years ago. For, out of the myriad species that have existed, only humans managed to develop a rational interpretation of nature. Why that was so is seen here too where the main events leading up to the birth of science are discussed through a brief history of humans.

PART II: THE PRE-SOCRATICS IN LIGHT OF MODERN PHYSICS

In an effort to trace the beginnings of science, this part introduces the most important scientific theories of the most famous pre-Socratics (including Pythagoras and Democritus) and analyzes them within the context of modern physics (or science in general). These philosophers and thinkers knew that nature is intelligible. It has secrets, to be sure, but like Promethean fire, their knowledge is discoverable and can be harnessed. They had ideas that were so phenomenal, fascinating, unique, strange, and daring that some actually anticipated various aspects of modern science that were well ahead of their time. And some of these ancient ideas, while defying common sense and apparent reality, have not yet been refuted—they remain still unsolved mysteries! These Greek theories are not as obsolete as is often assumed. They still spark the imagination. Hence our intellectual trip exposes the most beautiful and mind-blowing laws of nature and shows that, despite the two-and-a-half-millennia time difference, ancient and modern science share a fundamental qualitative similarity more often than is usually thought, and they complement each other's scientific uniqueness.

Modern physics is becoming more and more complex, both conceptually and mathematically, often hitting intellectual dead ends. But while the laws of physics are mathematical, its mathematics emerges from its concepts. And the pre-Socratics were undoubtedly masters of profound conceptual analysis (of course, some were also excellent mathematicians who conceptualized their physical theories using the mathematics known at the time). Therefore, ancient rationales will at times be used to reexamine and reassess some of the fundamental premises of current theories of physics. And while one of our basic questions is how ancient science measures up to modern physics, we may often find that the question can be reversed: how does modern physics measure up to ancient science?

If the Greek philosophers (i.e., Socrates, Plato, and Aristotle) are correct and philosophy truly does begin in wonder, and if natural philosophy is the precursor to what we now know as science, then it makes perfect sense that the urge to ask questions and to seek answers is a critical part of the scientific state of mind and its rise from the darkness of superstition to the light of knowledge.

Throughout its history science has asked many questions: What is the nature of nature? Does nature obey laws that can be understood? If yes, how do these laws work?

What are the things (stuff, matter) all around us made of? Is there a single primary substance of matter? What is the nature of matter? Does matter's behavior depend on space and time, and vice versa? What is the nature of space and time?

Is the universe finite (in size and age) or infinite? Is it static and fixed, expanding, contracting, or vacillating? Does it go through cycles? Is there some one thing that caused everything that now happens—a first cause? Was there a specific beginning of time? Will there be a definite end?

Where did the species, including humankind, come from? Is there a biological development process?

Is reality objective or subjective? Is this the only universe, or are there also others, each with its own unique reality? Can the intellect alone conceive a truer reality, or does it need the aid of the senses?

Is absolute knowledge ever attainable? Does God exist? Can science prove or disprove the existence of God, or will God always remain a matter of subjective belief? Why are humans intelligent? In fact, why are we the most intelligent of all the living or extinct species we know?

What is the nature of nature? It's a deceptively simple question but one that has packed within it the many additional questions outlined above—and many more besides. Two and a half millennia ago this fundamental question captivated the minds of the ancient Greeks and led them to strange but *rational* answers that demystified and demythologized nature and gave birth to science. From that time on, science has been influencing the world by guiding it out of the cave of ignorance and into the light of truth.

PART I

FROM CHAOS TO ORDER

CHAPTER 1

PLATO'S PARABLE OF THE CAVE

INTRODUCTION

The cave parable in Plato's *Republic* elegantly captures the essence of nature and the goal of scientists: nature has its secrets, and it is the role of scientists to discover, or "steal," them, as Prometheus stole divine fire from the gods and gave it to humanity.

Apparent reality is as incomplete and nebulous as the shadows of objects in the parable. The prisoners who live in the cave and see only shadows on the walls consider only these shadows to be real. They have no knowledge, none at all, of the objects that cast the shadows. But the prisoner who escapes from the cave has at last a better sense of reality and discovers for herself that the shadows are only a mere copy of the real objects. She sees that things are more than they appear. The parable helps us to realize that nature is much more beautiful and complex than the way the senses alone display it for us. And so for the scientist (as for the prisoner who escapes the illusive shadows of the cave) nature *is* comprehensible through both observation and rational contemplation. Logic unveils a much richer reality than the tangible one of mere sense perception alone. Ever since science opened the door to a rational understanding of nature, its discoveries have continued to amaze us by defying both common sense and apparent reality.

THE PARABLE

Imagine an underground cave where prisoners are held in chains since childhood and for many years thereafter, restricted from moving or even turning their heads, but are only able to look straight ahead and toward a wall in front of them. Behind and far from them is a fire the light of which projects on the wall shadows of various objects (including those of themselves, of passersby behind them, and of other things in general). Because of how they are bound, the prisoners see nothing else but these shadows. They believe that only these shadows are real, and the prisoners are utterly ignorant of the objects that project them, or generally of the greater truth about nature that is indeed accessible if only they could escape.

One day a prisoner does escape. Her instinctive curiosity guides her away from the darkness of the cave, first toward the fire and then upward and out of the cave and into the light of the sun. But being so long in the dark she is not accustomed to the light. And so she is frightened by her inability to see clearly. Gradually her eyes adjust, and she sees things she has never seen before, a panorama that is grand, boundless, magical. She is startled but also puzzled. Nature is so much different from the way she had imagined it. For the first time she sees mountains, flowers, people, animals; she sees the blue sky and the sea and the colors in the trees. She observes but also contemplates. She notices how similar all objects are to their shadows. And then for a moment she seems lost, anxious, absorbing all that is around her. While deep in thought she suddenly exclaims eureka! It is an exhilarating moment. She has discovered something. At last she has a better sense of reality because she realizes that the shadows in the cave are just mere copies of the real objects. Nature is more than it appears. But it *can* be understood!

Sad for her friends and morally obligated to free them from the deceptive darkness, she descends once more into the cave. But upon reentering she is frightened for a second time, for again she cannot see: her eyes have now adjusted to the light and are no longer accustomed to the dark. This does not go unnoticed by her friends, who think she is now different from them, that she has changed, that her vision is destroyed by the light. They think *she* is blinded by the *light*, not realizing it is *they* who are blinded by the *dark*. Her every effort to convince them otherwise and free them is futile and mocked. In fact any new knowledge she may impart is viewed as a threat, and they contend that anyone claiming to have such knowledge should be put to death.

MORALS

To know nature as it *really* is, we must use both our senses and logical reasoning. Sense perception alone is like the cave reality, incomplete but not absolutely false. What is false are the generalizations we make by relying solely on the experiences of the senses. Shadows *are* real, but not the only real things; shadows are *not* the truth, just *part* of it. Freeing oneself from the dark bondage of the cave in order to enter the light symbolizes the transition from a world of apparent reality, ignorance, and superstition to one of rational thought and potential enlightenment. In achieving such an escape, fear of learning (the pain in the eyes caused by the light) is eventually overcome, and superstition is replaced by reason.

The parable is also a story about ignorance and the cave-dwellers' failure to recognize their ignorance. Not only do the prisoners lack knowledge, but they do not know that they lack it. They were prisoners of time-honored prejudices and illusive sense perceptions. The cave may be seen as a metaphor for the incomplete knowledge we have about everything and the dangers associated with that partial understanding; especially when we do not realize that our knowledge is only partial, and therefore we think we know more than we do. Leaving the cave (to be educated) is as difficult as entering it to make others aware (to educate). The latter is far more advantageous because one is at least equipped with some knowledge that can be utilized to possibly anticipate forthcoming challenges.

CONCLUSION

Reading a symbolic version of a very complex story as if the symbols were real and the symbolic version were the actual story is an inherent risk in the very symbols used. Sense perception by itself is telling us just a symbolic story of nature—we see only shadows or imperfect representations of reality. But when combined with our intellect the two offer hope of providing us with a more realistic version of the story of nature. The pre-Socratics realized early on that appearances deceive, that the sensible world is like a cave, both illusive and incomplete. But even though our senses can be deceived, our rational intellect is perceptive and so offers a much-needed check on the veracity of what our senses report. This has helped us advance from a strictly sensual worldview to one in which our reason carefully assesses and analyzes what our senses convey about the world beyond us. And these ancients found that nature is a lot more intriguing, mysterious, and beautiful than it may at first appear, but also that it is intelligible to the willing mind only through the careful and deliberative approach of science.

CHAPTER 2

WHAT IS SCIENCE?

INTRODUCTION

Science (taken from the word for knowledge in Latin) is the systematic study of nature and the organization of acquired knowledge into timeless, universal, causal, and, most importantly, testable laws and theories that are derived from observation and rational consideration. A good scientific theory, therefore, makes experimentally verifiable predictions whose confirmation leads to the theory's acceptance as a true description of nature. The premise of science is the realization that with rational thought nature *is* comprehensible. Science holds that natural phenomena, while often appearing to be random and unpredictable, are actually orderly and to a certain degree predictable; they obey intrinsic causal laws that can be understood rationally without the need of invoking myths, superstition, supernatural forces, or the intervention of capricious and anthropomorphic gods.

CAUSALITY

Scientists are generally viewed as rational people, but like the rest of us they are passionate and willing to sacrifice much in the pursuit of their endeavors. Democritus once said that he would "rather discover one causal explanation than obtain the whole Persian empire."[1]

Causal explanations conform to the principle of causality, according to which later events are caused by earlier events. Causality is a relation between *causes* and *effects* that organizes knowledge in the hope of discovering general laws of nature. But the relation between cause and effect is still very much an open question. In a deterministic interpretation of nature a specific cause produces a specific effect. Say cause-1 produces effect-1. But cause-1 may also have a cause of its own—some previous cause, say, cause-2, which may itself be caused by cause-3, and so on. In fact the original effect-1 may also be the cause of some other effect, say, effect-2, which in turn may be the cause of effect-3, and so on. We can show this schematically as

$$\cdots \rightarrow \text{cause-3} \rightarrow \text{cause-2} \rightarrow \text{cause-1} \rightarrow \text{effect-1} \rightarrow \text{effect-2} \rightarrow \text{effect-3} \rightarrow \cdots$$

On the other hand, in a probabilistic interpretation of nature one of several probable causes could produce one of several probable effects. Therefore, it is obvious that neither causes nor effects are known with certainty; on the contrary, all events are expressible in terms of probabilities.

Classical physics (as understood by Isaac Newton's and Albert Einstein's theories) is deterministic, whereas quantum physics is probabilistic in its approach. The location, speed, direction of large objects (such as cars, billiard balls, and planets) can be described with accuracy by the laws of classical physics, but the same characteristics of tiny particles (such as electrons, protons, neutrons, neutrinos, quarks, etc.) are describable only in terms of probabilities through quantum physics. We can pinpoint the position of a car as well as its speed and direction, but this is not the case with an electron. Quantum theory can describe only where an electron is *likely* to be, not where it is or will be. The source of uncertainty and probability will be introduced in chapter 12, "Heraclitus and Change."

Causality only explains later effects by earlier causes, but it cannot explain the very first (or primary) cause; it can only assume the truth of such a cause and proceed from there to deduce its actual (or possible) effects. No cause can be assigned to the label "primary cause" since if it could, it would not be primary. Nevertheless, to find any causes and their effects, even the elusive initial cause, the phenomena of nature must be studied rationally with a method, the scientific method.

THE SCIENTIFIC METHOD

Isaac Newton (1642–1727) hypothesized that white light is a blend of all the colors of the rainbow. He tested his hypothesis by experimenting: he passed a beam of white light through a glass prism and observed how it disperses into a rainbow of color. Others, at Newton's time, thought that the colors came from the prism itself, that they were not a property of light. Newton proved them wrong by advancing his experiment. This time he passed just one color, say, the blue (from those dispersed by the first prism), through a second prism. Since only blue entered the second prism and only blue emerged from it (and not a rainbow of color), color, he concluded, was a property of light and not of the prism. He supported his hypothesis even further by mixing again the initially separated colors and noticing how once more they reproduced white light. The so-called Snell's law calculates the exact angle that each color of light refracts by crossing from the one medium, air, into the other, the glass prism. Hence what we must also emphasize here is that all scientific laws are quantified through mathematics. Without such quantification modern technology (of computers, cell phones, etc.) would not have been possible. Chapter 11, "Pythagoras and Numbers," will discuss how one may model nature mathematically.

As seen through the above example, the scientific method consists of observation, reason, and experimentation. By observing a natural phenomenon and thinking upon how it may have come about, its explanation is attempted by means of a rational hypothesis, a proposed explanation offered as a possibility. The validity of the hypothesis must be tested experimentally. A set of conditions is predicted such that if the hypothesis is confirmed, that set of conditions would in fact occur. A conclusion is reached about whether or not a hypothesis is verified (confirmed) or falsified (disconfirmed) by analyzing the predictions of the hypothesis in light of what is found to be the case when the hypothesis is tested. If the hypothesis is consistently verified by numerous individuals who attempt to test it, then a law (or a general theory of nature) is discovered and our comprehension of nature expands a bit more. Of course, often a hypothesis on how some aspect of nature works can be formed by pure theorization (reason) without first having to observe some particular phenomenon (as is the case of Einstein's theory of special relativity), although it is still likely that various prior observations would guide a scientist's speculations, for, in truth, no person is disconnected entirely from the influences of nature. The fewer assumptions a law makes and the more phenomena it explains, the more powerful it becomes. Furthermore, regardless how abstract and complex a law may be, it must be able to explain the natural phenomena as we see them, as we perceive them through our five senses. It is always possible, as any good scientist knows, that any law might later be shown with additional testing to be just a mere copy, a shadow, of some greater truth yet to be uncovered. In fact, this has been so throughout the history of science. Nevertheless while our scientific models may in the end prove to be nothing more than shadows of yet undiscovered truths, they still play a vital role in the development of human thought and understanding about the world and our place in it.

CONCLUSION

Science has not always been part of human civilization. In fact, even civilization has not always been part of human existence. So what critical event in human history created the potential for us humans to develop abstract thought and eventually science?

CHAPTER 3

URBANIZATION

INTRODUCTION

With perhaps the exception of the mastery of fire, the most momentous event in the roughly two-hundred-thousand-year history of *Homo sapiens* was urbanization, which emerged about ten thousand years ago. Had it not occurred, we would still be hunting and gathering. But it did, and that led to the birth of a new and complex way of life called civilization with everything this term entails. This development was the result of two great ideas that were beginning to be implemented at roughly the same time by our hunter-gatherer ancestors: namely, the domestication of animals, which led to herding, and the domestication of plants, which led to agriculture. Both activities, but particularly agriculture, promoted the need of a settled life, created food surpluses, and with these surpluses came an expanding population. From small villages of mere hundreds to crowded city-states of many thousands, life became increasingly more communal and diversified, an unprecedented transformation that by about six thousand years ago found the cultural phenomenon of urbanization spreading worldwide. To put these ideas in perspective, this chapter will narrate a brief history of *Homo sapiens* (covering our primitive beginnings and the main changes that occurred thereafter) from the time we first evolved two hundred thousand years ago to May 28, 585 BCE, the day of the solar eclipse predicted by Thales,[1] the first natural philosopher. Because Thales, a pre-Socratic, flourished around the time of this eclipse, this day may be regarded as the birthday of science. In fact, the entire sixth century BCE was an intellectually explosive period globally. But let's start from the "beginning."

THE ERA OF PURE SURVIVAL: 200,000–30,000 YEARS AGO[2]

Homo Genus[3]

The *Homo* genus consists of several member species all closely related to us, each of which is also unique. Its first member, *Homo habilis* (a distinct species having some humanlike characteristics), evolved more than two million years ago and became extinct about 1.4 million years ago. Its only extant member, *Homo sapiens*, evolved about two hundred thousand years ago in Africa and is the species to which every living human belongs—so each one of us can be traced back to a common African ancestor. *Homo sapiens'* evolution was happening just as our older evolutionary relative *Homo heidelbergensis*, who had evolved about seven hundred thousand years ago, was becoming extinct. It is at this point that our species might well have gotten its chance to evolve as a consequence of the eventual extinction of another species, in general not an uncommon phenomenon in the evolution of life. Sadly, the longest-living member of the genus *Homo*, *Homo erectus*, whose members evolved approximately 1.9 million years ago, also became extinct about 143,000 years ago.

Since their earliest days and down to about the end of the Paleolithic Age (the "Old Stone Age") roughly ten thousand years ago, *Homo sapiens* were basically hunter-gatherer nomads with remarkably few cultural changes. They constantly searched for food by following migrating wild animals and by seeking new areas of fresh edible plants. They lived and traveled in small social groups, including perhaps tens or hundreds of individuals from an extended family. Their diet included plants, nuts, eggs, honey, fruit, and meat from animals, fishes, and birds. Survival was a day-to-day affair, but they got by despite the dangers and uncertainties of their rough and wandering lifestyle. It is not easy to hunt large wild animals (horses, reindeer, mammoths, cattle, woolly rhinoceros, bears, and the like) with primitive weapons/tools. And food had to be found daily, but there was no guarantee of that. In fact *Homo sapiens* came near to extinction sometime between ninety thousand to seventy thousand years ago. During this period the climate had fluctuated significantly, and it was around this time, 74,000 years ago, that super volcano Toba erupted, complicating further the conditions for survival.[4] In spite of these challenging circumstances, *Homo sapiens* have been managing to survive.

Bottleneck Effect

It is speculated, however, that the human population may have suddenly decreased dramatically (to only a few thousand) at this time,[5] and as a consequence so may have the gene variety among the survived members—one of the reasons humans today are genetically so similar, hence our differences are truly skin-deep. To explain this genetic similarity biologists often employ the so-called bottleneck analogy. Imagine a narrow-neck bottle filled with marbles of many different colors. The marbles represent the initial large genetic variety in the population of a species. And the narrow neck represents a potential challenge that a species might have to experience—such as severely cold or hot weather, earthquakes, volcanic eruptions, droughts, earth-asteroid collisions. Now, just as the color variety of the few marbles that make it through the narrow neck (that "survive") is likely to be reduced when the bottle is turned over, the gene variety of the few members of a species that survive a harsh situation or catastrophe is also likely to be reduced.

The Peculiar Water

While extreme coldness is one of the toughest bottleneck passages a species might have to endure, fascinatingly, a life-preserving peculiar behavior of water eases the passage (lessens the difficulty). What is it?

Oceans, speculated to have been the habitat where primitive microscopic life-forms first evolved, begin to freeze from the surface downward, but they don't freeze all the way to the bottom when the temperature drops to freezing. The water at the bottom of an ice-covered ocean never freezes; it remains always around 4 degrees Celsius (about 39 degrees Fahrenheit). This is because of an unusual behavior of water: when cooled below 4 degrees it expands rather than contracts like most other materials. That's why a water bottle forgotten in a freezer will break—as the water in it cools below 4 degrees it expands (especially so as it turns into ice at 0 degrees Celsius) and cracks the bottle. Similarly, ice floats because its expansion at lower temperatures also lowers its density below that of liquid water, causing it to rise to the surface.

Now the water at the bottom of an ice-covered ocean can't lower its temperature below 4 degrees because it literally does not have the energy to do so. For it to cool below 4 degrees water must expand, that is, it must lift up all that heavy icy sheet over it, but it can't. Similarly, we can't lift something extremely heavy over our heads—we don't have the strength or the energy to do it. The water at the bottom of an ice-covered ocean remains nice and cool at 4 degrees, fluid, unfrozen, keeping the fish swimming, preserved, and pleased. Unlike deep oceans or lakes, however, water in shallow street pools does freeze completely because the quantity of the water in it is small; the water at the bottom of a pool can freeze as it easily lifts up the small weight of the water over it.

If water didn't possess this peculiar property, the evolution of life on earth might not have been possible at all because oceans would freeze *completely*, killing all life-forms within them. If water were contracting, not expanding, below 4 degrees, it wouldn't have to lift up any water or ice positioned above it in order to cool down during freezing weather. Instead it would easily become colder and transform into frozen solid ice first at the ocean's surface. Then, being denser than the water below (because of the assumed contraction) this ice would naturally sink to the bottom of the ocean (like a rock does because it's denser than water), and the water below would in turn be displaced upward toward the ocean surface. The new surface water would follow the same sequence of events (it would freeze, condense, sink, and displace the water below it to the surface) until all ocean water would freeze completely crushing to death all marine life. Even worse, in persistently cold weather life might not have gotten the chance to even evolve in the first place; the crushing forces of solid ice spreading everywhere in a frozen ocean would have been destroying the molecules of life, killing any chance they might have had to coalesce and develop. Thankfully, water is peculiar, and so life was able to get its start on earth.

Primitive Technology

The dawn of technology, marking the beginning of the Paleolithic Age, occurred more than two million years ago with the use of simple stone tools by *Homo habilis*, "handy man," or possibly by other species before him from the genus *Australopithecus*. *Homo sapiens'* tools, at different periods during the years covered in this section, included stone projectiles, blades, flakes, stone or bone hammers, long pointed wooden spears, stone or bone awls, needles and scrapers to perforate and turn hides into clothing, and in general anything made of stone, wood, or bone that could be useful for a hunting-gathering lifestyle amid varying and challenging climates. Metal tools and pottery did not yet exist.

Technology is often associated with science, but technology itself is not science. Tools and implements—primitive forms of technology—can exist without science. But usually, especially nowadays, technology is a consequence of science; it is the application of the laws of science for various practical purposes. Technology can also lead to science (as we will see in chapter 6, "The Birth of Science"), but it is not itself science.

As nomads who lived in small social groups, humans shared food and other resources, lived mostly outdoors but sometimes avoided the wind and cold weather of the last glacial period (110,000–10,000 years ago) by seeking refuge in natural shelters such as caves or human-made simple huts constructed by piling up stones or branches or by hanging animal hides on poles, or by gathering around a campfire.

Fire

The use of fire was a skill first practiced by *Homo erectus* (another of our extinct ancestors known for standing upright rather than crouched or bent over) possibly five hundred thousand years ago. Whether *Homo sapiens* discovered such skill independently or learned it by observing and imitating *Homo erectus'* example is uncertain. Also uncertain is when fire actually came into being as a human tool (see next section). An interest in fire may have been sparked in a way by something like this: after a naturally caused fire (e.g., as a result of lightning), a keen and courageous observer might have retrieved a burning branch, kept it ablaze by feeding it with new branches, and experimented with its properties.

Fire was used for warmth, for light at night, to keep insects and dangerous animals away, for hunting animals, and for cooking food. Cooking killed parasites (especially those on meat), made food softer and tastier, and increased the types of food that could be consumed. Cooking has aided in safer food consumption and digestion and has improved the survival chances of our species, and in so doing it has contributed to our general evolution and in particular to the evolution of the human brain. Eating cooked food delivers more energy than eating the same food raw.[6] This fact has been especially significant for the evolution of our energy-hungry complex brain, which although comprising only 2 percent of our body weight requires 20 percent of the energy from the food we consume.

Gathered around a campfire to relax, socialize, and bond, Homo sapiens might have even "told" their first stories, an activity that has certainly influenced the development of human language. I imagine these stories were very simple ones, language being in its most primitive form at that time. Regrettably, sounds do not fossilize, so before the first written record, left sometime during the fourth millennium BCE, the level and extent of language used by humans is uncertain. It is hypothesized that a form of basic vocal language must have evolved by one hundred thousand years ago.

The warmth of fire might have also helped parents to care for their helpless babies. Remarkably, among the primates, human babies require the longest period of parental care. By comparison, the childhood of chimpanzees, which genetically are our closest relatives from the living species, is roughly half as long as the childhood of humans. Caring for each other, conveying stories, and exchanging information undoubtedly enhanced Homo sapiens' survival skills and their evolutionary path. Moreover, learning from each other along with the long period of parental care contributed to the overall bond between humans. Parents who care for their own offspring, especially when this is done for a period of several years, instill in their children the tendency of caring as they are growing. As this process continues throughout the group and through generations, chances are that offspring and parents alike will also begin caring for their close family as well as their extended family. Such feeling could in turn inspire the desire for a more communal lifestyle, which is the basis for civilization. Directly or indirectly the phenomenon of fire has contributed to all aspects of the evolving human culture.

Migration

Between eighty thousand to sixty thousand years ago, *Homo sapiens* migrated from the African savannah first to Asia, then to Australia (ca. fifty thousand years ago), to Europe (ca. forty thousand years ago), to the Americas and to the Pacific islands (ca. fifteen thousand years ago), and by about ten thousand years ago even to the Arctic, but not to Antarctica. Migration was possible because the icy-cold temperatures of the last glacial period bound the water as ice on land, and so it could not flow into the sea. As a result, the sea level then was lower than that of today by about 100 meters, exposing more parts of the continental crust, which in turn functioned as natural land bridges between continents. These bridges lasted until about ten thousand years ago. Then the glacial period ended. The melted water flowed into the oceans and caused the sea levels to rise and these bridges to be submerged under the water. Consequently, the new islands that were created kept their inhabitants isolated.

So, by thirty thousand years ago *Homo sapiens* had developed better tools, built primitive huts, advanced their hunting and carving techniques, made and wore clothing, socialized around a hearth, buried their dead, possibly even played music (using flutes made from bones). But perhaps the most significant of their activities, which was not related to sheer survival, was painting! It was the first form of pure self-expression.

FROM PAINTING TO HERDING: 30,000–10,000 YEARS AGO

Painting

Art begins in prehistory more than thirty thousand years ago by our Paleolithic ancestors, the Cro-Magnons, who decorated their cave dwellings with splendid, colorful, and expressive animal paintings. Cro-Magnons, named so after the site in France where their remains were first discovered, are the *Homo sapiens* who arrived in Europe about forty thousand years ago and are believed to have looked and behaved more like modern-day humans. With their beautiful art they demonstrated their sensitivity and sensibility. Cave painting is "great art, manifested in works of subtlety and power that have never been surpassed."[7] While the true significance of these paintings may never be known, it has been speculated that in addition to painting for merely the sake of art, the animals of the paintings might have also had a sacred function.[8] As if life would imitate art and, as depicted in the cave walls, humans could become better hunters and wild animals captured and tamed. But if humans could express themselves so well through painting, they could probably speak as well, too. And so by now their stories around their hearths would have been more descriptive. Consequently they understood and learned from each other, making stronger both the human bond and their social lifestyle. Urbanization and the birth of civilization were thereafter naturally expected. Painting is a significant activity in human culture, for after painting comes writing—recorded information! This sequence of events is normal. For in painting one draws to represent things that already have a shape, for example, a bison. On the other hand, in writing, especially with an alphabet, one draws (e.g., letters) to represent things that do not have shape, that is, the sounds of a language. Comparatively, then, an alphabet is a more abstract invention than painting or a pictographic type of writing, and so it naturally followed both of them. How the Greek alphabet aided the pre-Socratics in conceiving the scientific view of nature is a topic to be contemplated in the chapter "The Birth of Science." A culture, even an urbanized one, cannot advance without recorded information. But between painting and writing there was a long time period and a few other significant cultural events, the first of which might have been lighting fire at will.

Fire Lit at Will

One of the most consequential events since our species evolved in Africa some two hundred thousand years ago was the lighting of fire at will. How and when this happened is unknown. But heat from friction is certainly what caused it. In their attempt, for example, to sharpen their tools by rubbing or striking them together (e.g., stone with stone or wood with wood), our curious ancestors could have easily started a fire. By rotating back and forth a pointed stick in a concavity of another in an effort to sharpen it further, a spark might have been ignited and fire lighted. An event this remarkable would not have gone unnoticed. It would have prompted the realization that fire can be lit and controlled at will! And although quite possibly that was an accidental discovery, lighting fire from that moment on became a favorite intentional human activity without which the notion of civilization would be nonexistent.

Fire has ever since been playing a critical role in our cultural and biological evolution. We gather around fire to warm up as well as to bond, socialize, learn, imagine, and cook our food. We use fire to make tools, thereby improving our lifestyle and increasing even further our chances for survival. No doubt humans decreased their chances for survival by using some tools and/or fire itself as weapons against each other. Much, much later, by about six thousand years ago, fire was used to melt hard copper alloys and mold them into refined tools. Today every technological achievement can be connected to fire—fuel is somewhere burning, creating electricity and powering our many devices, which to be made in the first place require fuel as well. Fire has been culturally so impressive and revolutionary that it became part of many rich ancient mythologies. For instance, in ancient Greek mythology the titan Prometheus defied the orders of the Olympian gods, stole the fire that was their exclusive secret, and brought it to humankind (more about this myth in chapter 4, "The Mythological Era"). Like Prometheus, scientists today aspire to "steal" (or discover) the other secrets of nature. Fire has also had a significant role in our scientific theories of nature (see, for example, chapter 12, "Heraclitus and Change" or chapter 15, "Em ped ocles and Elements").

It was the light of fire that dispelled the darkness of a cave, warmed the body and soul of our cave ancestors, and inspired them to paint their early wall paintings. If we accept that necessity (e.g., the instinct to survive) precedes luxury (e.g., of painting), then lighting fire at will might have been discovered by *Homo sapiens* before cave painting, or at least around the same time. For, as a result of the necessity to survive, it is possible that our ancestors foresaw the numerous benefits of fire and thought long and hard on how they could control and implement it in their daily lives.

Primitive Religion

Homo neanderthalensis (commonly, Neanderthals) might have been the first species who buried their dead. *Homo sapiens* practiced simple burial probably as early one hundred thousand years ago. But there is some evidence that their inauguration of painting coincided with their practice of more elaborate burials, which included burying of various grave artifacts (such as stone tools, animal parts, food) that might have been viewed as significant in the life of the deceased and quite possibly might have also been considered significant in an afterlife. Sculpture was another form of art that was enjoyed around the period of painting. Findings include various types of figurines, such as the "Venus": women with exaggerated bellies, buttocks, and breasts signifying perhaps fertility or other beliefs, carved from mammoth ivory or stone. Burials, painting, and sculpture are activities arguably indicative to a sort of primitive religious outlook, even though organized and systematic religions began with urbanization and the consequent development of specialized professions such as priests or group leaders. In the subsection "King-Priests, God-Kings, and Theocratic States" (of the next section), we will see how a priest and a group leader often were the same person, the king-priest.

Hunting

The hunting techniques eventually improved because in addition to spear throwing, which facilitated killing large dangerous prey from a distance without the risky physical contact, there was also the invention of the bow and arrow. Cave art depictions suggest that the bow and arrow had been in use by about twenty thousand years ago. But our evolutionary cousins, the Neanderthals (with whom we share a direct common genetic ancestor), did not invent such weapons. This, as we will see in chapter 6, "The Birth of Science," might have been one from a combination of reasons that caused their extinction between thirty thousand to twenty-five thousand years ago. *Homo floresiensis*, the most recently discovered type of human, emerged ninety-five thousand years ago in the island of Flores in Indonesia, stood about a meter tall, had a small brain size, but finally it, too, became extinct less than twenty thousand years ago (although its status as a separate *Homo* species is still controversial).[9] Thereafter *Homo sapiens* remained the sole survivor from the once diverse human family tree. But to be able to advance the species and better their quality of life they had to be inventive. Their first great idea directly involved the life of other species.

Herding

On the road to civilization a critical step toward a settled life was achieved through domesticating animals, an idea that soon led to herding. This happened about ten thousand years ago, at the end of the glacial period. The two important consequences of herding were reaching a surplus in available food and evolving toward a seminomadic way of life. The latter was a transitional step between the lifestyle of the once strictly nomadic hunter-gatherer and that of a fully settled farmer.

Herding and agriculture evolved about the same time, approximately ten thousand years ago, and their implementation was a turning point in the development of human culture. I believe herding most probably preceded farming, since herding would be a natural outgrowth of the hunting and gathering way of life, whereas farming is the result of a longer, tougher, and more organized type of work, which requires a far more thoughtful state of mind, since one must recognize the value of a crop and the hard work it takes to till the soil for months to achieve a harvest that can sustain oneself, one's family, and one's community. Thus if we assume things progress gradually, then the spontaneous nomad hunter-gatherer would naturally evolve first into the semi-spontaneous seminomad shepherd and then, eventually, the settled farmer.

Herding might have evolved by realizing that by capturing a herd of animals and keeping it alive (so the potential meat does not spoil) the daily meat could be secured for weeks, possibly months on end, making it much easier to survive without the uncertainty of daily hunting. Group members could kill an animal or two per day from the captured herd as their need required, but by the time they killed all captured animals of a herd, new ones were being born. So as long as people cared for the animals of the herd (e.g., fed and watered them, aided their reproduction), the supply of meat, milk, fur, hides, bones, and the like could be limitless.

Moreover, herding not only secured the daily meat on an ongoing basis but also provided extra resources such as new types of food (cheese, butter) and created the need for new occupations. Since not all animals of the herd had to be slaughtered at once, some were always available for milking. Some milk was consumed fresh and some was processed into butter and cheese. Animals could also be used as beasts of burden to aid humans. Ways are found to preserve the surplus food for the future. Preserving surpluses and planning for the future is a good thing and a sign of the human tendency to become urbanized. With a flock of animals under his control, a shepherd did not have to hunt every day to feed himself and his family. Thus as long as he maintained a healthy flock, food was plentiful for long periods of time. The first domesticated types of animals were goats, sheep, and cows. Animals of this kind do not eat meat, thus they do not have the natural drive to kill and are more peaceful compared to predators. Thus they were tamed easier. In addition, they eat what humans do not: grass and other types of plants, and besides meat they also give us milk. Goats, sheep, and cows were abundant in the Fertile Crescent and had been hunted before their domestication. The Fertile Crescent, which is the region from Egypt to Mesopotamia bounded by the Syrian Desert from the south and the highlands of Anatolia from the north, was among the earliest regions to develop a settled lifestyle and agriculture.

Semi-Settled Life

In addition to providing a food surplus, herding created the need for a seminomadic lifestyle. Group members no longer had to follow the migration of their prey. They could decide where their home would be, at least for an extended period of time. Animal domestication was therefore an important first step toward the settled life of community living. While a shepherd still had to be on the move in search of virgin grazing grounds, he was also required to have several temporary home bases (perhaps seasonal) where he could carry out various new activities: milk the animals, process the extra milk into butter and cheese, smoke and dry the extra meat in order to preserve it, care for the animals (e.g., in breeding), and develop specialized tools. Their semipermanent home bases might have been in places with rich grazing grounds such as fertile riverbanks.

The flocks were led to grazing fields during the day. But to limit the straying, by night they were brought back to a fenced base, thus a home. Fences were probably built by piling wood or rocks. Some seasonal dwellings might have existed even before herding, as long the natural resources of a region could sustain the nomads for some period of time. The benefits of seminomadic lifestyle might well have created the desire for an even more settled way of life and hence might have triggered the ingenuity in humans to search for ways to achieve it.

Idiosyncrasy

This transition from the nomadic to the seminomadic lifestyle was heralding a critical change in human behavior. Specifically, with herding, the once spontaneous hunter-gatherer who was basically searching for food whenever he was hungry gradually transformed into a shepherd, who, by domesticating animals, figured out a way to have easily accessible and abundant food for the future! Planning for the future cannot be a bad thing for civilization. But by far, this uniquely human approach to survival became even more refined, particularly with the domestication of plants, which led to farming and an even more settled lifestyle. The biology of the human brain was ready for it, for by this time the central part of that organ had evolved the ability to plan for future activities and wait for their resulting rewards to be fulfilled months and even years later.

FROM AGRICULTURE TO CIVILIZATION: 10,000–6,000 YEARS AGO

Agriculture

The decision to domesticate plants was one of the most consequential events in all of human history. This culturally explosive phenomenon, which occurred about ten thousand years ago, marked the end of the Paleolithic Age and caused the start of the Neolithic Age ("New Stone Age"), which lasted until about six thousand years ago. Plant domestication led to agriculture, which required a settled way of life, hence the development of urban living in communities that grew and in turn led to the rise of civilization.

This critical transition in lifestyle, from the nomadic hunting-gathering to farming villages, which gradually evolved to technologically advanced and densely populated cities with social and political structures in place, is known as the Neolithic Revolution. This was not some singular event; rather, it generally occurred independently at several locations worldwide during the Neolithic Age (though roughly simultaneously).

Although the rate of development differed from place to place, agriculture and urbanization soon became widespread. New ideas evolved from within specific communities, but ideas were also exchanged by people from different regions through their various contacts. Because of the need for fresh water, the first urbanized centers tended to develop near the fertile banks of permanently flowing great rivers. These were the Fertile Crescent (extending from Egypt by the Nile to Mesopotamia by the Tigris and Euphrates Rivers); south of Sahara; the Indus Valley; the Yellow River and Yangtze River valleys in northern and central China, respectively; and Central and South America. With an abundance of fresh water farmers did not have to depend on inconsistent rainfall to water their fields, their flocks, and themselves.

A factor contributing to this transition appears to have been significant climatic changes that not only made the weather in general more comfortable but also allowed for the evolution of abundant and edible new types of vegetation in certain areas of the planet. These areas later evolved into the first agricultural communities. Specifically, by twelve thousand years ago the ice of the glacial period began to melt. The climate was then becoming warmer and the landscape changing considerably. By ten thousand years ago the glacial period ended, and thereafter the climate has generally stabilized into the familiar one we know today. By about the same time huge quantities of wild wheat were growing in the Middle East (the region between the eastern Mediterranean, the Persian Gulf, and the Caspian Sea). Wheat was the first domesticated plant, followed by barley.

Through their adventures in the wilderness the hunter-gatherers must have observed so many times how seeds that fall on the ground from plants, in time, grow to become the same type of plants as those they had fallen from. At the right time, such observation must have inspired the deliberate planting and watering of seeds and in general the systematic cultivation of land. But the right time occurred when people came to first appreciate the benefits of food surpluses. So while they were still only hunter-gatherers or shepherds, semi-settled seasonably, before deliberate cultivation had begun, they might have taken advantage of naturally occurring surpluses of food (e.g., wheat). They harvested and preserved these extras for consumption in later days or months. Because the surplus of food was used so beneficially, it occurred to people to control and enhance food production further by deliberately planting extra seeds and aiding their growth. When harvest season came and abundance was everywhere, people became convinced that agriculture was a vital alternative to hunting-gathering or herding alone. Of course these groups continued to hunt and keep flocks. But since plants added to the animal food source and created even more of a surplus within a short time, agriculture would be a major occupation. Gradually, during the Neolithic Age the nomad hunter-gatherer transformed into a settled farmer, and the idea of systematic human-made cultivation of the land had been fully established. Everything that pertains to the notion of civilization is a consequence of agriculture!

Settled Life

Agriculture ushered in a more fully settled lifestyle. Thus with it we have the inauguration of urbanization. The farmer adapted to a domestic lifestyle by living permanently in settlements close to his fields. To put food on the table each day, the farmer had to commit to and endure hard work and careful planning for several months: clean the fields from unwanted plant growth, plow the soil, find seeds (e.g., of wheat, barley, peas, lentils, rice, beans, avocados, potatoes, squash, dates), plant them, redirect water from a nearby river to irrigate the growing crops, harvest them, eat some fresh and process some of the harvest into new types of food (grind grains to make flour, moisten it, then dry it with the heat of fire and make a nutritious bread), preserve and even trade the surplus, save seeds for next year, rotate the crops, keep his beasts of burden healthy (horses, donkeys, mules, oxen, camels), attend to his flocks, make/fix tools, and so on. But this hard work was rewarding. Agriculture increased food productivity, created and exploited surpluses, raised the confidence of the people, and in general brought them prosperity and especially security, a feeling we all understand even today once the daily food of future months is secured way ahead of time. Furthermore, since successful agriculture was requiring the collaborative effort of many people (e.g., in harvesting the crops or constructing specialized tools) the benefits of coexistence were being appreciated and pursued. The settled lifestyle of the farmer promoted human collaborations but also conflicts. Conflicts could be avoided and collaborations could be enhanced if farmers chose common leaders who would develop a set of rules fair to all. This gradually led to organized communities called city-states.

From a Village to a City-State

The first villages were simple and extended for just a few acres. They had anywhere from twenty to fifty square or oval mud-brick multiroom houses. The village population was perhaps a hundred to a few hundred people, most probably close and distant relatives. Rooms were used for sleeping, cooking, and socializing but also to store tools or the surpluses of food. The daily life in the village was basically agricultural and pastoral, thus all villagers had similar skills. Family and perhaps village leaders would have been the elders since these were the people with the most experiences.

With time, about six thousand to five thousand years ago, both the village and human culture underwent an extraordinary transformation. The village grew to become a city-state: a complex, self-governing, organized community of a few thousand (thus no longer made of just blood relatives), that included a city and its surrounding farmland and grazing pastures. Life in a city-state created complex human interactions. It promoted innovative ideas and rapidly accelerated their dissemination, and it set the stage for the rise of human civilization. With time the city-state would become more systematic, more advanced, safer, and its people more sophisticated and with a better understanding of the world around them. The settled agricultural economy provided new challenges but also opportunities and encouraged the flourishing of numerous specialized professions. In fact the successful operation of the city relied on the continuous emergence of innovative, specialized pursuits, as is the case today. It also necessitated that people collaborate on just about everything. One depended on the skills of another. The farmer depended on the craftsman for tools just as the craftsman depended on the farmer for bread and milk.

As a result of a predictable food supply and the safety that communal living provided, urbanization spawned an increase in population. Labor-intensive agriculture made it practical for a farming family to have many children (a common practice even today in agricultural villages). When the boundaries of city-states began overlapping, the cities could decide to unite into one nation or fight with each other. The city-states by the Nile followed the first path and so by about 3100 BCE, Egypt was the first nation of the world. Later the many city-states in the Yellow River Valley became one unified and centralized Chinese Empire around 221 BCE. On the other hand, the Mesopotamian city-states and later the Greek city-states generally tended to follow the second path.

The inhabitants of the first villages must have developed a feeling of group identity, which evolved into what we today call nationality as the settlement size increased from a village to a city and then to a nation. Now, what kind of a ruler and political system did a city-state have and why?

King-Priests, God-Kings, and Theocratic States

Before we urbanized ourselves in communities large and small the law of the people was to be found in their customs. In growing urban environments customs became increasingly more complex, thus peaceful coexistence in a city and its effective operation required a common vision and a leader. Such vision was often provided by religion, which itself became a custom, for organized religion (as we will elaborate in chapter 4, "The Mythological Era") was born with the onset of structured communities. Consequently priesthood was among the first type of specialized professions everywhere. But so was the profession of village or city leader. And because religion and politics were evolving together to accommodate the changing cultural, religious, and political needs of the city-state, the first states were theocratic. This means that the ruler of the people was both a religious leader (a priest) but also a political one (a king)—that is, there was no separation between church and state; religion and politics had from the beginning been closely intertwined. Thus the institution of king-priest and of a hierarchical society (a stratified society in which various priests would have different amounts of power in the governance of a city) developed almost simultaneously with the first city-states. But why?

Priests were the first type of people who contemplated the world in an abstract way; they were looking for what they viewed as hidden meaning and purpose in nature that needed to be unearthed and explored, hence to the other members of the community they appeared to demonstrate certain wisdom. As a result of the insights they voiced, many in the community regarded them as among the most knowledgeable of people. Now, as will be argued in chapter 4, the earliest religions were cults focused on natural phenomena that influenced the crops people grew, so a priest had to master the right rituals in order to encourage the spirits believed to be in nature to produce rich yields. Consequently, priests were socially significant because of their efforts to secure people's daily food, their paramount concern in the struggle of life. But these priests were also politically significant because they were perceived to be the ones most connected with the honoring of powerful gods or spirits. This is so because priests, as specialized religious professionals, were those who devised and implemented the various religious rituals of daily life in the village or city, and were thus considered *inspired*, *literally*, as if a divine spirit entered their body during a ritual, advised and generally endowed them with valuable knowledge. Why this was thought so will be discussed in the section titled "From Dreams to Spirits" in chapter 4. People imagined, for example, that during the rituals, the laws of the city and of people were given to the priests *directly* by the gods (through divine revelations or inspirations)—in fact it is in this sense that the political system was theocratic, "rule of god" in Greek. Hence, not only priests were regarded as divinely endorsed (thereby creating a link between ordinary people and their gods), but so were city laws (including the moral code of the community). Since both city law and priesthood represented the will of the gods, both earned immeasurable authority. And so priesthood status was elevated to prestigious, mighty king-priest status: a priest was in charge of both the religious as well as the secular/public matters of a city-state, thus he was also a king. There were in fact instances (for example, in Egypt) in which a king-priest himself was regarded as a god by his people, thus a god-king.

Temple offerings by the people—anything from food and personal items to flocks of sheep, land, even people—that were meant for the gods were in actuality owned and managed by the ruling priests, making them wealthy and powerful. And to preserve the special family privileges, kingship naturally became hereditary (or in general restricted to the socially privileged). All political, military, religious, and economic powers were centralized and controlled by the king's entourage and ultimately the king himself, who answered to no one. Democracy, which was born in ancient Greece (and played a significant role in the birth and development of science), had not been an option for the first ten thousand years of civilization (see next section). A good king-priest was respected for promoting peaceful coexistence and organizing the effective functioning of his own city. This was a challenging task, especially during early human history when rules and regulations were still in their infancy and thousands of unrelated people were living in relatively new social settings: living in harmony with one another in the close spaces of a large village, town, or city often meant that some of one's freedom had to be sacrificed for the greater good and stability of the community. A common religious vision probably eased such problems. At the same time, however, while a common vision and belief in a shared pantheon of gods or spirits could unify one city, it could also cause conflict between cities with different visions and pantheons, an age-old challenge that persists even today. Still, the benefits of urban life were immeasurable whether they were cultural, technological, political, religious, artistic, economic, or of some other form. But the single most important consequence of urbanization was the creation of free time!

Leisure

Food surpluses and specialized professions meant that at least some people (usually the more well-off) had certain leisure time to devote to anything their heart desired, including the arts, one-to-one personal and warm socialization with fellow humans, even contemplating abstract matters that at one time were the province only of the priests. Religion and the priesthood had been the first of these. With time philosophy, literature, history, science, and mathematics followed (though after the era covered in this section), as well as many other things that the notion of culture entails. All of these actual and potential benefits of communal living have been the joyous pursuit of all types of peoples all over the earth as a result of urbanization. Settled life fostered the evolution of the human intellect. The improvement of spoken languages as well as the invention of writing were the results of the evolving needs of an urban existence. They accommodated more efficient communication for trade and commerce, tool refinement, and techniques of farming.

FROM WRITING TO THE BIRTH OF SCIENCE: 6,000–2,600 YEARS AGO

The discovery of copper probably occurred as early as 6500 BCE, but its use was becoming widespread by about six thousand years ago, causing the transition from the Neolithic to the Chalcolithic Age ("Copper Age"). With the use of copper, tools improved significantly. The island of "Cyprus gave its name to copper"[10] because it was rich in it and among its earliest sources. Copper and tin form a hard alloy called bronze, which revolutionized not only everyday tools but also warfare. About the thirteenth or twelfth century BCE, the Greek and Trojan heroes fought with bronze weapons (Homer's *Iliad*). Nonetheless, bronze was no match for tougher iron. Formed in the cores of supermassive supergiant stars toward the end of their life before they become supernovae, iron makes stronger-edged tools. The Iron Age began around 1200 BCE. The combination of fire and iron has ever since been shaping both our technology and the general direction of our civilization.

Sumerians were probably the first to develop writing sometime in the fourth millennium BCE. Egyptians followed soon thereafter (if not simultaneously or prior). Writing is a momentous invention that accelerates the rate of progress because recorded information helps future generations to learn faster, more accurately, and more systematically all the accumulated wisdom of their ancestors. The need to keep track of the quantities of items accumulated and traded might have necessitated the invention of writing, including primitive forms of mathematics. With writing we have the beginning of recorded history and the accounting of what went on before (prehistory). The critical role of language (both in its oral and written forms) in the intellectual evolution of humankind (in fact even in our physical survival) will be elaborated in detail in chapter 6, "The Birth of Science." We will see, for example, how significant the Greek language was in the evolution of ideas and the birth of science.

Egyptians were obsessed with the afterlife, and around 2686 BCE they built their first of several elaborate stone burial tombs, the monumental pyramids. At the end of the third millennium BCE we find the first inhabitants of Greece. They were obsessed with life—Hades or the underworld for them was a gloomy place. Phoenicians invented the first type of written alphabet around 1050 BCE. And by 776 BCE the first Olympic Games were held. The social and political reforms implemented in Athens in 594 BCE by the Athenian leader Solon were the first crucial steps on the road to a democratic system of governance. In fact, the sixth century BCE was intellectually explosive globally.

Since sacredness is not easily challenged, theocracy proved generally repressive to religious or other innovations. So for millennia there was no significant evolution in religion. But sixth century BCE proved to be astonishingly different. The world at that time saw an outburst of religious and secular worldview reforms. This was seen in China with Lao-tzu (Taoism) and K'ung Fu-tzu (Confucius); in India with Buddha; in Persia with Zoroaster; in Israel with the Jewish prophets; and in various Greek city-states, too. Greece was a haven for the mystery religions whose subtle influence on the birth of science is explored in chapter 5, "Religion and Science." In southern Italy we find Pythagoras, whose school combined religious, mathematical, and scientific studies, and in Ionia (the Hellenic region of Asia Minor with the islands nearby) emerged the first natural philosophers whose rationalistic approach about nature marked the transition from mythology to science. Among several factors (to be introduced in chapter 6, "The Birth of Science") that contributed to the various reforms in ancient Greece was that Greek city-states (those that began forming after the fall of the Mycenaean civilization, around 1200 BCE) were not theocratic— "Greek gods do not give laws."[11] After thousands of years of no substantial progress in religion, any religious transition in itself, such as those that took place in sixth century BCE, was a promising prospect.

This chapter's brief human history finishes with the total solar eclipse of May 28, 585 BCE, foretold by Thales of Miletus in Asia Minor. We know the exact date of this phenomenon by using the modern calendar that predicts eclipses to calculate those that occurred in the past. The eclipse occurred the day of a great battle between the Lydians and the Medes. It suddenly brought a semblance of night to day and shocked soldiers and kings from both fighting armies. They interpreted the natural phenomenon as a divine omen against their continued hostilities. Frightened by the possibility of having angered their gods, they agreed to end both the battle and their six-year war. Thales was the first to use purely natural causes to explain the phenomena of nature. Others before him explained them supernaturally through myths and superstition. Since Thales flourished around the time of the famous eclipse, that era marks the birth of science. Parenthetically and interestingly, whether or not the eclipse was a divine sign is arguable among believers. But what is equally interesting is that this or any other eclipse also has a natural cause and a rational explanation.

Religion, philosophy, and science were all consequences of urbanization, a phenomenon so culturally impressive that stories from when it was first implemented seem to have survived until today: the story of the Golden and Silver Ages can be traced back to those early days and so can the legend of the lost city of Atlantis.

THE GOLDEN AND SILVER AGES

Hunting is risky but exciting. Farming is safe but dull. To hunt you get up and go, you are spontaneous and carefree. You have adventures with dangerous, magnificent, and strange animals in the wilderness. You move to be near the game that you hunt. The stories you tell your kids are thrilling, and you are the hero of the family. The gratification of the daily kill is instant though uncertain: but the food of the day must be found every day. However, with farming you settle down, plan for months ahead, and commit to a life that finds you tilling the soil. You have the safety of a home and a community, but your daily stories are ordinary. The gratification of farming is gradual but secure: planting occurs during one season, harvesting during another, but a rich harvest guarantees food not just for a day but for months ahead! This lifestyle difference has been a subject of ancient oral folklore but also of ancient written tradition, such as Hesiod's poem *Works and Days*.

Hesiod is an ancient Greek poet but also a farmer who lived around the late eighth century BCE. Among other things, his *Works and Days* describes the five Ages of Men (the stages of humanity since its emergence). The first was the best, the so-called Golden Age, a prehistoric utopia that, according to the poem, appears to have been the preagricultural, pre-urbanized era of carefree wanderers, the hunter-gatherers who lived spontaneously day by day. Incidentally, it has been speculated that the story of the Garden of Eden is from the Golden Age.[12] The era of agriculture and of the farmer was the second, the so-called Silver Age. Farming was committed, hard, tedious work, and farmer Hesiod had firsthand experience of it. So after a long laborious day in the fields, from dawn to dusk, farmers would go home looking forward to a well-deserved plate of food, worrying however about the constant challenges of tomorrow (the weather, their crops, their animals, their tools), and during their restful hours they would, I am quite certain, recollect with nostalgic envy on the carefree, uncommitted lifestyle of their not-so-distant ancestors, the hunter-gatherers (or in general those who had never committed to farming). Calling the hunting-gathering era golden seems to have been the reminiscing words of a tired, worried, reflective farmer.

Interestingly however, for Hesiod only carefree hunter-gatherers, from the Golden Age, or shepherds in general, were good enough to be "loved by the blessed gods,"[13] whereas hardworking farmers (like himself) from the Silver Age were not loved because they were "less noble by far."[14] Perhaps the challenges of coexistence in an urbanized environment were too great for the settled farmers. And so "they could not keep from sinning and from wronging one another, nor would they serve the immortals, nor sacrifice on the holy altars of the blessed ones as it is right for men to do wherever they dwell. Then Zeus the son of Cronus was angry and put them away [annihilated them]."[15] Incidentally, in the Old Testament book of Genesis, Cain was a farmer, and his brother Abel, whom he killed, was a shepherd.[16] Now who were those people who were annihilated by Zeus? Where were they from?

THE LOST CITY OF ATLANTIS

Hesiod's sinners, from the Silver Age, might have been a general reference to city people, in particular those corrupted by the newly found temptations of urban lifestyle. Or they might have been a specific reference to the corrupted

Atlanteans, citizens of the lost city of Atlantis, who, by Plato's account (from his dialogues *Timaeus* and *Critias*), were defeated by the prehistoric Athenians with Zeus's help, and their legendary city was afterward sunk by the gods in a single day and night.

Plato places chronologically the destruction of Atlantis about nine thousand years before the time of the Athenian leader Solon, approximately 11,652 years ago, an era that roughly coincides with the onset of urbanization. He describes Atlanteans to have initially been people of nothing but virtue, only to later become of nothing but corruption. Since both groups of people, Hesiod's Silver generation and Plato's corrupted Atlanteans, were, according to each author, punished by the wrath of Zeus, and also since both groups are placed chronologically from around the era of the onset of agriculture and urbanization, then I think is plausible that they are one and the same group. If so, then Hesiod's "golden race of mortal men . . . loved by the blessed gods,"[17] the hunter-gatherers/shepherds of his Golden Age, might have been Plato's virtuous Atlanteans before their fully settled agricultural lifestyle or during its early transitional stages when things were still simpler and people's minds were purer. Hesiod's "silver [race] and less noble by far,"[18] of the farmers of his Silver Age might have been Plato's corrupted Atlanteans well into their urbanized lifestyle, both unable to deal effectively with the new, ever-changing, and demanding challenges of peaceful coexistence, such as increasing populations, limited resources, territorial disputes, land ownership, new codes of conduct, different overall philosophical outlooks on life, and ultimately driven to obliteration. Are we, the modern humans, managing better?

So among humankind's earliest cities, the most prosperous and successful might have been Atlantis, whose sudden violent destruction by possibly a natural cause (such as flooding or earthquake), and/or by human-made ones (such as greed, deceit, slavery, internal conflicts, external wars, and generally corruption), left such long-lasting and legendary images in the minds of those early and impressionable urbanites (either Atlanteans who survived, or others who visited the great city before its obliteration), that the tales they had told their children, exaggerated, have since then been traveling through the continuum of spacetime and stinging the imagination of all those who have been hearing them. If Atlantis is discovered to be a real city destroyed by war (and not just some imaginary city of an allegory), then Plato's account will be a written reference of the oldest human war, which ironically would coincide with the onset of civilization. It would mean that, as we were getting civilized we were also preparing for war, a still-persistent irony of civilization.

CONCLUSION

Agriculture led to an urbanized culture and eventually to developed civilization. In turn, the coexistence and interaction of the many groups nurtured the development of the human intellect and precipitated the pursuit of a multitude of innovative and specialized professions. Thus hunting and gathering were no longer the only occupations that could secure one's daily food. But by far the most revolutionary effect of civilization was the creation of leisure time! For free time created the potential for abstract thought and the development of sophisticated religion, philosophy, and ultimately science. But the scientific view of nature is the newest, only about 2,600 years old, whereas the mythological is the oldest, at least ten thousand years old—as old as civilization itself. How did we view nature before the advent of science? Why did the mythology develop? When was religion born and what triggered it?

CHAPTER 4

THE MYTHOLOGICAL ERA

INTRODUCTION

The worldview of the first 7,400 years of civilization was purely mytho-logical. It was an intellectually unrefined era, nostalgically simplistic, and superstitiously phobic. Nature was imagined to be animated; natural phenomena were considered random, unpredictable, and the work of mysterious supernatural (not subject to any physical law) forces (e.g., capri-cious, anthropomorphic spirits or gods). There were nature deities (e.g., the sun-god and the sky-god), deified ancestors, plant and animal spirits, idolatry, totems (animals believed to have been the ancestors of tribes), king-priests, god-kings, and spirit-based political systems. Hence belief in the supernatural is literally as old as civilization—for what we today refer to as ancient mythology constituted humankind's earliest form of religion. While influencing each other since their birth ten thousand years ago, religion and civilization have been evolving from their crudest forms to their more sophisticated ones today.

DEFINITION OF RELIGION

Religion, at least the primitive kind, is the belief in the existence of beings with powers far greater than that of humans, with whom humankind wants to open communication and cultivate a relationship. All of the natural forces that humans did not control themselves were placed under the direct control and supervision of a spirit or god, which had to be understood and placated to ensure that these forces could be managed and made more predictable to the service of humans. With this in mind, the two culturally remarkable practices of Paleolithic humans, the possibly one-hundred-thousand-year-old practice of burying one's dead and the thirty-thousand-year-old practice of cave painting, are activities that while their mere practice in itself does not constitute religion, it may be a vague indication of an early tendency for religion.

BIRTH AND ENDURANCE OF RELIGION

Religion was born after humankind first concluded that such beings with higher powers—what came to be called gods—existed, and specifically at the moment humans felt the need to establish and nurture a relationship with them. This birth is generally believed to have occurred about ten thousand years ago with the onset of urbanization. The factors that caused this birth and the nature of the first gods are introduced later. What caused religion to endure, evolve, and become an essential part of human culture was when the idea of a human-god relationship developed over time into a great human need and thus a lifelong habit, passing systematically down through the generations. Specifically, this need triggered the development of various tribal rites and rituals through which people believed they could open communication with their gods, thereby developing, preserving, strengthening, and renewing it. As time went by, the rituals became more complex and constituted a significant part of communal life. This established and sustained the religious outlook as a way of life.

THE CHARACTER OF EARLY RELIGION

The goal of the religious outlook was twofold, initially selfless but later egoistical. It was selfless in the sense that through their rituals our ancestors aimed partly to secure the welfare of the tribe (and, consequently, of humankind in its struggle for life) by appeasing, befriending, encouraging, persuading, even trying to control and manipulate the gods to care for the people: to tend to the tribe's practical needs regarding food, shelter, health, and fertility. The character of early religion was tribal[1] not personal—in a sense, just as the search for food or shelter was a tribal activity or, to say the least, a family effort. It aimed to guarantee tribal well-being. Religiously speaking, initially the individual cared more for the welfare of his community than for himself! Little by little, when religion's character evolved to be personal (egoistical) as well (see the section under "Idolatry"), the gods were expected to tend even to the more abstract and personal needs of the individual, such as those regarding personal grief, happiness, knowledge, comfort, and hope, as well as meaning or purpose for this life and even in the afterlife. Humans worshipped and idolized the gods, built them shrines, and brought them offerings in the hope of receiving favors in return. "By gifts are the gods persuaded, by gifts [are the] great kings" says an ancient Greek epigram.[2]

But the goal of the religious outlook was also selfless on another level, for the rituals were not always aimed at securing what individuals (or the tribe) wanted from the gods, but also at what they could secure for the gods. Our ancestors literally sympathized with them in what they perceived to be their own divine struggles, for gods were imagined anthropomorphically—in our own image (with similar needs and challenges as humans). They felt or shared their gods' passions, their sufferings, their needs and challenges; they showed affection for them, expressed admiration and respect for them, even sacrificed for them (i.e., gave the gods something humans valued, such as food and drink, material valuables, or in extreme cases human sacrifice). (The act of sacrifice was indicative of the belief that not only did humans need the gods but also that the gods needed humans.) Some people devoted their lives to the gods (e.g., by becoming a priest in a temple in order to better serve the god who was thought to be literally living there). Finally, through the rituals, people wished merely and humbly to give thanks to these gods for their previous help (e.g., for making it rain or causing a rich harvest). Devotion to the gods quite possibly was even intended to promote harmony in the coexistence of humans and these divine entities.

CAUSES OF THE BIRTH OF RELIGION

While it is generally accepted that organized religion (with gods, their temples, priests, and rituals) was born with the advent of urbanization, more speculative is what might have sparked its birth and aided its development. No doubt a variety of factors might have played a role. These form three general groups: (1) theological, (2) biological, and (3) cultural.

The Theological Factor

Is religion (and, in particular, the belief in the existence of a god or gods) knowledge that came about via divine revelation, is it a human discovery, or is it a human invention that humankind created? If it is a revelation or discovery, then a god (or gods) is presumed to exist. If it is an invention, then a god (or gods) cannot necessarily be presumed to not exist, for arguably any human invention/ creation might be, directly or indirectly, caused by a god. This is so because, while science can determine various properties of the universe causally, it cannot determine its first cause—what caused the universe (recall the section titled "Causality" in chapter 2 "What Is Science?"). In science there will always exist an initial unexplained axiom (a primary cause)—and as we will emphasize in the section "Unborn and Imperishable" in chapter 13, "Parmenides and Oneness," there is not, and there can never be, any scientific explanation of how something (e.g., the universe) can come to be from absolute nothingness. Hence the *why* of this first cause, whether it might be some god or related to some god (and thus, why the universe is what it is and why it exists), will always remain a subjective matter, and consequently so will the belief in the existence of a god and the cause of the birth and development of religion. A god of *absolute* powers (e.g., wisdom) is unprovable, for the only way to recognize, understand, and prove the absolute is if we ourselves had the same absolute powers. But we do not. So even if an absolute god were to reveal itself, the only thing we could logically be certain of is that the being revealed has great powers but not necessarily absolute ones— hence we would not know if it were the true god or just another being with powers merely greater than ours. The most central feature of religion, therefore, is faith, not causal knowledge. But if religion is a divine revelation, why shouldn't everything else that we come to know, such as science, be a divine revelation also? And if we were to accept that things are divinely revealed to us, then a question begs for an answer: what is the role of our mind?

Having said that (and leaving the nature of the primary cause of the universe to be a subjective issue), if we are still interested in pursuing a scientific understanding of the universe, we must remain of the conviction that the universe is inherently rational, that it obeys understandable causal laws (something proven repeatedly by science). And so everything that happens in it, including the birth and development of religion, has a rational explanation. Human biology and culture are part of a rational explanation concerning the birth and development of religion.

The Biological Factor

(a) A Biologically Evolved Brain but an Ignorant Mind

A requirement for critical inquiry of nature is a biologically evolved brain. Now, while a critical examination of the phenomena of nature has always been an impossibility for all other species—since their brains have not yet evolved to that crucial level required for abstract thought—for humans, it has not. By about ten thousand years ago, our prehistoric early Neolithic ancestors already had a brain that was as biologically evolved as our own. This early species was therefore as capable of critical analysis as we are now. And so for the first time they became profoundly curious about their surroundings and developed the desire to know the nature of the world they lived in. But without much prior knowledge as a point of reference on how natural phenomena worked, and without an advanced language to express themselves clearly, these first explanations were childish, irrational, dogmatic, mystical, and overly simplistic. For, while Neolithic humans were intelligent, they were like young children who have the biologically advanced organ of knowledge—the brain—yet an ignorant mind. Hence, naturally, early Neolithic humans' raw intelligence led them to model natural phenomena after the only thing they knew best: themselves. In their own image they modeled these phenomena with feelings, needs, motives, challenges, desires, pleasures, passions, and powers that were in fact not only greater than their own, but justifiably they were also imagined to be supernatural in their extension and capabilities because nature is impressively powerful (e.g., the thunder is unpredictable and loud, the lightning is fast and bright, and the sky is huge and ever-present).

Initially it was, for example, the *object* sun that was imagined to be animated and a god—the sun-god. That is, there was no reason to see an object, such as the sun, and think that it is something different (e.g., an anthropomorphic being) than what it appears to be. But with time, the attributes of the animated phenomena were imagined so much to be like and possess the attributes of humans (and/or of other animals), that the phenomena became tangible and anthropomorphic (or, in general, zoomorphic), and so in addition to humanlike behavior, they also acquired humanlike form. Consequently, object and god were gradually separated: for example, the object sun-god became the anthropomorphic sun-god, who, among other things, could control the rising and setting of the object sun. As in human culture, the gods were imagined to have specialized professions. The moon, planets, and stellar constellations of the zodiac are additional examples of objects that initially were object-gods that gradually evolved in the human mind to become gods of a zoomorphic or anthropomorphic nature (or a mixture of the two). In ancient Greece the father god of all, the heaven-god Uranus ("heaven," in Greek), and the mother goddess of all, the earth-goddess Gaia ("earth," in Greek) had similar evolutions. Uranus and Gaia initially were the physical sky and physical earth, but with time they were imagined as two anthropomorphic gods. Their son Cronus had a son of his own, Zeus. Uranus and Gaia gradually decayed to lesser gods, and mere grandson Zeus evolved to become "Father of men and gods."[3] Even more impressively, as Greek religion evolved from Homer's and Hesiod's Olympian to the mystery religions (the Eleusinian, Dionysian, and especially the Orphic), so was the status of Zeus attaining a sole and absolute divine supremacy as an all-encompassing father of all. So a deity who once was the natural object/phenomenon itself was gradually becoming increasingly disconnected from the object ultimately acquiring a divine existence of its own with new roles and attributes. Now, since capricious supernatural gods controlled nature and since humans could not know their divine minds, early humans thought that nature was random, unpredictable, and at the absolute mercy of these divinities. But humans hoped that their rituals could persuade the gods to act favorably toward them, and, so in a way, indirectly, humans hoped they could have a certain amount of control over nature.

Motion and sound, which are among the important characteristics of a living person (although not just of a living person), were also characteristics of the phenomena of nature; trees shake and appear to cause the wind (although what happens is the reverse; the wind shakes the trees), the sun and moon rise and set, the thunder is loud, the rain comes down from the sky. And so for our Neolithic ancestors various things and phenomena appeared as alive as they themselves, only more powerful. Their deification and personification followed naturally thereafter.

Because humankind recognized the benefits of communicating and cultivating a relationship with the animated superpowerful phenomena of nature, to achieve it rituals were devised and implemented. The phenomena became instantiated as worshipped gods, and the stories about them emerged as myths. Some myths were only about gods, others about humans and gods. By one account of ancient Greek mythology, for example, Orion was once a handsome, skillful hunter who hunted with Artemis, goddess of the hunt.[4] But he was killed by a scorpion sent by Mother Earth because he was arrogant and threatened to kill all the animals in order to impress Artemis. At the request of Artemis, Orion was placed by Zeus among the stars in order to be remembered. Orion therefore was to the ancient Greeks the personification of a particular grouping of stars. To modern astronomers this grouping is *Orion the Hunter*, one of the most recognizable constellations from a total of eighty-eight that divide the sky—like the fifty states divide the country of the United States. In the evening it's easily spotted high in the sky and toward the south during the winters of the Northern Hemisphere (and visible during the summers of the Southern Hemisphere by looking north).

At first the myths were primitive, for although the human brain had the biological potential for progress, the mind was still imprisoned in absolute darkness. While awake, humans regarded nature as animated, zoomorphic, anthropomorphic, powerful, supernatural, and random; while asleep, it was seen as enigmatic, spirit-filled (the primitive interpretation of dreams, as will be seen in section "From Dreams to Spirits"), and often frightening. As civilization gradually evolved, its new challenges and needs made its myths more diverse, imaginative, and rich, and a number of these myths became the religion of some. So what we today refer to as ancient mythology constituted humankind's earliest form of religion. Note that to the ancients a myth was not necessarily a fairy tale as the word "myth" has evolved to often imply nowadays, because many ancient myths have with time proven to be just that, fairy tales.

The first myths and attributes of the gods arose through early human attempts to interpret the phenomena of nature. The loudness of thunder and the brightness of lightning might have, for example, been associated with the wrath of the sky-god, and the wind with this god's breath (the latter, for example, was a belief of some North American Indians of relatively recent times). And so religion was humanity's first serious attempt to understand how the phenomena of nature work. Interestingly, therefore, our curiosity and desire to make sense of nature were among the stimuli for the birth and development of both religion and science. Nonetheless, the application of causality in each approach is fundamentally different: in religion the cause of the phenomena was divine (supernatural), in science it is naturalistic.

(b) Long Parental Care

Religion might have also been born as a consequence of yet another reason of biological origin, namely, the need to be cared for and the need to care for. The human species has evolved biologically in such a way that human babies lack the ability of self-reliance, so their self-preservation and general survival depend on the unconditional and uncommon long period of parental care. Ever since their birth and for several years thereafter, human babies need their parents' care and love. Without such long parental care, the babies would die. Now, if our physical attributes (e.g., brown or blue eyes) have a biological origin and explanation, why not our emotional ones (e.g., the need to be cared for)? If yes, then the biological need of the human baby to be cared for might exist in our genes and endure (subconsciously) throughout adulthood, driving humans to discover/invent their god through which they can continue to be cared for and to be loved. And so as we are growing older we are also naturally searching to supplant the gradually lesser parental care with a gradually greater divine care in order to fulfill the genetic need of being cared for. That is, in a similar manner that the genetic (biological) need to eat and drink (or to have sex) drives us to sources of food and drink (and to seek a sexual partner), the biological need to be cared for and to be loved (in fact even the need to care for and love, for parents have such need for their babies) might have driven us to seek God and religion. Through religion not only do we continue to be cared for (by the divine), but we continue to care for (the divine)—for, as we saw earlier, religious rituals have an egoistical and selfless objective.

Note that while the young of various other species have the biological need to be cared for, too, still such need is not as evolved as or as extensive as it is in humans. Even more importantly, unlike humans, all other species still lack the critical mind that could stimulate such biological need to more abstract endeavors (such as religion). My point is that we need to remember that the cause of something is often a complex interplay of several factors. Concerning the birth of religion the biological factor is coupled with a cultural one.

The Cultural Factor

With its evolving urbanized culture, humankind came to realize that its survival and overall well-being depended on nature's animated powers: the sun for light and warmth, the earth for sustaining plant and animal life (for food), the rain and rivers for water, the sea for fish. Humans could not control the rain or the light from the sun, but the rain-god and sun-god could. Intelligently our ancestors sought to establish communication with these natural powers and to cultivate a relationship in the hope of appeasing and befriending them, and through offerings to ask in return for their help in the human struggle to survive. So early humans deified and worshipped these natural phenomena. Modeling the human-god relationship after the human-human one was a consequence of the human culture: since individuals and groups befriended and formed alliances with other persons and groups for mutual benefit, by exchanging goods, a custom practiced for millennia before religion, that's how these humans thought they should behave toward their gods if humanity was to be successful in its relationship with them.

Fear of the phenomena (the personified powers of nature) was yet another culturally related reason that could lead to religion, for such fear could now be conquered through communication with these powers. Even the admiration, respect, or envy for such magnificent powers could lead to religion, for such feelings could trigger humanity's humbleness toward them, its desire to communicate with them in order to understand them better, perhaps even imitate them or sympathize with them in the hope of becoming like them. Nature was worshipped not only in terms of its particular objects (the sun, the stars, the rain, etc.) but also in terms of its natural processes, such as the cycle of moon's phases; day and night; the changing seasons; the birth, growth, decay, and rebirth of plants—for example, night was a goddess and so were the seasons. Humans' rituals were often an attempt to act out and imitate such processes and in general sympathize with nature (the gods), believing that through sympathetic magic life would imitate the ritual. And thus we humans could, at least in some aspects, be like the gods. *Enthusiasm*, for example, which etymologically meant that a god who was honored literally entered the body of a worshipper and influenced him in strange ways (e.g., spoke to him or made him move a certain way), was a state of mind hoped to be achieved by the worshipper in ancient Greece during the Mysteries (ancient Greek religious rituals of the historic era), in order that the person would feel godlike, not only physically but also intellectually—that is, he wished to know the mind of the god. The latter was such an ambitious goal that the desire to fulfill it, as will be discussed in chapter 5, "Religion and Science," might have been among the factors that sparked the rationalist viewpoint of nature and consequently the birth of science. For it had at some point been realized that achieving divine knowledge (knowledge supposedly exclusive only to the gods) ritualistically was really unattainable but could be attainable through one's rational analysis of nature.

There were as many gods as the vast array of phenomena in nature would permit—it was difficult to settle for only one supreme god just yet, and monotheism was millennia away. And since nature is ever changing, so were our human preferences for god. During a full moon for example, the moon-god is more important than the sun-god, but in the morning, the preference is reversed. Worshipping the natural powers as gods, coupled with the changing human culture, had gradually led to the belief that the phenomena of nature had even more attributes than initially imagined. It also caused human understanding of nature to be reversed, for with time, it was not nature's phenomena (the gods) who were imagined as taking human form; rather, humans began to imagine themselves as having the form of gods. In ancient times, for example, the mythical titan Prometheus created humans in the image of the gods by molding into human shapes a mixture of soil and water.[5] And, according to the book of Genesis, "God created man in his own image. In the image of God he created him."[6]

In general, as humans became socially more responsible and humane toward one another—that is, more conscious of their actions and the consequences of those actions—so did their gods and religion. Hence contributing to the development of religion was humanity's changing needs, challenges, and overall understanding of the world, all of which were coming about through the evolving complexity in lifestyle required by urbanization. The birth and development of science was yet another contribution to the evolution of the religious outlook, although that influence came much later as science was born about seven and a half millennia after urbanization and religion. Interestingly, however, unlike the birth of science, which occurred once and in one specific place (as will be seen in chapter 6, "The Birth of Science"), the birth of religion was universal.

UNIVERSALITY

Religion was generally born independently but universally (in all major urbanized centers of early civilization) and relatively simultaneously. It had no founder and evolved and became established through diverse custom and traditions. This universality is no surprise because the factors that caused the birth of religion were themselves universal: we all have about the same mental abilities to address our basic needs, challenges, and experiences, which themselves are generally also about the same for all, especially so in the early stages of civilization, because nature itself, which presents us with these challenges, obeys universal laws. The first types of gods were universal, as were the first types of religions.

FIRST GODS AND FIRST RELIGIONS

The prevailing view is that the first gods were those associated with the great powers of nature: things that were so powerful, admirable, impressive, fearful, wondrous, that appear so *alive*, such as the light-bringing bright, warm sun; the moon of many phases that illuminates the night; the strong whistling wind; the fire-carrying lightning; the vital rain; and the loud thunder. The concept of the divine might have extended to something huge and intangible, such as the ever-present thus immortal sky, which is strikingly so high above all, appearing to blanket and look over all, that from it the rain falls. Or something semi-tangible but still immensely powerful, also ever present and seemingly alive, such as the earth from which everything is born, nourished, grows, and on which humanity itself so much depends for food, shelter, and care. For this life-giving property earth was female, thus the earth-*goddess* was the mother of all life, and the sky, who rained onto her the fertile rain, was male, thus sky-*god* was the father of all, a religious notion common to almost every primitive religion. With this in mind, the first religions were cults aiming to warrant fertility. This is a reasonable development because people's main concern has always been their daily food, which after urbanization was coming mostly from farming, and so naturally people wished to have their priests, through correct rituals, properly encourage the gods to produce rich harvests. Furthermore, since these great powers of nature are themselves universal (experienced by all people in all lands alike) and timeless, likewise the first gods were the same universally as well as usually immortal.

But indigenous gods differed from place to place for various groups also made a god out of something minor and local to them like a stone, a fountain, a well, a river, a tree, animal fur, even a dead ancestor, by attributing to them supernatural powers that they did not in the first place appear to possess. Now, how did such minor and local things manage to become gods?

FROM DREAMS TO SPIRITS

The intellectually undeveloped humans of early civilization were in no position to understand the meaning of either death or dreams. But they had the mental ability to wonder about them and to seek explanations. So when the dead appeared in one's dreams, moving and speaking, it was imagined as if they were somehow still alive, as if something from them, let us call it spirit (psyche or soul), endured despite the physical death and decay of the corporeal body, and had the supernatural ability to *enter* the body of those who were asleep (the etymology of the term "in-spire"). For something that could be seen, even in a dream, was thought to be there where it is seen, thus in the body of the sleeping person. The images of a dream were as real to the unrefined but intensely curious mind of Neolithic humans as the shadows were to the prisoners in the parable of the cave. Now, to have the ability of entering the body, the spirit was imagined to be like a shadow or a breath or air, incorporeal and form-changing. Air or a cloud appears to have such properties, for example, they change their form and seem to fit everywhere. The spirit could also then move through solid walls (for, how else could it enter the body of one who is asleep inside a house with closed doors and windows), squeeze through small openings, even appear instantly and spontaneously here or there (e.g., from the body of the sleeping person to the place dreamt of). It could bring a message to the one who is asleep, give advice, endow the sleeper with knowledge, or take the sleeper's own spirit someplace far away. After the sleeping person wakes up and is reassured (by others in the same room with him) that his body never left his bed, a simple way to explain how he had dreamt to be in a place far away was to assume that he, too, had a spirit. Now such superhuman, supernatural mystical abilities of the spirit of someone who died could naturally lead to his deification, especially when the individual was already influential and respectful during his life. The burial and subsequent memorial rituals for his sake were elaborate to begin with but also were glorified and mythologized further with the passing of generations and time. Since the spirit of an important dead human ancestor could be imagined as a god, his children were also gods, and perhaps their children, too. Consequently, the origin of the human race could be traced back to some mythical, deified first parents. And the glorified life stories about some of them, those that endured time and stung the human imagination the most, had gradually become the stories about great gods.

There was, for example, stormy weather once upon a time. And lightning started a fire. Foreseeing its benefits, a brave man retrieved a flaming stick and brought the gift of fire to his group. But an angered thundering, blazing sky-god punished the unsuspecting curious man by burning his hands, for the man had defied god and stole the exclusively divine fire. Still to his children he was a hero because the knowledge of fire changed their luck. And to their children he was a creator since they could trace their life back to him. Some of those children were called Greeks, and such a man was called by them Prometheus ("foreseer"); rightly then he was their favorite and most admired titan god, who, by one account, created man from water and mud.[7] As implied by Hesiod's *Works and Days*, Prometheus reigned during the fabled Golden Age, the period of first-generation humans when all—people and gods—lived once in harmony together. But his further gift of fire brought his own downfall and also ended the Golden Age for his precious creation; people began transgressing and thus were separated from their gods. Still, through rituals and ascetics a fallen soul could once again be purified and reunite the race of gods, the Orphic (a mystery religion) held (as we will see in chapter 5, "Religion and Science"). Imagining gods forming families and living initially with humans (or humans to be the children of gods) is not difficult, since some gods, it appears, had once themselves been merely humans who had been enviably admired in live heroic actions, had been zealously dreamt about in inspiring great dreams, and had been eagerly imitated in daily life.

Both Hesiod in his *Theogony* (on the origin of the gods) and Aeschylus (ca. 525/524–ca. 456/455 BCE) in his *Prometheus Bound* imagined for example that Zeus, the king-god of the Olympian gods, god of thunder and lightning, punished his cousin Prometheus for stealing fire from him and bringing it to humankind. Fire was supposed to be the exclusive knowledge of the gods. So he tied Prometheus to a rock on a high mountain, naked and helpless. By day a wild giant bird would devour his liver. But Prometheus's suffering could not end, for he was an immortal god who could not die. His liver would regenerate at night, only to be devoured by the bird again the next day, in a continuous, eternal, and painful cycle. But at the end, after thousands of years, Prometheus is liberated by Hercules. Prometheus's punishment captures the fear of inquiry (and of the gods in general). But his liberation captures the human hope, if not the desire, for godly knowledge (e.g., of fire, of search and discovery), and the admiration (even envy) for one who dared defy the gods and steal it at will. So while defiance of the divine for the sake of knowledge is horrifically punished, it is also admired and worthy of being pursued, the story implies, for ultimately Prometheus is redeemed.

FROM SPIRITS TO IDOLATRY

Of course we dream not only of the dead but also of the living and of animals and inanimate objects. Analogously, the spirit of an inanimate object—say, of a mountain or an impressively large rock, or a lion's teeth, or a bear's fur—could lead to the object's idolization. Idolatry (in Greek, the worship of an image or an object) is the worship of inanimate objects imagined to be occupied by spirits and to have mystical powers.

Spirits were envisioned to move into other bodies or objects and assume a different kind of form or life than what they came from—an idea that forms the basis of reincarnation. So a bear's or a lion's spirit could move into a man's body, enriching him with its unique animal characteristics such as strength, courage, even new knowledge. Humans grew to admire and respect various animals as rivals in the struggles for survival, and so the deification and worshipping of animal spirits were common. Idols such as stones, animal teeth or fur, bird feathers or claws, even human-made artifacts (e.g., statues such as Venus figurines and generally various images of gods) could be regarded as embodying a spirit, too and thus considered alive and worthy of religious reverence. All these constituted idolatry.

Hence an idol was worshipped because it was imagined to be occupied by a spirit—although often it was also worshipped solely for the sake of the object itself disconnected from a spirit, for with time the distinction between object and spirit might have been lost. For instance, a lion's fur might have been considered embodied (in-spired) by the spirit of a dead lion. Wearing it, a proud hunter might have believed, could empower him with the same strength and courage of the lion. So while the fur itself does not have any powers, the lion's spirit that might reside in it can bring it power. Hunters and warriors prayed to the inspired object, extolled its virtues, encouraged it to help them, and brought offerings to it for all previous times it came through. But should this idol-god fail, the worship might quickly turn to a hatred; the deity then could be threatened. Incidentally, a Greek saying "Even a saint needs a threat," which is used allegorically in the modern times, probably has its root in prehistoric practices of idolatry. The idol was threatened and beaten, with the belief it could be persuaded to serve man better in fulfilling his ambitious dreams. But if this idol-god did not comply, it was regarded as useless, soon forgotten, and replaced with a new one. When the inspired fur did not live up to the hunter's expectations, it seemed as if the spirit decided to abandon it, in which case the hunter could also decide to abandon the fur and search for another mystical idol.

Trees themselves were initially worshipped as more than mere plants.

With time, however, they were imagined as the abodes of spirits that possessed various powers such as commanding the wind or rain, even causing fertility in plants and animals. Present-day maypole festivals during May harvest, which celebrate the end of uncultivable winter and the beginning of farmable spring in the Northern Hemisphere through dancing around a tall wooden pole raised from the ground, arguably attest to tree worship in the distant past. Wells were also worshipped for themselves, but with time a fairy (a spirit) was often imagined to be living in them. Throwing precious articles as offerings and making a wish expecting that it will be granted by the fairy was a common belief in early local religions. The modern custom of making a wish before tossing coins in a fountain certainly has its origin in the worship of wells, lakes, and rivers.

Now there were times when humans worshipped things, such as various animals, simply because of some certain quality that was admired or desired. To be successful in hunting and to survive the endeavor hunters had to be acquainted with the skills of the animals being hunted: for example, the quickness of the hare, the wiliness of the fox, or the strength of buffalo. Hunters respected and often envied such skills, feelings that led them to deify animals or their spirits. What was worshipped, however, was the species as a whole and not the individual members, attributing perhaps the species' unique skills to its mythical first parents. Early humans often regarded themselves as having a certain kinship with an animal or viewed themselves as the creature's descendants. This is known as totemism. While the hypotheses that triggered totemism are several, possibilities include the realization that humans and animals possess similar skills (even physical attributes, such as eyes, legs, etc.) and face similar challenges in the daily struggle for survival. Thus by associating with an animal humans might have thought they would begin acquiring its desirable skills and manage their lives better. A sacred animal often was not eaten.

So humans and many other things in nature were thought to have a spirit. Hylozoism (the view that all matter is alive) was a characteristic of humankind's developing religious worldview. Because *inspired* objects (the idols) were of various types, including small and human made that could be carried easily around by individuals, they could become personal gods. Without a doubt this must have contributed to the birth of personal religion. Before that, religion was tribal. Prayers, rituals, and offerings were practiced by the tribe (and for the sake of the tribe) and not by the individual (nor for the satisfaction of his individual needs). Of course, another factor related to the evolution of personal religion might be merely the desire of humans to have the gods attend to their personal needs and/or ambitions: if gods could be appeased and befriended by the tribe and for the sake of the tribe, why not also by the individual and solely for his sake? This is not something unlikely to have crossed the Neolithic's mind. The development of personal religion had lessened the priesthood role in religion.

Dreams had undoubtedly played a significant role in the birth and development of religion—an idea that was speculated also by Democritus[8]—even in the explanation of death and of an afterlife. The body dies (*expires*, i.e., the opposite of *in-spire*) when its spirit decides to permanently abandon it, but the spirit lives on. Interestingly, the interpretation of dreams as divine revelations or inspirations still remains a common religious viewpoint. In general though, how were minor, local religions converted into major and powerful state or kingdom religions?

STATE RELIGION

As different tribes, villages, and cities united (either voluntarily or by force), the gods that dominated were those imposed by the victor in battle, by the strongest culture, and/or those with the most appealing stories/myths. Some gods faded away, others fused into one another, and new ones emerged. Gradually, with advancing civilization, humankind's unique regional experiences caused local religions to evolve into great and diverse state religions. For example, the most powerful city's main god becomes a kingdom's most important god, too. Because a kingdom is geographically large, the main god's principal shrine might be either in the capital city of the kingdom or at the center of each of its major cities. Such a god (much like the king of a kingdom) is therefore literally at a greater distance from the worshipper compared to local gods (or leaders) in a small village. Consequently, that god's (or a king's) tangible familiarity is reduced, but being the most important and powerful god from all the local ones (as is a king compared to local chiefs), that god's (or king's) admiration and adoration become deeper, more mystical and abstract, and so does religion as a notion.

For example, a city, with its many and diverse citizens who are no longer blood relatives, can be more successful if the sentiment of social responsibility begins to evolve. People do not help just the blood relatives in a city but one another, too, as fellow city dwellers. In the evolving complexity of an urbanized lifestyle citizens realize their codependency and the need for mutual help (encouragement, sympathy, altruism, and communal good over the individual), and so they actively pursue innovative ways to make it work, for only then can the city and thus themselves benefit and progress. Since humans were becoming socially more responsible, so were their anthropomorphic gods (who are modeled after them). And since city life promotes the collective good over the individual, a city god is imagined to be as unbiased to any one city group of people. Such a god is imagined to be more universal, as the god of all the people of the city or state. This type of thinking can be applied to large empires. Like the king of an empire who looks after all his subjects, so, too, is the empire's main god/s imagined to be. However, local gods are also worshipped, and each city has its own individual guardian god. But again, with evolving civilization, the new personal religious sentiment is now defined by the more powerful state religion for which the main god (or a smaller group of main gods) has a more universal (state) appeal. This type of universality was heralding a religious transition, from polytheism to monotheism.

POLYTHEISM AND MONOTHEISM

Since humankind modeled its gods after its own experiences and social changes, then, just as a group of people had a recognized leader, a tribal chief, or later a king, in time so, too, was the case for a group of deities. Thus the preferred god from a group of many had its status elevated, and the notion of a supreme, or father god, or mother goddess of all is born. When with time the lesser gods were forgotten or willingly abandoned, polytheism had in some cases been replaced by monotheism.

The first time this was attempted was long after the birth of religion, during the historic times when the Egyptian pharaoh Amenhotep IV (also known as Akhenaton), who ruled Egypt from 1379 to 1362 BCE, believed that there was only one god, the sun-god Aton. His view, however, did not become popular. The Egyptian priesthood and consequently their followers adhered to their age-old polytheistic views. In Judaic tradition monotheism is attributed to Abraham at an even earlier era, a few centuries before Amenhotep, but this reference is found only in the Bible.[9]

Nonetheless, for Ulrich von Wilamowitz-Moellendorff (1848–1931), a renowned scholar of his time on ancient Greek matters, he who "upheld the only real monotheism that has ever existed upon earth,"[10] was Xenophanes (ca. 570–ca. 475 BCE), a pre-Socratic philosopher (and natural scientist) who believed that "there is one god, among gods and men the greatest, not at all like mortals in body or in mind. He sees as a whole, thinks as a whole, and hears as a whole. But without toil he sets everything in motion, by the thought of his mind. And he always remains in the same place, not moving at all, nor is it fitting for him to change his position at different times."[11]

Xenophanes has been called the first philosopher of religion. How philosophy was born of religion in Greece around his time and how the ancient Greek religion proved catalytic for the birth of science, will be the topic of chapter 5, "Religion and Science." Xenophanes criticized both approaches to religion—the popular polytheistic view as well as the anthropomorphic one.

ANTHROPOMORPHISM

Xenophanes wrote: "Both Homer and Hesiod have attributed to the gods all those things that are shameful and a reproach among mankind: theft, adultery, and mutual deception."[12] "But mortals think that gods are begotten, and have the clothing, voice, and body of mortals. Now if cattle, horses or lions had hands and were able to draw with their hands and perform works like men, horses like horses and cattle like cattle would draw the forms of gods, and make their bodies just like the body each of them had. Africans [it is Ethiopians in the actual Greek text] say their gods are snub-nosed and black, Thracians blue-eyed and red-haired."[13] Actually Greek philosophers did not "eliminate divinity from the world. They preferred to depersonalize their gods."[14]

In Xenophanes's time science was in its infancy. It was a transitional period from the purely mythological and superstitious worldview to the rational and scientific one. Did the mythological worldview aid in any way the birth of science?

RELIGION AND THE BIRTH OF SCIENCE

Greeks were seafaring people, and their travels put them in contact with diverse traditions and mythological worldviews. But when one is exposed to a variety of conflicting and inconsistent mythological explanations of nature, one must decide which one from such explanations is right or whether all are wrong. For example, what god is in charge of rain, the Ethiopians', the Thracians', both, or neither? Greeks (starting with the pre-Socratics) thought none were and in general that all mythological explanations of nature were illogical, aesthetically unappealing, and plain wrong. Their next step was the search for a more universal, naturalistic, and causative worldview. They managed to do this through science by examining nature rationally.

CONCLUSION

Religion was born after humans had first concluded that the phenomena of nature were controlled by beings with higher powers, in other words, gods, and specifically at the moment that humans sought to open and cultivate a relationship with such gods. Religion became firmly established and turned into a way of life when the idea of a human–god relationship developed into a huge human need and thus a lifelong habit, and passed systematically down the generations through rituals and traditions. With this in mind, and emphasizing again that what we today refer to as ancient mythology constituted humans' earliest form of religion, how could science evolve then, from within an age-old, time-honored, sacred, often frightening mythological (antiscientific, antirational) establishment in which myths, superstition, and the supernatural were the dominant worldview for at least the first 7,400 years of civilization? Could religion have been a stimulus for the birth of science?

CHAPTER 5

RELIGION AND SCIENCE

INTRODUCTION

In the history of the world, it was religion that came first (about ten thousand years ago with the onset of urban living), followed by science much later (about 2,600 years ago). These two fields of inquiry have always shared a very intimate connection; they have been inspired by the need to understand the phenomena of nature in terms of abstract thinking. Now if "science must begin with myths, and with the criticism of myths"[1] then from some general point of view religion—the ancient mythology—may be regarded as the first and most basic type of science that had gradually been forming in the mind of the intellectually evolving human species as a means to understand all the unfolding phenomena in nature. Ultimately true science was born, but it was given birth from within a well-established and time-honored religious outlook. And such challenge, though formidable (for sacredness is not easily questioned or opposed), was remarkably overcome. Since people's religion, especially its evolution, tells a lot about the way they think—their aspirations, endeavors, hopes, passions, needs, desires, daily life challenges—anyone wishing to understand the success of this transition must search for possible hidden scientific tendencies and signs that might have existed within what appeared to be a purely religious outlook but really was not.

THE GREEKS OF MYSTERIES

At least this (the presence of subtle scientific tendencies in a society where the main outlook was the religious/mythological) was the case in sixth century BCE Greece, the era of the birth of science. The popular religion then was no longer the simplistic, placid, and happy Olympian[2] of Homer and Hesiod, where death nonetheless was a terrifying end, but rather it was the emotionally moving, ritualistically rich, and intellectually intriguing mystery religions[3] (such as the Eleusinian, Dionysian, and Orphic, called so because details of the rites were kept secret), in which the afterlife was a hopeful beginning. The ancient Greeks were generally religiously oriented, a fact evident from their art (statues, vases, and wall paintings), diverse and imaginative myths, rich pantheon of deities, famous oracles, temples, as well as their religious ceremonies, especially those of the mystery religions. But their continuously evolving religious outlook contained subtle elements indicative of their forthcoming transition from mythology to science. What exactly were these elements?

On the one hand, in the Olympian religion, nature was the playground of capricious, often immoral gods (although moral ones as well), with people and nature completely at their mercy. Human knowledge, and even actions, were believed to be decidable and controllable by the gods. Zeus, for example, could choose to strike with a thunderbolt, Eros could cause someone to fall in love, Apollo could heal as well as bring on a plague, and Artemis could teach hunting skills. Demeter could instruct in agriculture while the Muses would inspire people with the knowledge of the arts and sciences. Like humans, gods, too, had specialized professions. So nature and people's own future were entirely up to the goddess Fate and all the gods in general. Excluding immorality, it was in a way like a Disney Tinker Bell movie: different processes in nature (e.g., the changing seasons) are carried out by different types of fairies of specialized professions, like the tinker fairies, the winter, warm, water, garden, light, frost, plant, animal fairies, and so on.

To the contrary, the mystery religions (especially the Orphic) promoted an entirely different outlook. The worshipper was an intellectually and ethically evolving individual with greater personal responsibility for his own future including the *afterlife*. He held a deep conviction that he had certain control over his knowledge and actions; he was hopeful that, through mystic rituals and asceticism, divine immortality and wisdom were also humanly achievable and thus not an exclusive privilege of the gods. Such change of religious attitude was indicative of an unsettled, curious, open mind, one unsatisfied with the passive and strictly dogmatic mythological worldview of the Olympian religion as a means to understand nature, life, even death, and in search of something more profound and meaningful—something rational, natural, objective, universal, even humanly controllable.

In particular, during the Mysteries the worshipper ate, drank wine, danced, rejoiced, sorrowed, and felt a divinely inspired madness; he went to the extremes of frenzy hoping to experience *passion* (a physical and an intellectual suffering), *ecstasy* (etymologically, the release of the soul from the dependence of the body—recall a spirit/soul could enter a body, i.e., in-spire it, but also escape from a body, that is, undergo ecstasy), and ultimately *enthusiasm* (unification with the honoring god). With such intense emotional arousal the worshipper behaved in ways different from those of the everyday: free from the daily inhibitions and oppressions he was his true self. Simply put, during the Mysteries the believer took matters in his own hands and tried ritualistically to feel and act just as he thought his god did and hoped that life would imitate the ritual, an idea as old as religion itself but now with a new twist, an intellectual expectation by the worshipper. This ritualistic emotional enthusiasm (this potential unification with a god) led to the belief that everything divine—immortality, omnipotence, bliss, even the godly mystic (secret) knowledge—was humanly attainable, *yes, even the godly knowledge* (the secrets of nature)! The believer, of course, first had to achieve absolute purification of his fallen soul through a system of complex sacraments. Once, according to Hesiod's *Works and Days*, during the Golden Age, the first-generation people had pure souls and were allowed by gods to live with them in bliss. But people sinned and were separated from their gods. Still, until they join them again, divine aspects, such as godly knowledge, could begin to be experienced by the worshipper, so he believed, at least ceremonially during the Mysteries, via proper soul purification, prayer, the reconciliation of the gods with offerings and sacrifices, and perhaps through sympathetic magic: "If I manage to feel and act the way I think my god does, I hope I will then begin to become like one." Anthropomorphism, recall, evolved from imagining merely the *gods* in the image of man, to imagining zealously *man* in the image of the gods, a radical reversal in human psychology indeed: that is, man aspired to be like the gods: almighty, all-knowing, and immortal.

PHILOSOPHY BORN OF RELIGION

But such belief *did not* remain only ceremonial; to the contrary, it was gradually affecting even the daily life of the Greeks of Mysteries. In particular, the Orphic, who aimed for spiritual drunkenness (so wine was used only symbolically), believed that the exercise of a proper ascetic way of life could ultimately purify one's soul, elevate it to the otherworldly heights of the gods, and thus release it from its cycle of constant deaths and rebirths (thought to occur through metempsychosis). He believed, that is, that he could reach the state of apotheosis (become godlike in every aspect: power, wisdom, happiness, immortality) and thus be allowed once again the honor of eternal bliss and absolute knowledge alongside his gods—the goal of the mystery religions, which were after all religions of salvation. And so death was no longer the terrifying, hopeless, and gloomy place of Hades (etymologically, of the "Invisible," thus supposedly unknowable, where the soul is powerless and in oblivion), as in the Olympian religion, but rather a hopeful state of existence at the Elysium (the Island of the Blessed, of heroes and gods, where the soul could be immortal, conscious, free, and with divine wisdom). Interestingly, modern scientists, through their efforts to figure out the laws of nature—of creation, one might say—have in a sense similar one of the Orphic aspirations: to know the mind of the Divine.

That offered the Orphic, and generally the Greeks of Mysteries, hope for the future (including the afterlife), and in their searches for ways to satisfy their changing religious needs and fulfill their goal, they were stimulated for a philosophy of life (for the search of a deeper truth, about existence, death, nature); a philosophy born of religion. *That* was an intellectual turning point for if "philosophy . . . is something intermediate between theology and science,"[4] then the discovery of science was the expected logical next step.

This however took place only when the religiously, philosophically, and morally evolving ancient Greek, who desired the longed-for divine knowledge at any cost (through the risk of a Promethean-like retribution in the Olympian religion or through rituals and/or asceticism later in the mystery religions), grew impatient waiting for the gods to decide to (literally) *in-spire* him. And he had, at some point, realized that mystic rituals, sympathetic magic, reconciliations of anthropomorphic erratic gods in hopes for an *inspiration* (of knowledge "handed out" via godly revelation), and generally religious dogmatism or asceticism, were not working out. But what could work was simply the free and rational critique of both, nature *and the worldview of one another*—that is, learning, he thought, should come from thinking persons themselves, not from Fate or rituals, and so finally and without fear he began to imitate the actions of his favorite cultural hero Prometheus, who stole the godly secret of fire at will. And when the ancient Greek tried it he found out it was the only thing that worked. *That* gave birth to science. This "rational critique" attitude, in search of the truth, is of course useful not just as a way to do science but as a general way of life as Socrates would have attested.

Additionally, there was neither an official religion in any of the Greek city-states, nor was there a written religious corpus to which a city or an individual had to conform, nor an organized priesthood to impose particular dogmas, rituals, or a lifestyle. Consequently (and unlike the analogous case of other civilizations), Greek religion not only did not interfere with attempts for a non-mythological worldview but in subtle ways it promoted them. To understand the nature of the world they lived in humans had to think for themselves, an action, which I think, unavoidably forced them to study nature rationally. In fact, because there was no theocracy in Greece, neither political law nor morality came from the gods (or priests or kings) but from the people themselves, the type of people whose idiosyncrasies are studied in the next chapter, and who also invented democracy and pursued moral philosophy.

CONCLUSION

Human knowledge was initially thought decidable only by the gods, then hoped for through rituals and/or asceticism, but at the end proved obtainable only through one's own reason. And so in the vastness of existence, earth was only another planet, the sun just another star, neither was a god or the center of the universe. All things (including human beings) were composed of the same primary substance/s and obeyed common natural laws (or one grand law), which should be describable mathematically. Advanced life-forms evolved from simpler ones, and neither illness was caused by demons nor eclipses by gods.

The idea of intellectual progress captivated the Greeks so intensely that for the sake of knowledge they risked angering their gods, "stole" their fire and their (nature's) other secrets, and gave birth to science. But why were they able to do so?

CHAPTER 6

THE BIRTH OF SCIENCE

INTRODUCTION

What led to the intellectual transition from mythology to science 2,600 years ago, and why did this cultural phenomenon begin to unfold first in ancient Greece? The factors that are generally accepted as having created favorable conditions for such a transition were geographic, economic, religious, and political. In this chapter I add to the usual list of factors three new ones: the power of the Greek language, the effect of making a habit of scientific thinking, and the ancient Greek idiosyncrasy. The study of these factors will help us understand why science may be regarded as being born out of the Greek civilization. We begin by reviewing first the commonly accepted factors.

GEOGRAPHY, ECONOMY, RELIGION, POLITICS

1. Geographic: Locally, a landscape of natural boundaries—mountains separating cities, and the sea separating islands—helped in the formation of relatively isolated city-states (a thousand or so) and promoted intellectual diversity. Diverse ideas were ultimately shared and improved when people moved and interacted. Globally, the crossroads location of Greece exposed its people to ideas of other great civilizations from Europe, Asia, and Africa. Moreover, Greece's long coastline and many surrounding islands resulted in the establishment of coastal- and island-cities and made Greeks seafaring people. But their sea adventures aided them in demythologizing the phenomena of nature and stimulated them in conceiving rational explanations.

2. Economic: Average people became technologically inventive to better

their lives. And even though technology is not science but the application of it, technology can lead to abstract theorization about how it can be improved and consequently the discovery of laws of nature upon which technology is based. On the other end of the economic spectrum, well-to-do people used their leisure to philosophize and theorize.

3. Religious: Contrary to theocracy and hierarchy, which impose dogmatic thinking, restrict inquiry, and impede progress, religious freedom in Greece allowed for contemplation of diverse views and created a potential for betterment.

4. Political: Social freedom and democracy prompted free debates on just about everything, resulting in the conception and exchange of new and improved ideas.

Because these factors have been contemplated extensively in the literature,[1] my focus in this chapter will be on the three influences that have not been sufficiently appreciated.

The first influence is the Greek language itself. The notably communicative nature of ancient Greek helped in the conception and diffusion of knowledge in the most efficient way possible. While the first alphabet was Phoenician, the first alphabet to contain vowels was the Greek. With this innovation Greek became the first easily read and written language of the world, and the facility of written Greek became significant in the evolution of ideas and the birth of science. The second influence is simply the force of intellectual habits. Using ideas from the theory of biological evolution, I will argue that the good habit of the pre-Socratics to practice science imposed an epistemological kind of natural selection by promoting intellectually favorable environments where learning science could continue to happen and new scientists could exist, thrive, and become abundant, contributing therefore to the constant development of the scientific outlook at the expense of the mythological one. The third influence is the idiosyncrasy of the ancient Greeks: they were rational, passionate, and excessive. But it was the proper moderation between passion and logic that allowed them to become *creatively* excessive.

LANGUAGE

What first interested me in investigating the language factor was a brief statement by Nobel laureate Bertrand Russell (1872–1970): "The Greeks, borrowing from the Phoenicians, altered the alphabet to suit their language, and made the important innovation of adding vowels instead of having only consonants. There can be no doubt that the acquisition of this convenient method of writing greatly hastened the rise of Greek civilization."[2] Although the Greek language is usually not regarded as a factor that created favorable conditions for the birth of science, I will argue that its influence was subtle but profound and thus cannot be overlooked.

I will first lay the groundwork, in the next two subsections, by contemplating the general effectiveness of language in human survival and intellectual evolution. Then, in the third subsection, I will link directly the influence of the ancient Greek language on the birth of science.

The Sound of the Fittest

From the family tree of biological evolution the more anthropomorphic primates (the hominids, species that are more human than ape) are a family of species whose first member is believed to have evolved some seven million years ago.[3] Its two most recent members, who are relevant to our consideration of the effect of language on both our physical survival as well as our intellectual evolution, are the evolutionary cousins *Homo neanderthalensis* (Neanderthals) and *Homo sapiens*. Both species are thought to have evolved only about two hundred thousand years ago, with Neanderthals preceding. So at one time the two cousin species shared the earth and possibly interacted.

Neanderthals are our closest genetic relative. Physically, in some very general terms, the two species were not that different—a visit to either the American Museum of Natural History in New York City or the Smithsonian National Museum of Natural History in Washington, DC, where artists and scientists reconstructed Neanderthals' possible appearance from various findings including fossilized bones, will convince anyone of this.[4] Neanderthals were short and stocky with a more elongated skull, and *Homo sapiens* were taller and thinner with our characteristic high-dome skull. Furthermore, because the two cousin species share several brain similarities, it has been speculated that they were of comparable intelligence. This hypothesis, however, is the subject of current contention.[5]

With such general similarities, both species would have been expected to survive, but only *Homo sapiens* have managed. Unfortunately, between twenty-five thousand to thirty thousand years ago, Neanderthals became extinct. The theories for their extinction vary and are hotly debated. The cause might be just one or a combination of several, such as climate change or an isolated existence in clans, which might have resulted in limited exchange of ideas and thus a slower rate of intellectual progress than needed for surviving life's constantly changing challenges.[6]

One theory of extinction relevant to our discussion on the importance of language in survival is Neanderthal-human competition. Such competition might have been destined to be biologically unequal. For through a mutation (a purely chance change in the genome, the hereditary substance) *Homo sapiens* were accidentally gifted by nature with an anatomy comprising a more efficient larynx that could produce a richer variety of sounds, creating the potential to develop a relatively more advanced language than that of Neanderthals. This must have aided in the general survival of *Homo sapiens*. But some experts hypothesize that in a more specific way, this also might have been a contributing factor in our survival at the expense and general extinction of Neanderthals, by giving us a competitive advantage. It is probable that a better capacity for language enabled *Homo sapiens* to communicate essential survival skills such as hunting and gathering, making and refining tools, finding shelter, making friends, living together in extended social groups, forming alliances, trading, and generally learning from each other.[7]

Consequently, *Homo sapiens* developed a better understanding of the world around them and achieved an intellectual edge over their cousins the Neanderthals in all aspects of their competition. But during the early competitive environment of predators, limited resources, and in general a nature where survival was of the fittest, such intellectual advantage achieved through language skills (regardless of how primitive initially) made a difference between life and death. Thus, this theory holds, *Homo sapiens* secured their survival by overpowering and driving their own cousins to extinction.[8]

Language is a useful skill, possibly the most powerful of humankind, not only in the struggle to survive but also in our efforts to thrive and live fully. Language controls the flow of information and creates the potential for knowledge. But how rapidly does intellect evolve with the influence of language, especially an evolving language?

Biological versus Intellectual Evolution

The effectiveness of language can be appreciated further by comparing the time required for the extremely slow biological evolution of the anthropomorphic family of species with that of the immeasurably faster intellectual evolution of the only species that managed it, *Homo sapiens*, and trying to explain the reason for such a huge time difference.

Specifically, on the one hand, the biological evolution of this family describes a seven-million-year process (from its first member species, the *Sahelanthropus tchadensis*, believed to have evolved about seven million years ago, to its last and only extant member, *Homo sapiens*, who evolved about two hundred thousand years ago), but on the other hand the incredible intellectual evolution of this entire anthropomorphic family is due exclusively to the achievements of just this last member species. And depending on what might be regarded as advanced knowledge, such evolution can be condensed to an unbelievably small time interval. It could be thirty thousand years (since splendid art was painted on cave walls by Ice Age cave dwellers); or ten thousand years (since the end of the last glacial period, which roughly coincided with the transition from the lifestyle of hunter-gatherer to farmer, urbanization, and consequently the birth of civilization); or about five thousand years (since the beginning of written history when Sumerians in Mesopotamia invented the first type of writing in the world at 3100 BCE); or 2,600 years (since the birth of science); or some five hundred years (since the rebirth of science with the contributions of Renaissance astronomer Nicolaus Copernicus [1473–1543]); or three hundred years (since the Industrial Revolution); or, even more impressively, a mere few decades (since the discovery of the computer)!

To emphasize the unprecedentedly rapid cultural and intellectual evolution of the last few decades, I recall a comment by noted science fiction and popular science writer Isaac Asimov concerning the conclusion of his *Chronology of the World*: that, while his initial intention was to write the entire history of the world, from the big bang (the event that gave birth to the universe, as we will learn in subsequent chapters) to the date his book would be completed (a fifteen-billion-year period for him then), he was finally forced to conclude it with the events of 1945 instead of 1989, the book's completion date, falling short of his initial goal by a mere forty-four years.[9] And he explained that the reason was that the changes brought about by the evolving human culture between 1945 and 1989 were so many, rapid, and universal that to be effectively described would require their own book as extensive as *Chronology of the World*! He said this in 1989. Can you imagine what he might have said today, especially after the explosive evolution of the Internet? In the 1980s the Internet was just being born.

I concur with Asimov's assessment and base my understanding on the evolving notion of language itself. For from the simple sounds and symbolic cave art of the distant past to the rich languages, modern mathematical symbols, and sophisticated electronic communications of the present, language has been evolving diversified and creative new modes that allow for better conception, dissemination, and improvement of knowledge and have thus been transforming our species intellectually faster than ever before.

More precisely, with time and as a consequence of advances in mathematics, science, and technology, the notion of language has been broadened. Mathematics has added a versatile variation in symbolic and quantitative communication, while science has enhanced our imagination and invented naturalistic and rational interpretations of nature, and technology, mainly after the invention of computers (especially their interconnection via the sociologically innovative Internet), has enriched communication through myriad modes, including ones that affect all people of this planet and potentially intelligent beings of other star systems. Traveling at the speed of light, a radio signal transmitted from the Arecibo Observatory in Puerto Rico in 1974 has as its destination the globular cluster M13, a group of some three hundred thousand stars in the constellation of Hercules twenty-five thousand light-years away from us.[10] The signal's coded information about us can be easily decoded if intercepted by an intelligent alien life-form.

For millennia, the idea of language has included more than gestures and sounds. Knowledge can be recorded many different ways and in places other than the human brain. Thus while we no longer need to remember everything, everything can still be remembered because the knowledge of the past is readily available and therefore accelerates the rate of progress. One can learn the accumulated knowledge of millennia by simply reading a book!

And all of this can take place because we are anatomically able to speak sounds, are instinctively curious to develop them into coherent language, and are intellectually successful in habitually passing on such great skill to our offspring. And such is the power of language: it is a skill for rapid and extraordinary intellectual bursts! Unquestionably language has been aiding in the advancement of science. But did it aid in its birth?

Ancient Greek Language and the Birth of Science

The evolution of the Greek language has been a huge topic for scholarly research. While I admit ignorance on such an immense linguistic field, I also know the generally accepted facts about Greek's extraordinary richness, such as a plentiful vocabulary, thorough and rigorous grammar, diverse phonology, and successful orthography (i.e., spelling), all of which contribute to the language's highly expressive and communicative nature. This distinct nature leads me to contemplate the connection of the language and the birth of science. But first some history.

Spoken since at least 2000 BCE and written since at least 1400 BCE (not yet with the Greek alphabet, which evolved a few centuries later), Greek is one of the world's oldest recorded living languages and the longest documented from the Indo-European family of languages where it belongs. Phoenicians invented their alphabet around 1050 BCE. Modeled after that, the first true alphabet containing vowels was invented by the Greeks around the eighth century BCE. It was rapidly diffused throughout ancient Greece. With this innovation Greek became the first most easily read and written language of the world. This is so because alphabets are phonetic: each different sound of a language can be represented with a unique symbol, and thereafter symbols can be combined to write and sound all the words of the language. Therefore, with an alphabet every language can be written and read relatively easily. In contrast, a pictographic writing system, in which a picture represents a word or phrase, is more complex. The success of the Greek alphabet is also indicated by the fact that after some three thousand years, Greek is still written with the same letters that served as a basis for the Latin letters, and which, in turn, have been the basis of several modern languages. While Greek has been evolving, its overall identity has been basically preserved. Greek has remained relatively the same language until today, a rather rare but not accidental linguistic phenomenon. Parenthetically, part of the explanation for this might be that the Greeks value so highly the written works of their ancestors that their references to them kept the essence of their language unchanged.

Because of its simplicity, the Greek alphabet assisted in making the good habit of literacy accessible to all in ancient Greece. By the fifth century BCE every male citizen was expected to know how to read and write.[11] "The Greeks founded such an eminently literary culture."[12] Such widespread literacy undoubtedly accelerated progress. In contrast, the complexity of some other cultures' writing systems, often combined with their theocratic (and hierarchical) political systems, made writing the nearly exclusive privilege of priests and professional scribes and not the populace, a situation arguably unfavorable for developing science. Greek literature begins with Homer's monumental epic poems the *Iliad* and the *Odyssey*, dated by consensus from around eighth century BCE. However, their surviving present form is at latest from sixth century BCE, the century when Greek philosophy, science, and mathematics began. From around 700 BCE are Hesiod's poems *Works and Days* and *Theogony*. All four works were significant in educating Greek youth.

These chronological facts indicate that Greek's relatively early growing richness was present by the time of the birth of science in early sixth century BCE. This evidence, together with the fact that Homer was from Ionia—which was also the birthplace of the first scientists (the natural philosophers Thales, Anaximander, Anaximenes, Xenophanes, Heraclitus, Pythagoras, and Anaxagoras)—proves that science was born at a place and time where language was already advanced enough to aid the evolving scientists in the clear articulation of their theories.

This is a significant conclusion, for it links directly the positive influence that the ancient Greek language had on the birth of science. Greek had equipped the early philosophers with the skills for conceiving and formulating their abstract thoughts, clearly expressing their minds, and efficiently converting their raw intelligence to systematic, rational, transferable, and debatable knowledge. Without such a productively expressive language, their scientific theories would have remained unrefined, perhaps not even conceived in the first place.

A poor language reduces not only the ability to express oneself but also the potential to learn from others, for if neither we nor others can think and communicate clearly, we can neither influence nor be influenced. And the poorer the speech and writing acquisition are, the more inadequate the cognitive process becomes.

It seems no accident that the Greek language had been maturing roughly simultaneously with Greek thought in philosophy, science, and mathematics. The sounds and symbols of a communicative language could create clearer thoughts, which could then refine further the language in a continuous interactive cycle of the evolution of both. But mathematics is also a form of language, most particularly the language of science. So while by language we usually mean the communication in terms of sounds and written words, mathematics has tremendously empowered such new ideas by utilizing numbers, equations, complex diagrams, and abstract concepts. Mathematics helped to develop abstract thinking and to quantify science. In turn, science enhanced technology, which in turn enhanced both science and mathematics, in a mutually productive process. Now since mathematics adds a valuable extension to the definition of language, can we find yet another link between language (specifically the mathematical) and the birth of science?

During the rise of Greek civilization, science and mathematics were driving each other and evolving simultaneously. The first natural philosophers were both scientists and mathematicians. Russell has said, "The preeminence of the Greeks appears more clearly in mathematics and astronomy than in anything else."[13] Mathematics was a skill that enabled them to conceptualize and more easily make their scientific theories rational; but equally important, their unprecedented physical intuition concerning the workings of nature aided them in advancing mathematics and thus the language of science.

Thales (who flourished in early sixth century BCE) was also a geometer. After him, the Pythagoreans were superb mathematicians and the first to implement the mathematical analysis of nature, a practice of vital significance in modern theoretical physics. Physicist and Nobel laureate Erwin Schrödinger (1887–1961) argues that what guided Democritus (the last of the pre-Socratics) in conceiving his atomic theory of matter was his deep insight of mathematics.[14] In fact, the most enduring discoveries from Greek science of antiquity were by natural philosophers who were also accomplished mathematicians.

The mathematical knowledge that was a common characteristic among most of the pre-Socratics seems to indicate that science could not have been born by persons who did not know the language of mathematics. This is yet another conclusion that links directly the positive influence of the ancient Greek language, which in its broader definition includes mathematics, with the birth of science. Without a doubt, the clear conception and coherent expression of complex ideas were made easier by the communicative nature of the prolific ancient Greek language. But could the scientific birth have survived and matured without good habits?

HABITS

A combination of factors aided in the emergence of the first natural philosophers and in the transition from mythology to science. This unfolding new knowledge gradually advanced, and spread, grew popular, respectable, and of practical value but also abstractly meaningful and satisfying. Among the Greeks generally, seeking knowledge became a way of life, a scientific habit that characterized the culture. And even though acquired properties such as knowledge and skills are not biologically inherited, habits (such as practicing science) and behaviors (such as a desire to advance the scientific outlook) associated with such properties are transmittable culturally through teaching and can still change the environment in complex and subtle ways. And in turn, through the process of natural selection from biological evolution, the environment can influence a species by controlling the direction of its evolution.

Specifically, the good habit of the first natural philosophers to practice science imposed an epistemological kind of natural selection by promoting scientifically favorable environments where learning could take place and new scientists could exist, thrive, and contribute to the constant development of the scientific outlook at the expense of the mythological one.

But since my goal is to explain the critical role that habits play in our intellectual evolution from the point of view of biological evolution, I first need to discuss further the notion of natural selection imposed by a habit.

Imposed Natural Selection

The process of biological evolution of the species begins with a mutation (a random alteration in the genome that can result in a new hereditary characteristic) and continues with the mechanism of natural selection (which says that inheritable characteristics that are also environmentally favorable become more common in successive generations; hence it describes the role of nature in the preservation or extinction of a species). Natural selection can proceed as a consequence of a variety of environmental influences such as chronic periods of coldness, hotness, dryness, wetness; the eruption of a super volcano; changes in atmospheric composition; an asteroid-earth collision; radiation from the sun; or a supernova explosion.

But natural selection can also be imposed by the habits of a species; after all, species are part of nature and their actions affect it. In this case, if some members of a species already have or develop an inheritable trait (a mutation) that is favorable to a kind of environment created by a habit—either their own or another species'—then they will be naturally selected. This means that these members will begin growing up more easily, prospering, adapting, preferentially reproducing, and becoming more abundant in such an environment that is friendly to their rare trait. Assuming the habit persists, in time the species will gradually evolve to the point that most of its members possess the genetic trait favorable to the environment created by the habit. Let's look at two specific examples:

1. Microbes: While on the one hand a moderate use of antibiotics can be beneficial to our survival by killing myriad common but still harmful microbes, on the other hand a habit of thoughtless overuse of antibiotics can promote the evolution of rare but much more dangerous microbes (superbugs) that are resistant to the antibiotics we use. Natural selection, in this case imposed by the habit of overuse of antibiotics, can make common population characteristics

rare (common microbes get killed) and rare ones common (mutant microbes resistant to our antibiotics get multiplied).

In the microbes example, the habits of one species, humans, can impose natural selection onto another species, microbes. Humans actually have since urbanization been imposing natural selection onto several species; the plants we have been domesticating, the animals, even species with which for one or another reason we must systematically interact—like the aforementioned microbes.

Equally fascinating is another fact of evolution: that the habits of a species can also impose natural selection onto itself—for example, humans can impose natural selection onto themselves! Before I discuss humans, let me first discuss briefly the birds.

2. Birds: With the desirable genetic trait of wings, birds avoided many predators only when they began habitually using their wings for flying and building their nests high up in trees. Such a habit imposed natural selection by creating an environment that selected and promoted even further the evolution of birds that could fly the best. With time these skilled high fliers became more abundant, while birds that could not fly proficiently became rarer.

The mythological worldview was once popular and the scientific rare, but since the birth of science their status has been gradually reversing, a fact that is contributing to the overall intellectual evolution of the human species. This observation brings me to a hypothesis, to be introduced in detail in the subsection below, that the good habit of doing science imposes an epistemological kind of natural selection that gradually selects people with scientific and, in general, intellectual tendencies. Such a habit not only secured the safe birth of science during the critical early stages, 2,600 years ago, but also has since then been contributing to the overall evolution of the scientific outlook at the expense of the mythological.

Erwin Schrödinger in his *What Is Life? & Mind and Matter* gives a detailed analysis of (a) how behavior in general influences natural selection and thus the process of biological evolution and (b) how our invaluable characteristic of intelligence allows us to conceive and implement incalculable choices and so both our behavior and consequently our evolution depend on us, at least to a certain degree.[15] Thus, he argues, our evolution does not depend solely on chance mutations. This is an encouraging but also challenging prospect. Based on these two points, he speculates on the potential of intellectual degeneration in our species. Below I will focus on an analysis exploring the opposite: how practicing science habitually has imposed an epistemological natural selection and has been influencing positively the evolution of the human intellect. (This is not to say, of course, that it could not influence it negatively.)

Habits Influence Evolution

Since habits can impose natural selection and cause biological evolution, they can also cause intellectual evolution, for our organ of intelligence, the human brain, is just one of many body organs known in biology to have been evolving. So good human habits can cause a biological evolution of the brain and consequently create the potential for intellectual evolution. Just as birds that could fly the best were selected in the environment where flying became a bird habit, it is not unreasonable to suppose that the developing good habits of the pre-Socratics to understand nature scientifically instead of mythically imposed natural selection by creating favorable environments for new scientists to flourish, multiply, and evolve. In short, the good habit of doing science set up an epistemological environment where the scientific man, in general the intellectual man, or, to say the least, the man of scientific appreciation, is favored and thus naturally selected.

The pre-Socratic era was the first habit-forming period for science. Specifically, several good habits of pre-Socratics—people who were keenly observant; curious; skeptical; investigative; unconventional; open-minded; free-spirited; innovative; rational; passionate; critical; philosophical; eager to speak, write, and debate; truly scientific; and generally epistemological (interested in knowledge of diverse fields)—have been inherited by succeeding generations, from their place to another, from the few to the many, from then to now, from ancient Greece to the rest of the world and seem to have been imposing an epistemological kind of natural selection by promoting scientifically favorable environments.

These good habits have therefore contributed systematically to the formation of an ever-improving scientific worldview at the expense of the mythological one and consequently advancing our overall intellectual evolution. For truly epistemological individuals have found such environments intellectually appealing, rewarding, welcoming, and increasingly more adaptive, so much so that today's humans have evolved to become intellectually superior to our ancestors, in fact to any other species known. Hence the kind of environment set up by the good habit of learning (or flying, in the bird example) favors, through imposed natural selection, the increase of those interested in learning (or flying).

A Good Genetic Trait and a Good Habit

So, having an environmentally desirable genetic trait (for example, a larynx, a complex brain, legs, wings) from which a good habit can develop (language, learning, walking, flying) is only one required element in the struggle for survival. Using the trait systematically and habitually is the second required element. For only then can the trait influence the environment via imposed natural selection so that the members who have it can be naturally selected even further and consequently increase their chances for survival and betterment by becoming environmentally more fit. In the evolution of birds, for example, those that did not take up the good habit of flying, despite their anatomic ability to do so, generally have less chance to survive attacks by predators. On the other hand, the expert fliers that use their wings proficiently tend to flourish. Varieties that are not environmentally favorable (such as birds that despite having wings are not using them) can become rare and perhaps extinct. But even if they do manage, not following good habits makes their existence much more vulnerable.

Extending the logic of the bird example into the realm of humans and their intellectual habits, it can be argued that a greater chance to flourish belongs to those who use their brains intelligently and try to develop good learning habits, such as attending school, in order to keep up with new challenges and opportunities of a fast-changing environment. In this bird-human analogy there is, however, an important difference in favor of humans. We have a far greater level of intelligence. We have a choice of how to behave, and since behavior influences evolution, through our choices we, too, contribute significantly to our own evolution. Specifically, through chance mutations we were endowed by nature with the raw intelligence of an anatomically complex brain, but what also plays a critical role in its development is our conscious choice of using it productively. Again, I assert that practicing science habitually has imposed an epistemological kind of natural selection, has changed the intellectual environment, and has allowed us to realize our potential to live up to our name and become truly *sapiens*: wise. Starting around the sixth century BCE, science, philosophy, and mathematics were gradually becoming a way of life in ancient Greece, increasingly systematic and habitual, not just for a few individuals in a few places but for whole populations in many cities, especially in the education of the young, creating therefore a better chance for this way of life to be passed on to future generations and to people in new places. Since then, because learning science has gradually become a significant skill in life, the numbers of those with the mythological worldview have been decreasing, while those who are scientific (rational) have been multiplying. This development is comparable to the declining numbers of the rarer birds that cannot fly proficiently and the growing numbers of the numerous expert high fliers that flourish in an environment where flying became an important skill for survival.

Practicing Science Habitually

The notion of a habit was crucial in the development of Greek civilization. There is absolutely no reason to believe that, before or after the Greeks and independently of them, others in the world would not have conceived a scientific idea about nature or a mathematical demonstration of some theorem. In fact, the proof of my opinion is that all kinds of people from all over the world do, or can learn to do, science and mathematics. But if something profound like practicing science had not happened habitually, such good skill would not have spread and perhaps soon would have vanished without significant effect on society. But evolutions occur when a phenomenon leaves a mark on the environment. A significant reason for the rise of Greek civilization was that philosophy, science, mathematics, and the love of free thinking—and consequently democracy, which aided in the preservation and continuation of the good habit of practicing science—all evolved into a good habit that has been influencing the world ever since.

Hence, it must be acknowledged that in addition to a variety of more commonly understood factors, an important element in the intellectual transition from mythology to science in ancient Greece was that the Greeks, starting with the pre-Socratics, pursued their new ideas in a systematic, persistent, and habitual manner. And in their explanations of how nature works, the pre-Socratics applied exclusively the scientific outlook for all phenomena of nature. For them every phenomenon had a natural cause; thus supernatural interventions were ruled out. This is the reason why Greece, from roughly the sixth century BCE, may be considered the birthplace of science. Had the pre-Socratics explained some phenomena naturalistically but others supernaturally, this birth of science would not have occurred. From the Greeks the scientific outlook spread and today is a way of life and a culture, a human culture.

IDIOSYNCRASY

Was it geography that broadened the Greeks' intellectual horizons, or was it their curious, adventurous *open* mind that led them to see the world (nature, their own culture, and that of others, as well as human itself) with a critical eye—freely questioning and analyzing it for, like Socrates, they thought "the unexamined life is not worth living" and so they sought for more objective, coherent, universal, and timeless truths about nature (including humans)? Was it tools (or free time) that sharpened their intellect, or their intellect that sharpened the tools and pursued leisure time to philosophize? Was it the absence of theocracy (which imposes dogmatic and mythological thinking) that promoted free and rational thought among the Greeks, or was it their tendency for free and rational thought that opposed the formation of theocracy? Was it democracy that encouraged free public debates and open dialogues, or the Greeks' critical mind and love for freedom of speech and independent thinking that contributed to the invention of democracy? Was it the advanced Greek language that aided their thoughts (to be clearly conceived, expressed, and disseminated), or their thoughts (formed out of love for self-expression, rhetoric, dialectic, literature, and inquiry—for *any* subject was to them open for contemplation, debate, and criticism) that aided the advance of their language? Could causes be mistaken as the effects? Although not absolutely, to some degree, yes, they could. If so, then in the rise of Greek civilization the idiosyncrasy of the Greeks cannot be overlooked. Well, then, what kind was it?

Rational, Passionate, Excessive

Plato's dramatic chariot allegory (from his dialogue *Phaedrus*) captures the essence of the ancient Greek idiosyncrasy.[16] According to the allegory, a charioteer, who represents (personifies) the human soul, tries to drive a two-winged-horse chariot to the proper destination, that of truth. But the journey is not easy because the horses, which represent the complex twofold nature of the soul, have excessive energies but competing tendencies and pull forcefully toward their separate ways. One horse represents the tendency of the soul to conform to orderly reason, the other to surrender to defiant passion. Reason can tame passion, but passion can cloud reason. Reason alone may lead the soul down a safe but dull path, while passion alone may lead it down a risky but uncommon one. Nonetheless, it is not the one or the other tendency that is the good or the bad for the soul but their combination. Destination Truth (the discovery of something extraordinary) can be reached only by harnessing properly both horses' excessive energy. For not only reason is natural and needed but so is passion. The ancient Greek idiosyncrasy, like the charioteer, was a synthesis of the rational and the passionate nature. And each nature was excessive and at war with the other, like the mighty horses that pull vehemently against each other. But their delicate harmonization, whenever it was managed, is what steered the Greeks to creativity.

To eliminate the backward passions of magic and superstition and to liberate humankind from its dependence on irrational supernatural forces, the ancient Greeks rationalized nature and gave birth to science and intellectual freedom. To fight the corrupt passions of tyranny and autocracy they figured out the law and gave birth to democracy and political freedom. To face the blinding natural passions of human soul they contemplated morality and criticized themselves, each other (in fact, welcomed the latter), and the policies of their cities—and all in public—while giving birth to moral philosophy and philosophical freedom. It took a rational nature for them to philosophize, politicize, and do science; as well as to, or actually try to, "know thyself" (to have self-knowledge, to be conscious of their limitations and potentials) in order to take responsibility and control over their own understanding, actions, and future (away from tyrannies, hierarchies, superstitions, dogmas, and even their own passions). But also it took a passionate nature for them to aspire and choose freely in spite of consequent suffering, even death (as the heroes in their tragedies); to choose to die for freedom by battling internal oppressing tyrants and external despotic kings (as the heroes of their wars); and to despise powerful capricious gods and age-old mythical traditions and mysticism for the sake of religion based on philosophy and of knowledge based on rational inquiry (as the heroes of their everyday life). But none of these pursuits, neither religion nor philosophy nor science nor democracy nor freedom, were achieved through reason or passion alone. It was not just the one or the other that was the good but their combination. The Greeks were passionately rational but also rationally passionate.

Russell has written of the Greeks, "Without the Bacchic element [passion, ecstasy, enthusiasm, frenzy, impulse, spontaneity, suffering, sorrow, joy, a divinely inspired madness, the liberation from the constraints, agonies, and pressures of the everyday, and the expression of one's true self, all feelings aroused at the Bacchic, the Dionysian Mysteries], life would be uninteresting; with it, it is dangerous. Prudence versus passion is a conflict that runs through history. It is not a conflict in which we ought to side wholly with either party."[17] Plato (in his dialogue *Phaedrus*) believed something similar: "He who, having no touch of the Muses' [divine] madness [passion] in his soul, comes to the door and thinks that he will get into the temple [become a prophet] by the help of art [by reason alone, by mere knowledge of the art of prophecy]—he, I say, and his poetry [sane prophecy] are not admitted [for they lack a touch of a divinely inspired madness, passion]."[18] Russell continues, "A large proportion of them [Greeks], were passionate, unhappy, at war with themselves, driven along one road by the intellect and along another by the passions. . . . They had a maxim 'nothing too much,' but they were in fact excessive in everything—in pure thought, in poetry, in religion, and in sin. It was the combination of passion and intellect that made them great, while they were great. Neither alone would have transformed the world for all future time as they have transformed it."[19] By the way, passion was one of the "evils" sealed in Pandora's box, but Pandora's curiosity—*Greeks'* curiosity—freed it, for reason without passion is dull. At the end, apathy (etymologically, not suffering) was the evil not passion (suffering, physically and intellectually).

Their life's philosophy was everything in moderation, self-control. But it was inspired because they were excessive and not of moderation for "noble self-restraint [reason] must have something [excessiveness, passion] to restrain."[20] One wants to make self-control his philosophy because one is excessive in his psychology. Hence philosophy's goal was to warn them of the potential dangers of excessiveness, and their hope was to keep their excessive nature under control through self-control. The result, the proper moderation between passion and logic, allowed them to become *creatively* excessive, and in their search for a new worldview they invented democracy, science, and philosophy. Passion drove them away from the ordinary, but logic controlled their passion and allowed them to embrace the extraordinary. It is their zealous and continuous search for such moderation that shaped their adventurous path in history, leading to Western civilization and those non-Western that aspire to Greek ideals—great liberties such as the personal, the civic, the political, the religious, the scientific, and in general the intellectual. Incidentally, had we not had these great liberties (those of us lucky enough to have them), would we be willing to sacrifice ourselves in order to get them?

The ancient Greeks did. Nonetheless their critics say that the Greeks did not live up to their own ideals. Bruce Thornton responds that this failure, whenever it happened, "reflects only the banal truth that humans [*everywhere*] rarely live up to their own aspirations"[21]—Greeks, in other words, were humans first. And this is a consequence of the fact that the war between reason and passion—the challenge to steer successfully the "chariot," that is, our self, to the Good—is a universal human condition, a great truth that points toward yet another great truth of Greek origin—particularly a Stoic philosophy—that of our common humanity transcending our uniqueness in individuality: we are all rational beings with passions by nature, all other differences are superficial and/or accidental. Commonality, and in fact oneness, as a general law of nature (of which man, too, is part), appears (as we will see) also in Greek natural philosophy as early as Thales—all things for him are made from the same stuff, water—and attains an astounding uniqueness in the worldview of Parmenides—for only what is, is, only the Being exists for him, one and unchangeable.

Stephen Bertman asked why the Greeks were the ones who invented science. "Because," he argued, "the seminal principles of Greek civilization—humanism, rationalism, curiosity, individualism, the pursuit of excellence, and the love of freedom—were uniquely compatible with science's own essence."[22]

CONCLUSION

Geography, economy, religion, politics, language, and the practice of good habits, all appear to have an interwoven and critical role in the creation of favorable conditions for the rise of Greek civilization and the birth of science. However, it is likely that even these do not tell the whole story. For example, I believe that to understand the rise of a civilization we must also attempt to understand the idiosyncrasy of the people who caused such a rise. With this in mind, the ancient Greeks were passionate, rational, excessive, original, critical, political, religious, philosophical, scientific, and brilliant. They debated freely, zealously and with no fear, any theory to the end despite its implications. They took chances, made mistakes but rose victorious. And one of those rises was the birth of science. Their scientific theories, although the very first, were extraordinary! What were those theories and how do they measure up with our sophisticated mind-bending modern science after two and a half millennia of scientific progress? The answer will be surprising.

PART II

THE PRE-SOCRATICS IN
LIGHT OF MODERN PHYSICS

CHAPTER 7

CLOSE ENCOUNTER OF THE TENTH KIND

It is the beautiful season of summer. I have been reading this great story for hours, since dusk and during the absolutely moonless night. It is now almost dawn. I am very tired but do not want to put the book down. It is unusually original and wonderfully profound. The pages are one by one turning. I am really exhausted and am fighting sleep because the story is so good.

All of a sudden I hear a splash. The day is hot and bright, the sky blue, the sun yellowish-white. I look up and see Thales falling into the fresh, cool, flowing waters of a river. "Not very wise," laughingly remarks the atomic Democritus, staring curiously in the void with undivided attention. But Thales is carefree and zestful, having the matter of water primarily under control, getting up but purposely jumping back in again, with novel, childlike, passionate, and playful curiosity. Off from the center of the action, revolving around a Central Fire, carefully preparing his food while singing in a harmonious but almost secretive whisper, is the legendary Pythagoras. Inspired by the moment, he stops the song and begins counting the proportionally spaced ripples of the water from each splash Thales makes. He is a prominent mathematician, able to face squarely all mathematical oddities, but he is also a cosmopolitan musician who enjoys masterlike attention from orderly and exclusive gatherings of crowds. "Cosmic justice, conserve and save the phenomena," thirstily shouts the infinitely abstract but also practical Anaximander, the genius of antitheses, feeling the heat of dry air and expecting that it will soon be neutralized by the opposite coolness of wet water.

It is really a beautiful and hot day, but Anaximenes finds a creative and concrete way to moderate the heat and thereby cool down. With his lips nearly closed, he blows air out onto his body, noting that it emerges colder than when his mouth was wide open, causing his condensed sweat to rarefy and evaporate. Sitting at a distance, away from the many, boldly being where no one has been before, is the enigmatic Heraclitus, who skeptically observes the process of the constantly changing events, going through conspicuous but also subtle changes. He is quite certain he has previously taken a bath in this river's fresh waters, but then again, strangely, everything looks new and changed. What is the Logos (cause) of all these changes? To the contrary, judging all of the sense-perceived reality to be deceptive, there is the one and only Parmenides the ontologist, proud and relieved. For journeying during the darkness of night and into the light of day, through the unknown, from afar, he found the true way here by intentionally avoiding the known and opinionated way of all others.

Anaxagoras's *nous* (or intellect) finds everything to be a puzzle: "How is it that all these people from different eras of time and different places are here?" He wonders by skillfully placing his hands over his head. "Indeed a paradox, a paradox of space and time," adds the argumentative and prolific Zeno, who, through dialectic (the method of reductio ad absurdum, or reducing to the absurd), is trying to prove that motion is an illusion of the senses, and so no one really moves, despite that all appears to so do. "Are you sure space and time are the only elements in the puzzle?" melancholy Empedocles challenges, while, in the name of episteme (knowledge) and his love for strife, holding tight onto his clepsydra (a water clock), he risks a dangerous experimental leap through the air and over the flames of fire but lands safely on earth, in fact in the water, just beside me.

"And who might you be, young fellow—the modern physicist?" he asks. As I respectfully nod in awe, I feel all eyes curiously staring at me as if I'd been expected. And immediately the brightness of the day surprisingly turns into a mysterious twilight. "I have been predicting an eclipse at your arrival," Thales says, while nostalgically shaking off primarily the substance of water from his wet, muddy, ripped, and unfashionable clothes. "We have been longing to know your story," he adds. Moments later, the bright daylight is pleasantly restored. It is now noon. "It is a beautiful day indeed," I say humbly, "for I am learning yours. And I will tell you mine, too, but under *your* sunlight, for eclipses are ephemeral and pass, but your knowledge is timeless. You *still* bring fire to modern science." The day is still young, and who knows of the morrow?

Everyone's senses are keen, observing the changing sights, listening to curious sounds, smelling soul-awakening aromas, tasting the sweet air, touching the cosmic elements. But so is everyone's intellect contemplating it all. What a beautiful day! What a beautiful nature! What is her nature?[1]

CHAPTER 8

THALES AND SAMENESS

INTRODUCTION

Thales (ca. 624–ca. 545 BCE) was interested in how nature works. He was the first to ask what things are made of and what the properties of matter are. These are still the most fundamental and difficult questions of science. His answers were based on solely rational arguments, uncluttered by myths, superstition, rituals, or the actions of capricious gods. His approach was therefore the same as that of modern science.

He reasoned that in spite of the apparent diversity and complexity in nature, all things are made from the *same* stuff: water, and all things obey a common set of unchanging basic principles, water's transformations (e.g., its solidification, liquefaction, and evaporation). Thus for Thales nature is characterized by a certain sameness or unity between all things, however diverse they may be, an overall intrinsic simplicity.[1]

THE EARLIEST SCIENCE VERSUS THE LATEST SCIENCE

While the primary substance of matter is not water, what is the primary substance had not yet been discovered. Nonetheless, presently, according to the standard model of physics (introduced further in the section titled "Sameness"), the building blocks of matter are microscopic particles called quarks—constituents of protons and neutrons—and leptons—particles including electrons. And the plethora of diverse things is partly due to their transformations (from one type of particle into another), not to the transformations of water. Despite these new discoveries, still, Thales's notions about sameness—that all things are made from one and the same substance—and about the transformations of

103

matter, are of timeless scientific appeal. His idea about the transformations of matter, in particular, not only describes a fundamental property of the modern concept of energy (or matter, since, as Einstein's special relativity theory makes clear, they are equivalent and transmutable into each other), namely its ability to transform into various forms and cause change, but also employs causality, because for him the cause of all other things is the transformation of just one primary substance. But why was water the underlying principle/cause of such sameness and unity?

WHY WATER?

Several observations might have stimulated Thales in his speculation that all things are transient forms of water. Some ancient accounts such as Aristotle's[2] and Aëtius's[3] give us some insight. Water is required for the survival and development of all kinds of life. Primitive life exists in moist environments, and animal sperm is liquid. Also, since water transforms easily into the three forms of matter, the solid (as ice), the liquid, and the gaseous (as water vapor), and into a variety of shapes, it could, Thales might have thought, also transform into everything else, such as rocks or metals. Now, while all substances transform into the three states of matter (e.g., given enough heat, a solid piece of metal can melt and evaporate), water is the only substance of daily experience that does this before our eyes and on a regular basis through the changing seasons, something that observant Thales could not have missed. Furthermore, it transforms more easily: its evaporation temperature of about 100 degrees Celsius (at sea level) is smaller than that of, say, copper, bronze, or iron—materials that in antiquity were heated and melted to make tools—so one does not need a lot of heat to vaporize it; and in the cold winter, water is the only substance to turn to snowflakes and solid ice of all sorts of shapes. So its choice as a primary substance over other things appears logical. Thales might have reinforced his water hypothesis, I speculate, from another everyday observation, namely, that when heated or burned all things release (or so it seems, anyway) water vapor. One example that might have been an inspiring clue for his water doctrine might be observing the rising smoke from a burning piece of wood mixing with air and clouds, which in turn can be mixed with rain or snow and blend with the soil on earth. At first glance, smoke, air, and clouds, are like (or seem to be) water vapor; rain and snow are water, soil contains the water and snow of the rain or snowstorm, so it might actually be thought of as transformed water, and so might then be the plants (thus wood), since, starting as seeds, plants grow from the soil and "are nourished and bear fruit from moisture,"[4] and so might also be the animals since they eat plants or each other.

Since processes of this sort appear causal with water as the first cause, then it seemed logical to assume that everything is made from the same stuff—reconstructed from the same first principle—and that, in general, everything in nature is characterized by a certain subtle sameness.

SAMENESS

Sameness is a core concept in modern physics, not only because it emphasizes a universal, underlying, simple principle as a characteristic of all things in nature, but also because it points to a commonality in their ultimate origin. Unity (in the sense that everything can be derived from one and the same principle), Thales reasoned, is a subtle, intrinsic property of nature.[5] This idea inspired all other pre-Socratics (each creating his own special theory on unity, as we will see in subsequent chapters), and, in turn, they have inspired scientists of recent times.

James Clerk Maxwell (1831–1879) unified successfully the electric and magnetic forces by proving mathematically that they are really two manifestations of the same force, the electromagnetic. The electric force is caused by the electric charge: the positive and negative. Objects of opposite electric charge attract one another while objects of the same type charge repel. The magnetic force is caused by an electric charge in motion. A permanent magnet has two poles, the north and the south. Opposite magnetic poles attract, and similar magnetic poles repel.

Albert Einstein (1879–1955) from 1925 until 1955 attempted unsuccessfully to unify the electromagnetic force with gravity—which makes things with mass attract each other, such as the earth and the sun. Gravity, still the most puzzling of the forces, has not yet been unified with any of the other three known forces of nature (the aforesaid electromagnetic, and, coming up below, the nuclear weak and strong), despite the fact that through the work of Isaac Newton it was the first force to be described by a mathematical theory: Newton's law of universal gravitation.

Nonetheless, success struck another physics front with the combined efforts of Sheldon Glashow (1932–), Steven Weinberg (1933–), and Abdus Salam (1926–1996). In the 1960s the three physicists managed the unification of the electromagnetic force with the nuclear weak force in what is known as the electroweak force.[6] The nuclear weak force is responsible for the radioactive decay of unstable nuclei such as that of uranium, and the transformation from one type of material particle into another—Thales's notion on the transformation of matter is a significant process in modern science. The experimentally confirmed unification of the electromagnetic force with the weak force occurs at high energies and temperatures—where the two forces have the same strength and are indistinguishable, thus they are considered as one force. Whereas at lower energies/temperatures (generally those of everyday experiences) these two forces are two expressions of the same force: the electroweak.

The standard model of physics is the theory that combines the knowledge of the electroweak force and the nuclear strong force—which binds the quarks in the protons and neutrons and also the protons and neutrons in the nucleus of an atom. It is the best model so far because it combines successfully several theories to explain how particles interact and how the universe works. Quarks and leptons are among several experimentally confirmed predictions of the standard model. In fact, even more importantly with respect to Thales's view, according to the standard model, the materialness of quarks and leptons—in particular the source of their mass—is a single type of particle called a Higgs boson. On July 4, 2012, scientists working at the Large Hadron Collider, the most powerful atom smasher in the world, announced that they had discovered a new particle consistent with the predicted properties of the Higgs (the Higgs is a topic to be revisited in chapter 9, "Anaximander and the Infinite" and chapter 17, "Democritus and Atoms").

Through a Grand Unified Theory (GUT) physicists hope to extend the standard model by creating an experimentally verifiable theory in which the electroweak force and the nuclear strong force are unified. Several good candidates for a GUT do exist, making concrete testable predictions (such as the decay of a proton, not yet observed), though none has so far been experimentally verified.

Finally, a community of ambitious physicists is currently on the quest for the ultimate principle of sameness, that is, for the absolute unification of all four aforementioned fundamental forces of nature in what is termed the Theory of Everything (TOE). A TOE hopes to establish that everything in nature is explainable by a single overarching principle and its associated equation, that everything is truly a consequence of just one primary substance, as Thales initially claimed. A possible TOE that is still in its formative stages and thus highly speculative goes by the name of string theory. It seeks to describe nature in terms of vibrating strings of energy in ten or more dimensions. According to string theory and in agreement with the essence of Thales's idea, everything is made from the *same* stuff: absolutely identical strings! These strings, like violin strings, have different modes of vibrations that are speculated to manifest as different types of particles, which include the quarks and leptons.[7] In string theory, therefore, it is these exotic strings that are theorized to be the primary substance of the universe.

Now, we are aware of the three dimensions of space (think of them as three edges of a cube meeting at a vertex) and the one of time—thus with respect to what we can readily experience, we live in a four-dimensional universe. The other six spatial dimensions predicted by string theory are hypothesized to be curled up into unimaginably small ball-like geometrical shapes of 10^{-35} meters (as big as the vibrating strings themselves), and thus not easily detectable. When string theory considers an eleventh dimension it becomes a more complete theory (with fewer mathematical inconsistencies) that goes by the name of M-theory. Because in M-theory two-dimensional membranes vibrate in an eleven-dimensional universe (instead of one-dimensional strings vibrating in a ten-dimensional universe), M may stand for "membrane," though it has never been specified. The extra dimensions were required to make the equations of string theory consistent. The objects that resulted from the solution of those equations were not our familiar point-like particles (e.g., the electrons or the quarks) but rather were one-dimensional strings or two-dimensional membranes. Through a TOE we hope to reduce all of nature into one fundamental substance with its transformations, an utter simplicity and sameness of Thalesian grandness.

The paramount challenge in finding a TOE is rooted in our inability so far to combine the rules of quantum theory with the rules of Einstein's theory of general relativity. On the one hand, quantum theory describes successfully the behavior of the microscopic world of subatomic particles (such as electrons, protons, neutrons, quarks, etc.), where laws of quantum probability—the central concept of the theory—replace the deterministic laws of classical physics—which includes Einstein's, Newton's, and Maxwell's theories. On the other hand, the theory of general relativity describes successfully the behavior of the macroscopic world of planets, stars, galaxies, and generally the large-scale universe by explaining how space, time, matter, and energy are all inextricably intertwined and how gravity works. (Aspects of both of these theories will be discussed in later chapters.) Quantum theory and general relativity are significant improvements over Newton's and Maxwell's physics. Nevertheless, as special cases of the former two, the latter two are still abundantly practical.

BLACK HOLES: CHALLENGES IN THE QUEST FOR SAMENESS

But since nature is one and beautiful, one and beautiful should also be the theory that would explain it, physicists think. Hence a TOE must be capable of accurately describing all the phenomena of nature from the microscopic to the macroscopic. Unfortunately, general relativity and quantum theory seem irreconcilable, a very displeasing situation in science. For example, by employing relativity we derive one set of properties for black holes and by combining relativity and quantum theory we get a different set of properties. General relativity and quantum theory don't see eye to eye when attempting to explain black holes, but why?

A black hole is a dense object with gravity so immense that, according to Einstein's general relativity (that first predicted such strange things), within a certain distance from the object, nothing, not even light, can escape. The invisible spherical boundary around a black hole that defines this distance is called the event horizon. It is appropriately named because matter and light can fall through the event horizon but cannot escape—and so whatever events might be occurring within a black hole cannot be seen by us. Consequently, if some kind of civilization existed there, its citizens could see us, but we couldn't see them. Could we ever find out what's really happening inside a black hole (within its event horizon), or will this type of information be a truth, a secret of the universe that will always be hidden from us? Since mythical time of Prometheus scientists have never liked the idea of nature eternally withholding its secrets from us.

To answer this question about black holes we need to first answer another more general and fundamental question: what happens to an object that falls into a black hole? What is its destiny? Specifically, could we still somehow recover all the information about everything that happens to it inside the black hole? If yes, then information is considered preserved. But if not, if after a certain point in time we cannot link any information to that particular object, then all information regarding the fate of that object is thereafter considered lost in the black hole. This is an open question, but how is its resolution related to a theory of everything?

Within the context of general relativity, information is lost forever. But according to quantum theory information should be preserved—as this is a fundamental quantum premise: unlike general relativity, the equations of quantum theory do not produce objects with event horizons that can hide or, as we will see immediately below, even destroy information. Let's analyze the destiny of information in a thought experiment by considering what happens to a book falling into a black hole.

According to relativity, first the book crosses easily the event horizon since the invisible boundary is supposed to be an energetically calm region and nothing special to pass through. But ultimately crushed by the immense gravity the book pieces condense at the infinitely dense point-like center of the black hole. What once was a distinguishable book is now just matter indistinguishable from all other matter already present there. And all information about it is thereafter lost forever.

But quantum theory (at least through the "firewall" approach, which is one of various in the study of black holes) sees the fate of that same book differently: although the event horizon is a highly energetic boundary (in the firewall approach), a wall of fire that burns the book as it goes through, still, some energy is radiated to us from the region just outside the black hole that amazingly contains subtle information about the fate of the book (about the changes it undergoes) inside the black hole. Information therefore is preserved. Now the notion of a firewall violates a fundamental premise of general relativity: the principle of equivalence. This principle partly asserts that an astronaut should feel no difference between free-falling into a black hole and accelerating in empty space (far from black holes). Similarly, an object crossing the event horizon should experience no sudden changes and nothing special, but that's not what happens to the book in the firewall.

These two proposed conclusions clearly clash: if information is lost, quantum theory is fundamentally wrong, but if information is preserved, it is general relativity that is fundamentally flawed. Consequently, depending on one's approach, the properties of black holes are contradictory and the question still persists: is information conserved or lost? This contradiction is known as the information paradox.

A possible reconciliation of these two major theories that claims to leave both intact (at least for now) in their battle for cosmic dominance came through a revolutionary proposal in early 2014 by world-renowned theoretical physicist Stephen Hawking (1942–).[8] His proposal shocked the scientific community: through a new set of arguments his educated guess determined that information *is* preserved, which is an encouraging result on the one hand since information preservation is a fundamental premise of quantum theory. But on the other hand, if we are to save the equivalence principle of relativity, information preservation must hold only if black holes do not exist! Hawking's conclusion is ironic since he has been the leading authority and proponent of black holes for the last forty years. But now he claims that matter falling into a black hole can finally be spewed out as radiation, which, when analyzed, can in principle tell us what happened to that matter inside the black hole. Thus information is preserved— though, he continues, this will in practice be as difficult as weather forecasting. But restoring the principle of information preservation for quantum theory and, at the same time, the principle of equivalence for general relativity comes at a high price. According to Hawking, an event horizon, which is the defining characteristic of black holes, cannot exist, and so neither can black holes (at least in the sense of objects with an event horizon from within, which nothing can escape). This of course remains to be seen. Hawking believes that the final verdict concerning the true properties of black holes will be rendered only after we first reconcile general relativity and quantum theory through a theory of everything—that is, one principle/equation that describes consistently all of nature. The horizon of Thalesian sameness has never been more eventful.

THE SAGE

Thales was regarded as one of the seven sages of the ancient Greek world and the wisest among them. For this he was offered a golden cup, which he respectfully declined by offering it to another of the sages, who offered it to another until the cup was again returned to Thales. He then dedicated it to the god Apollo at Delphi. Thales, a humble man, is credited (among several other Greeks) with the famous aphorism "Know thyself." He was a philosopher, scientist, astronomer, mathematician, politician, even a theologian known best for his belief in hylozoism, that "all things are full of gods."[9]

But he was also a practical man. As an engineer, for example, he aided the army of King Croesus of Lydia in crossing the Halys River by digging a deep trench in the shape of a crescent and diverting its waters. It was a Herculean feat indeed. The waters initially flowed by one side of the army but later diverted; they flowed by the opposite side. Through his observations of the night sky he discovered the stars of the Little Dipper (Ursa Minor), which includes the North Star Polaris, and used them to teach navigation. He also wrote treatises on various calendars such as on the spring and fall equinoxes (which by the modern calendar occur about March 20 and September 22), on the summer and winter solstices (about June 21 and December 21), on the phases of the moon, on solar eclipses, and on the rising and setting of certain stars such as the Pleiades.

While in Egypt Thales is said to have computed the height of a pyramid by first noticing that at a certain time of day his own shadow was as long as his height. He then concluded that the length of the pyramid's shadow at that same time of day was, according to the law of similar triangles, equal to the pyramid's actual height. While on land and through the use of geometry he was able to calculate his distance from a ship at sea. Furthermore, he estimated correctly the angular size of the sun and of the moon relatively to the angular size of their apparent orbit in the sky to be equal to 1/720.[10] Angular size of an object is the angle created from your eye to two diametrically opposite points on the object. Say the dot, •, represents the eye, letter I, the object, and that they are situated like so, • I, from each other. The angular size of I is the angle depicted in the following geometry: •<I. Today we know that the angular size of both the sun and of the moon is 0.5 degrees (a fact that is true because the sun is much farther away than the moon). Incidentally, had these angular sizes happened to be unequal, their apparent sizes would also have been unequal, and, consequently, a total solar eclipse, during which our view of the sun is completely covered by the presence of a new moon in between (like the one Thales predicted on May 28, 585 BCE), would not have been possible. Now the sun's and moon's angular size of 0.5 degrees divided by 360 degrees, which is the angular size of their apparent orbit around earth, is exactly equal to Thales's estimation, namely, $0.5/360 = 1/720$!

Thales was also known for his weather predictions, a skill proven valuable in teaching his fellow citizens an important lesson about life regarding their negative attitude toward philosophy. In spite of all his knowledge (practical and abstract) and all his wisdom Thales is said to have been poor. And because of his poverty some people criticized philosophy by calling it a useless and impractical way of life. According to one account, "As Thales was studying the stars and looking up . . . he fell into a well. A Thracian servant girl with a sense of humor . . . made fun of him for being so eager to find out what was in the sky that he was not aware of what was in front of him right at his feet."[11] But had the great Dante Alighieri (1265–1321) witnessed the incident he would not, I am certain, have made fun of Thales but would, I am still certain, have responded to the girl by saying:

> The heavens are calling you, and wheel around you,
> Displaying to you their eternal beauties,
> And still your eye is looking on the ground.[12]

Hands-on Thales responded similarly, not in words but through a practical action. "He perceived by studying the sky that there would be a good olive harvest. While it was yet winter and he had some money, he put down deposits on all the olive presses in Miletus [his hometown] and Chios [a neighboring island] for a small sum, paying little because no one bid against him [as it was way too early for anyone to worry about the next harvest that would occur during the next autumn and winter]. When harvest time came and everyone needed the presses right away, he charged whatever he wished and made a good deal of money—thus demonstrating that it is easy for philosophers to get rich if they wish, but that is not what they care about."[13] What they do care about is the rational critique of nature.

One phenomenon that was analyzed rationally was the annual overflow of the waters of the Nile River—an unexpected phenomenon within the context of the generally dry Mediterranean summers when it begins to occur—that puzzled several Greek thinkers including Anaxagoras, Democritus, Herodotus, and Euripides. The occurrence was explained in a naturalistic way first by Thales, who ascribed seasonal northerly winds as its cause that hindered the river from emptying into the Mediterranean Sea and forced its waters to spill over its banks. Ancient Egyptians attributed the flooding to the tears of their mourning goddess Isis over the loss of her husband, Osiris. Today we know that the Nile's overflow is due to seasonal precipitation (mainly rain) on the highlands of Ethiopia (south of Egypt), where one of the sources of the Nile can be traced. Incidentally, Democritus's explanation was similar to ours.[14] That Thales was wrong is not so important as is his attempt to offer a rational explanation for this natural occurrence. Similarly, Thales and the pre-Socratics in general treated eclipses as natural phenomena, whereas the Babylonians viewed them as omens, despite the fact that the latter kept fairly accurate records for their repeated cycles. Comets, too, were generally thought of as bad omens, but not by the pre-Socratics. Anaxagoras and Democritus, for example, thought that comets "are a conjunction of planets that, when coming near each other, create the illusion that they touch,"[15] an explanation, which although incorrect, is logical because it explains why a comet appears to be a strip-like light in the sky instead of point-like, as planets and stars are. Today we know that the strip-like appearance is due to a comet's tail. It is created from the evaporation of some of its ices, while a comet approaches the sun in its elliptical orbit around it.

Another example of a naturalistic explanation is found in Thales's cosmology. Although again wrong, it was reached by a rational analysis. According to it, everything is in essence water. Mud, and so, consequently, the earth formed via the solidification of water. Before Thales, the Babylonians had a deceptively similar idea, that nature had once been only water. But unlike Thales, who invoked only natural processes to explain dry land, the Babylonians invoked the supernatural: they imagined their god Marduk creating dry land and thus earth by first placing a carpet over the water and afterward adding mud on top of it—although how the carpet and mud were readily available to the god when nature was only water had not been explained. Naturalistic interpretations of nature were the approach of all the pre-Socratics and remain the approach of modern scientists.

CONCLUSION

By reasoning that all things are ephemeral transformations of *one* primary substance of matter, Thales attempted to attribute an all-encompassing, common, and unifying principle to all the phenomena of nature, the main goal of physicists today, as well as to understand a notion of great importance in science—namely, change. The concept of change (and the degree of change) has been hotly debated for centuries. Some have accepted it as self-evident, and others have flatly denied it as an illusion. Consensus has yet to be found. Every scientist, past and present, has looked to identify a permanent principle in all of the apparent changes. What that principle might be has varied from one scientific theory to another and from one epoch to the next.

Thales was more of a practical man who accepted change undeniably. His student Anaximander was practical but also an abstract thinker. His primary substance of matter was imperceptible and although he, too, accepted change as self-evident, he also required that change in nature obeys laws and happens with measure. But in all the conspicuous changes, he reasoned, something subtle must endure. He called it the *infinite*.

CHAPTER 9

ANAXIMANDER AND THE INFINITE

INTRODUCTION

Anaximander (ca. 610–ca. 540 BCE) taught that the fundamental substance of matter is the *infinite*: a limitless supply of undifferentiated, timeless substance encompassing all the world and manifesting itself as competing antitheses (e.g., hot versus cold). In modern physics energy has properties that are strikingly similar to those of the infinite and so does the much-sought Higgs boson particle of the standard model of physics (nicknamed the God Particle by Nobel laureate Leon Lederman [1929–] for its elusiveness as well as its significance for our understanding of the structure of matter[1]).

THE INFINITE

While itself intangible, the infinite transforms into all concrete things of everyday. Thus it is the true beginning of everything, animate and inanimate. It is also neutral, having no competing opposite. But it transmutes into opposites in struggle with one another—water versus fire, hot versus cold, wet versus dry, light versus darkness, sweet versus sour, and so on. The unjust dominance of one opposite over the other is ephemeral, for eventually it is rectified at annihilation; then, neutralized, both opposites transform again into the neutral infinite. And since eventually the effects of one opposite cancel those of the other, their endless creations and annihilations neither add anything to the infinite nor subtract. Thus even through its transformations, the infinite remains eternally conserved. In modern physics, it is energy that is conserved through its transformations into competing opposites, that of matter and antimatter, and, like the infinite, energy is also limitless and everywhere.

ENERGY AND THE INFINITE

In physics the notion of energy includes mass, too, since, according to Einstein's famous equation $E = mc^2$, from his theory of special relativity, energy (E) and mass (m) are equivalent and transmutable into each other—they are connected via the speed of light (c). Like the infinite, energy is limitless, timeless, indestructible, and omnipresent even in "empty" space (a challenging concept to be revisited in chapter 17, "Democritus and Atoms" when discussing his atoms in the void). Even more, energy causes change by continually transforming from one form to another (e.g., from light to heat) and from pure energy (e.g., light) into matter (e.g., electrons) and antimatter (e.g., antielectrons, also known as positrons). But even with these transformations the total energy content of the universe is always constant. This is known as the law of conservation of energy. One can neither add more energy to the universe nor subtract any from it. Conservation laws, of which there are several in physics, ensure that changes in nature occur with measure, as in the theory of Anaximander. Measure, in modern physics, means that in all of nature's changes some things/properties remain numerically equal (e.g., energy). This equality (this measure) is in basic agreement with the view of Anaximander, who, to save nature (as we will see), reasoned that neither of two opposites could ever dominate totally. Now, to appreciate further the notion of measure and the similarities between energy and the infinite, we need to first understand matter and antimatter since these, in modern physics, are opposites in struggle with each other, created from energy, and into energy once again they return.

Every particle of matter has a corresponding antiparticle of antimatter. A particle (like the negatively charged electron) and its antiparticle (the positron, which is really a positively charged electron) have the same mass and opposite electric charge (of equal magnitude). They are regarded as competing opposites since when they meet they annihilate each other by transforming completely into pure energy—like Anaximander's water and fire that neutralize each other and transform into the infinite. Furthermore, as opposites, not only do they compete—interact via the forces of nature they obey—but, since their effects cancel each other out, they compete with measure, by obeying conservation laws such as the conservation of energy.

For example, an electron and its competing opposite, a positron, can be created out of energy, interact—they initially move apart but after a brief time in existence they recombine—and ultimately annihilate (neutralize, cancel) each other by converting their masses *entirely* back into the energy from which they came. Just like Anaximander's opposites, which are created and annihilated from and into the infinite. Just as the infinite remains constant during such processes, so does the energy since, according to the law of conservation of energy, the energy content of the universe is the same before the creation of the electron-positron pair, during its existence, and after its annihilation. The energy content never changes, only the forms in which it manifests itself.

Other conservation laws are also obeyed, such as that of the electric charge. In this case, the electric charge of the energy from which the pair was created was zero—pure energy always has zero electric charge: the electric charge remains zero when the pair is in existence—for an electron has an electric charge of -1 (in some units) and the positron +1—and continues to be zero when the pair is annihilated as it once again becomes pure energy.

Let us take the conservation of the electric charge one step further. Since the net electric charge is always conserved, neither type of electric charge has an absolute dominance, as Anaximander would require. Nonetheless, one type of charge has a relative dominance over the other. The negative electric charge dominates temporarily at the vicinity of the electron, whereas the positive electric charge dominates temporarily at the vicinity of the positron. It is this temporary dominance that creates the electromagnetic force of interaction and overall competition between opposite charges. It is the cause of the opposites' becoming and decaying, their attractions, repulsions, motions, conversions to and from energy, and in general, such temporary dominance is a contributing cause of the phenomena of nature.

Competing opposites are necessary for Anaximander and modern physics, as they will also be for Heraclitus, if nature is to remain diverse, eventful, and beautiful. There is an electron here but a positron there. They, and other particles and antiparticles from similar processes, convert to light and to heat. The particles form atoms, molecules, composite objects like the sea, the trees, the breeze, the earth, the sky, and forms of life. It is summer here but winter there. It is warm now but will be cool later, night now but was day earlier. The unity of the world is preserved in harmony by the very competition between opposites. Temporary dominance and the resulting struggle of opposites produce the rich plethora of diverse phenomena while simultaneously, cosmically (universally) absolute dominance is not, *should not* be allowed, for conservation laws must be obeyed. Not only does Anaximander's worldview see the nature of nature as being cosmically just, but because of conservation laws so should modern physics. Curiously, however, it appears that nature is not cosmically just, for one of the most puzzling questions of science today is why there is more matter than antimatter in the observable universe, a question to be pondered in a section below after we first elaborate on the notions of opposites and neutrality a bit more.

NOT AN ORDINARY THING

Anaximander reasoned that the primary substance of the universe could not have been any one of the ordinary things, such as water or fire. For they have opposition with one another, and opposites destroy; they do not generate one another. If water were the infinite—that is, if everything in the universe were initially water—it would be impossible to have its opposite, fire, ever created, for water destroys fire; it does not generate it. And that would be terrible because in such a scenario eventful beautiful diversity would be absent from the cosmos. Thus something that has an opposite cannot be the primary substance of the universe for it presents a serious threat to the cosmic justice, to the unity and order of nature, to its diversity and in fact, to the very existence of nature itself—for such type of substance, with an opposite, would cancel itself out, and thus it would cancel nature itself!

Anaximander saved the phenomena—kept nature eternal, diverse, and eventful, and without the possibility of absolute dominance by any one of the opposites—by requiring that the primary substance, the infinite, be neutral—with no competing opposite. It must be neutral to itself and to the opposites that it creates. With such choice, neither opposite is a threat to nature any longer, since their effects cancel each other out, nor is the infinite, since it has no competing opposite to cancel out. Hence, unlike the opposites, the infinite is permanent and indestructible. And so then is nature itself, for nature's essence is the infinite. The neutrality of the infinite saves the phenomena, but the opposition of the opposites beautifies them. Both neutrality and opposition are central ideas in the world outlook of Anaximander and of modern physics.

The modern physicist's version of Anaximander's reasoning would be that the presently accepted primary particles of matter, the quarks and leptons, cannot really be primary, for they have opposites, the antiparticles of antimatter, the antiquarks and antileptons, and as opposites, a particle and its antiparticle annihilate, not generate, each other. Furthermore, there are six types of quarks and six types of leptons, and the pressing question is why there are so many building blocks of matter and why do they have different general characteristics (different electric charge, mass, spin, etc.)? Why not just have one primary substance, one infinite-like type particle?

On this issue, Nobel laureate Werner Heisenberg (1901–1976) has said, "All different elementary particles [particles that are not made of other things, thus have no substructure, such as the quarks and leptons] could be reduced to some universal substance which we may call energy or matter, but none of the different particles could be preferred to the others as being more fundamental. The latter view of course corresponds to the doctrine of Anaximander, and I am convinced that in modern physics this view is the correct one."[2]

We have seen how energy may be regarded as the infinite, but what kind of material particle of modern physics has infinite-like properties, including the key property of neutrality?

THE GOD PARTICLE

The search for a universal *neutral* substance that addresses Anaximander's concern and saves the phenomena has never been more intense. The particle with the most qualities required by such a substance is the Higgs boson. It has been mathematically predicted to exist by various physicists in the 1960s, including Peter Higgs (1929–), from whom it took its official name. In fact it is required to exist in order to save the phenomena as described by the standard model of physics: the Higgs particles are thought to give mass to all material particles (such as quarks and leptons) by pulling on them, thus forcing them to slow down, clump, and form all the composite objects in the universe, from nuclei, atoms, molecules, plants, animals, planets, and stars, to galaxies, and in general, all the complexity in the universe. The mass-giving mechanism of the Higgs and whether mass is truly an intrinsic property of particles or an acquired one through their interactions are more relevant topics for chapter 17. With the Higgs we can explain mass, and as a result the universe is diverse, beautiful, and saved. Without the Higgs, all particles would be massless, would fly around at light speed, and would not be able to come together and form atoms or in general composite objects, including us; the universe in such a case would exist in one boring, undiversified state, which of course would be in contradiction to the actual diversified universe we live in. Indeed, then, the Higgs saves the phenomena!

Like the infinite, the Higgs particle is intangible, neutral (in a few interesting ways, as we will explore in the next section), and the field that represents it permeates all of space. (Use of the term "field" means that something exists everywhere, whereas "particle" implies that something exists only somewhere. In quantum theory particles are manifestations of field fluctuations. Consider this analogy: if the sea is a field, a splash in the sea—a fluctuation, an excitation—is a particle that can be detected. Because of its all-pervasiveness, the Higgs field was related to Anaximander's infinite by Lederman.[3]) But while itself neutral, the Higgs also manifests itself as the competing opposites, as particles of matter and antiparticles of antimatter, and as some of the forces that matter and antimatter obey; in fact, it is not expected to be observed directly (analogously, neither was the infinite), instead its existence was confirmed indirectly by studying the behavior of various other particles that the Higgs decays into. Thus, the observed opposites in nature are in a sense different aspects of the same thing, the Higgs—or, analogously, the infinite. Anaximander proposed the infinite to put his opposites, and Peter Higgs (by modifying the standard model) proposed the Higgs field to put his opposites (the particles and antiparticles) in order to explain why they have mass.

Michelangelo's simile of a formed, asymmetric statue inside an unformed block of marble awaiting to be exposed by the sculptor can be applied to our discussion of the infinite and the Higgs. First, note that the unformed block of marble is symmetric hence boringly uniform. To the contrary, the formed statue is asymmetric hence beautifully irregular, with curves, lines, expression, and character. With this in mind, the asymmetric opposites (that is, formed matter, fire, water—the beautiful statue) are hidden in the unformed, symmetric nature—the dull but necessary raw marble—until the sculptor of nature, the infinite or the Higgs (or in general, the human mind), breaks her symmetry and exposes them.

Now this simile can be discussed within the context of a more literal scenario, that of the cosmological model known as the big bang, which speculates how the universe got started and how it has been evolving. According to this model, about 13.8 billion years ago the entire universe was unimaginably small, possibly a mere point, infinitely dense and hot. It then exploded in the absolutely most extraordinary event called the big bang and has ever since been expanding, cooling, rarefying, and creating the eventful universe we live in. Thus the universe was once, and for a minuscule moment during its early blazing stages of existence, much smaller, denser, hotter, perfectly (or more) symmetric than today (i.e., with all forces unified), and with none of the particles having any mass. An inconceivably small fraction of a second after the big bang, when the universe expanded and cooled, the Higgs field was activated, broke nature's initial symmetry by giving mass to only some particles, and has ever since been playing a crucial role in the plethora of beautiful and diverse phenomena we observe today. Analogously, Anaximander hypothesized that nature was once uninterestingly symmetric and unstructured until the infinite's eternal motion caused the eventful opposites to separate out from it and break the monotonous symmetry.

NEUTRALITY

To save the phenomena, neutrality must be an essential characteristic of a primary substance. The Higgs is neutral in various ways: (1) it is electrically neutral, thus it is its own antiparticle—it is both matter and antimatter, and in this respect it has no warring opposite to be destroyed by (note, even when the Higgs particles decay, the Higgs field still endures and is all-pervasive). (2) It is also color-neutral— "color" (or more precisely, color charge) here is a property of the quarks and of the nuclear strong force (like the electric charge that is a property of the electromagnetic force) and not the color of everyday sense. Parenthetically, color charge comes in three flavors: blue, green, and red. These color charges represent values that control the strength of the nuclear strong force between quarks. Where do their colorful names come from? The electric charge, which produces the electromagnetic force, has two values: the positive and negative electric charge. The color charge, which produces the nuclear strong force, has three values, which could have been called anything, such as A, B, C. But, because physicists are interesting people, they were fancifully named after the three primary colors of light, so blue, green, and red. (3) Moreover, its spin is zero, thus it is direction-neutral, an unusual notion that we need to elaborate on further.

Spin (like electric charge) is an intrinsic quantum property of elementary particles. In a simplistic view, imagine the spin of a particle to be like the spin of a top around its axis. However, unlike a spinning top, which may spin slow or fast and in every direction, an elementary particle spins with a fixed magnitude and only in certain directions. Now, the direction of a particle's spin is related to the direction of its motion through space. For example, a neutrino (an electrically neutral point-like particle, belonging in the family of leptons as electrons do) is observed to always be left-handed. This means that a neutrino moves through space like a left-handed screw: it advances (moves forward) by spinning counterclockwise. An antineutrino on the other hand is always observed to be right-handed. It moves like a right-handed screw: it advances (moves forward) by spinning clockwise. Unlike a neutrino, an electron can be ambidextrous— or move in space by spinning in either direction. The important point is that all particles are directional—the direction of their motion through space is restricted by how they spin.

But a primary substance must itself be free of such restriction; it must be nondirectional, direction-neutral, that is, isotropic (with no preferred direction of motion)—because a left-handed substance would, in Anaximander's terms be like, say, fire, and a right-handed substance like water. If a primary substance's own motion were restricted it would not have been able to generate the existing particles with all their various observed directionalities, which collectively are isotropic. In other words, if Anaximander's cosmic justice is to hold, a preferred (special) direction in the universe, toward which particles would be moving, should not exist. In fact, *on a grand scale*, the view of the universe is similar in all directions, thus the universe itself is isotropic (there is no special direction in it). One proof of this is the observation of the so-called cosmic microwave background, light that comes to us from every direction in the universe, showing that the matter that emitted it had almost exactly the same temperature, about 3 degrees above absolute zero.[4] Now, since the universe is made of galaxies and stars, which in essence are made of particles, then the isotropy of the universe must really be a consequence of the isotropy of its constituent particles. And if we believe that one day we will conceive a theory of everything, describing one primary substance, such a substance must have the property of isotropy, for only then it can generate the universal isotropy that we observe today. Well, to be isotropic, direction-neutral, the spin of such primary a substance must be zero since without spin the direction of its motion cannot be restricted. The Higgs boson particle is electrically neutral as well as color-neutral, and being the only particle of the standard model with zero spin, it is direction-neutral, too! Anaximander's notion of neutrality should be a vital property of the primary substance of the universe, and consequently of the universe itself. Nonetheless, it seems that the universe does not obey this fundamental notion of neutrality! For although it is isotropic and thus cosmically just (neutral) direction-wise, it is also unjust matter-wise; matter appears to dominate antimatter! What happened to the cosmic justice?

WHY IS THERE MORE MATTER THAN ANTIMATTER?

In modern cosmology, there is an open question: why is there more matter than antimatter in the observable universe? In Anaximander's terms, this problem might have been phrased: Why is there more water than fire in the observable universe? This observation makes no sense if indeed the universal substance is a kind of neutral, which transforms into equal amounts of opposites with properties that cancel each other out through the conservation laws they obey—so that the universal substance can remain neutral. Absolute dominance by any one opposite should not be allowed. Yet matter appears to have an absolute dominance in the universe. If true, and our observations are correct, where is Anaximander's cosmic justice? Are the laws of physics as we now know them really incorrect? The answer does not yet exist, but I will speculate cautiously.

Think of a creation process like that of an electron-positron pair. It obeys cosmic justice. During its existence, the region where the electron is located is dominated by matter, and the region where the positron is located is dominated by antimatter. But such dominance is relative and ephemeral. No law forbids equal and relative dominance by matter and antimatter in various regions of the universe. In fact this would be expected if indeed equal amounts of matter and antimatter are generated by a neutral-type universal substance. What *is* forbidden is *absolute* dominance, for which the amount of either matter or antimatter in the universe is absolutely more. Now, with relative dominance in mind we can speculate on why there is, or actually, why there *appears* to be more matter than antimatter in the observable universe.

First, the universe is immense, and we haven't observed all of it yet. So an unseen part of it might be composed of mostly antimatter that could balance out the matter we see, hence we have cosmic justice.

Second, what if our universe is not *the* universe but rather is only but a mere universe, a mere region, in a multiverse (which, by definition, is supposed to be composed of many universes, perhaps like ours, perhaps different)? In fact, we should not be rash in dismissing such a view, for not too long ago we thought the entire universe was what we now call the Milky Way galaxy—but we now think there are some 170 billion galaxies. If the multiverse hypothesis holds, the puzzle of the observed asymmetry between matter and antimatter might then be resolved. Matter might be dominating temporarily in our universe, but antimatter could be dominating temporarily in another universe, and in such a way that neither can claim absolute dominance in the world—the multiverse. Because such temporary dominance can be neutralized when such universes collide, converting their matter and antimatter to pure neutral energy. Anaximander's cosmic justice would then be restored.

Anaximander believed that there exist innumerable worlds, coming into being from the infinite and perishing back to it. Analogously, there are many modern cosmological models that speculate on many universes, all of which are part of a multiverse. Such hypotheses are a result, for example, of various interpretations of quantum theory, such as the so-called many-worlds interpretation (to be introduced in chapter 16, "Anaxagoras and Nous"), or attempts to construct a viable theory of everything.

COSMOLOGY

In addition to the hypothesis of the abstract infinite, Anaximander makes another conceptual leap by holding that the earth is motionless in space and without any physical support. This happens, he argues, because of its equal distance from everything, which is also uniformly distributed around it, and because of its equilibrium (its equal tendency to move in every direction).[5] This is in contrast with Thales's view of an earth floating on water and thus supported by it. Philosopher Karl Popper (1902–1994) remarked, "In my opinion this idea of Anaximander's is one of the boldest, most revolutionary, and most portentous ideas in the whole history of human thought. It made possible the theories of Aristarchus and Copernicus. But the step taken by Anaximander was even more difficult and audacious than the one taken by Aristarchus and Copernicus. To envisage the earth as freely poised in mid-space, and to say 'that it remains motionless because of its equidistance or equilibrium' (as Aristotle paraphrases Anaximander), is to anticipate to some extent even Newton's idea of immaterial and invisible gravitational forces."[6] Aristarchus of Samos (310–ca. 230 BCE), sometimes called the ancient Copernicus, was the first to advance the heliocentric model, revived in the sixteenth century by Copernicus (perhaps Copernicus could be viewed as the modern Aristarchus).

As commended by classicist John Burnet (1863–1928)[7] Anaximander's doctrine of innumerable worlds, each with its own earth, heaven, planets, stars, and especially its own center and diurnal rotation, is inconsistent with the existence of an absolute center or preferred direction of motion in the universe. With the lack of an absolute direction of motion, Burnet continues, Anaximander's argument that an earth that happened to be equidistant from everything in its world has no reason to move in any direction is quite sound. This is in fact a clever use of symmetry, a notion of central significance in modern physics. Symmetry, as in an unformed block of marble or a circle, implies a certain constancy, a similarity that persists without change. The unformed marble is the same throughout, for example, or the circle looks the same from its center at all angles, or, in Anaximander's case, the earth remains in equilibrium because of its equidistance. Symmetry in physics does not describe just appearance. It also underlies conserved abstract properties of nature such as the conservation of energy, momentum, and electric charge. For example, conservation of energy is a consequence of the hypothesis that the laws of nature are symmetric (invariant) with respect to time translations: they work the same today as they have in the past and are expected to continue so tomorrow. Because this hypothesis has been true so far, we accept conservation of energy as a law of nature.

The absence of a special direction in space was employed by Democritus in describing the motion of his atoms as random (see chapter 17). Though true, abandoning the notion of an absolute direction is still difficult today. We are tricked by the phenomena (such as falling objects or "the earth being under our feet and the sky up above us"), and so we often think of up above and down below as if they were really absolute up and absolute down. We don't realize that for those living on the opposite side of the earth, our relative up is really their relative down, and our relative down is really their relative up.

ON LIFE AND EVOLUTION

Thales, Anaximander, and (as we will see in the next chapter) Anaximenes (collectively known as the Milesian philosophers since they were born in Miletus, a Greek city in Asia Minor) explained nature in terms of the variations of *one* universal substance. Nature, of course, includes us and all life-forms. So an immediate consequence of their monistic theories is either (a) that there is no lifeless matter, to the contrary, everything is somehow alive; each philosopher's primary substance (the water, the infinite, or, in Anaximenes's case, the air) is somehow alive and so is everything that it transforms into. Or (b) that humans as well as all other species originated somehow from lifeless matter (the water, the infinite, or the air).

View (a), which is known as hylozoism, was held by Thales and Anaximenes, and although highly controversial, it is still an interesting notion, for, despite 2,600 years of advancements in science and philosophy, a clear-cut distinction between animate and inanimate matter cannot be made. An unambiguous definition of what is alive or dead does not exist as argued, for example, by four Nobel laureates: Charles Sherrington (1857–1952),[8] Erwin Schrödinger,[9] Werner Heisenberg,[10] and Richard Feynman (1918–1988).[11]

View (b) has a certain similarity with the premise of the modern theory of biological evolution—that regards the various species to have evolved gradually from a common ancestor (or two, possibly more) speculated to have arisen spontaneously from lifeless matter. By spontaneously I mean that the exact mechanism of life's origin is not yet known, although chemical reactions are generally the assumed cause. Now, compared to the other Milesians, Anaximander had a more concrete and extraordinary theory of the origin and evolution of the species, including humans, that captures four specific aspects of the modern theory of biological evolution: (1) life arose spontaneously from lifeless matter, (2) more complex life did not arise spontaneously but evolved from the less complex, (3) life's adaptation to its environment, and (4) survival of the fittest.[12]

Equally importantly, his theory was based on an accurately analyzed *observation*. Noticing that human babies are helpless at birth and for several years thereafter, Anaximander argued that humans could not have originated with the young of the species in their present form because they would have never survived. While newborns of other animals quickly support themselves, human babies cannot survive without long parental care. Therefore, he held that humans (and in general all animals) evolved from species, precisely fish, whose newborns were more self-reliant than human (or land-animal) babies.[13]

His general doctrine was that the most primitive forms of life were generated spontaneously in the moist element as it was evaporated by the sun—note that his notion of antithesis is present here, too, as the wetness of moisture versus the dryness of the sun. These living creatures had a protective spiny membrane and were the first kind of fish. With time, he speculated, they evolved to various other forms of fish. Then, some of their descendants abandoned the liquid element and moved to dry land, adapting to different conditions and evolving to new forms of life, including humans.[14] Modern theories of biological evolution are quite similar: primitive microscopic life is speculated to have appeared spontaneously initially in water, evolved to fish, then to the sea-land transitional amphibia, to mammals, to primates, then to the first hominids from which modern humans ultimately evolved.

Newborn self-reliance was the first state in the development of life. Newborn helplessness and long parental care have developed afterward and most probably simultaneously: as a newborn need arose an able parent addressed it. As if the evolution of a bad characteristic happens simultaneously with the evolution of a good characteristic, a notion, which if true, resonates well with Anaximander's doctrine of the simultaneous appearance of opposites: an inefficiency in some area, say, the inability to walk at birth and for several months subsequently, may evolve simultaneously with a corresponding efficiency in some other area, for example, an advanced brain, so that one may moderate the other—for example, the evolution of an empathetic brain allows parents to care for their helpless offspring for a long period. In general, the growing-up time increased with increasing brain size and complexity.

The selfless act of long parenting not only guaranteed the survival of the species but also, I believe, contributed to the overall bond between all people. Because once we began caring for our babies, we gradually began to care for our immediate and extended family, consequently increasing the chances to care more for our village, city, country, the human race, and ultimately life in general.

CONCLUSION

Anaximander's intellectual leap is marked by three of his theories: in cosmology of an earth motionless in space without the need of a physical support, in biology on the origin and evolution of living creatures including the human species, and certainly of his theory on the primary substance of matter. With the latter, he modeled change and diversity in terms of constant transformations of the intangible, neutral, and conserved infinite, into the concrete, competing, and transient opposites of everyday experience, and back and forth and with measure. Nonetheless, it was Anaximenes who formulated the first graspable theory of change—of how matter can transform between its various phases: the gas, the liquid, and the solid.

CHAPTER 10

ANAXIMENES AND DENSITY

INTRODUCTION

I n his search for the primary substance of matter, Anaximenes (who flourished ca. 545 BCE) returned to the tangible world and chose air. His way of
studying nature was economical and straightforward. Starting with a single
material (air) of unchangeable nature, he managed to explain the manifold of
natural phenomena quantitatively, in terms of condensation and rarefaction of
matter. For with these opposite processes in mind, it was no longer necessary
to ascribe all sorts of different properties to each object—such as rigidity, softness, hotness, coldness, wetness, dryness, fluidity, weight, color—just how dense
it was. This idea in itself has a certain truth. But from a grander point of view as
regards the evolution of science, this idea had surely been catalytic for the discovery of one of the most successful theories in science: the atomic theory.

CONDENSATION AND RAREFACTION

In Greek, "air" refers to any gas, and quite possibly in Anaximenes's view air was
vapor water. His main question, however, was how a single material, air, in its
gaseous state, could be transformed into all other forms of matter and account
for the overabundance of dissimilar things, while itself remaining unchanged.
What mechanism or processes could be applied to air, keep its substance
unchanged, yet convert air into all the different things—solids, liquids, and
gases? Change, he proposed, occurs via two opposite processes: condensation
and rarefaction of matter.[1] Successive condensations of gases transform them to
increasingly denser matter, the liquids and solids, but successive rarefactions of
solids transform them to increasingly rarefied matter and once more back to the

129

liquids and gases, an essentially accurate idea. These processes cause changes in the density of matter but do not alter the very nature of matter (its very substance). Hence, every object is really air—in general, made of the same material—condensed or rarefied.

WHY AIR?

Air is in various ways of simpler form than other everyday substances. It is highly mobile and can be found almost everywhere. It is invisible, thus apparently unstructured and symmetric, rarefied, thus quantitatively less. Symmetry, perceptible or subtle, was and still is a much-desired characteristic for nature in both ancient and modern scientific theories. In addition, starting from less (at least in a quantitative and visual sense, e.g., rarefied invisible air) and aiming to explain more (e.g., a denser, thus quantitatively more, visibly more structured, and thus in a way more complex substance), has always been the preferred approach in both science and mathematics; in mathematics, the fewer the assumptions (axioms), the more powerful a theorem is. Parenthetically, as regards religion, the reverse is true: polytheism preceded monotheism.

Now, although Anaximenes thought that fire is rarefied air and thus quantitatively less than air, still, as a primary substance fire seems to not have been adequate for him, for unlike air, fire is visible, has a variable form, and so is structured and asymmetric. Furthermore, air is needed for life through breathing, whereas fire destroys life. In fact air's traditional association with soul (from the pre-Homeric times) might have influenced Anaximenes, for he writes: "Just as our soul, being air, holds us together, so do breath and air encompass the whole world."[2] The significance of fire in the explanation of natural phenomena will be elevated in the philosophy of Heraclitus (see chapter 12).

Lastly, Anaximenes was an empiricist, thus he abstracted his theory as a consequence of careful observations of various meteorological phenomena for which air had (or so he thought, anyway) a significant role. "When it [air] is dilated so as to be rarer [more rarefied], it becomes fire; while winds, on the other hand, are condensed air. Cloud is formed from air by felting [due to condensation]; and this, still further condensed, becomes water. Water, condensed still more, turns to earth; and when condensed as much as it can be, to stones."[3] He imagined objects to be in either one distinct phase (the solid, liquid, or gas) or in a mixture of phases; "Hail is produced when water freezes in falling; snow, when there is some air imprisoned in the water."[4]

FROM RAREFACTION AND CONDENSATION TO THE ATOMIC THEORY OF MATTER

It will be argued that the discovery of the ancient atomic theory of Leucippus and Democritus—of atoms in the void—followed as a logical consequence of the ideas of rarefaction and condensation. And as we will see in chapter 17, modern atomic science has its roots in the atomic theory of antiquity.[5]

Softness and the Void, Rigidity and the Atoms

Softness occurs with rarefaction and rigidity with condensation, Anaximenes held, but how? Since everything is made of soft and penetrable air, why are some objects (e.g., the solids) rigid and impenetrable? Why is a piece of metal (which is supposed to be condensed air) incompressible and impenetrable, while air is compressible and penetrable? Why can we walk through air (so it seems, anyway) but not through a solid wall (which is supposed to also be air)? How do rarefaction and condensation really work, and how can we explain the varying degree of softness or rigidity in an object? Furthermore, what keeps matter together in a condensed or rarefied state?

First, let us take all four notions for granted—rarefaction, condensation, and the resulting softness and rigidity in an object. Then we ask: If we could imagine rarefaction and condensation to occur ad infinitum, what kind of an object would the absolutely most rarefied or condensed be? The most rarefied type of object would have zero density and would be *absolutely* soft, compressible, and penetrable; as if the object were void of matter; as if it were immaterial and did not exist; as if it were nothing! Now, an object void of matter is really a *void*, empty space. And so the most rarefied type of objects could be thought of as material-less gaps in space. On the other end of the limit, the most condensed type of objects would still be of the *same* substance, would have infinite density, would be *absolutely* rigid, incompressible, and impenetrable, and could be thought of as the matter that is filling up the nonempty space. These impenetrable pieces of matter that are also disconnected from each other by the void between them, are precisely the atoms of Leucippus and Democritus, and the philosophically controversial void is precisely what they invented to facilitate the motion of their atoms.

"There are but atoms and the void," said Democritus,[6] or equivalently, "the full"[7] and solid, and "the empty"[8] and rarefied. First note that Anaximander's opposites are present here, too, for the full and solid is the opposite of the empty and rarefied. Furthermore, "the full" seems to correspond to the aforesaid absolutely most condensed object, and "the empty" to the absolutely most rarefied object. Comparisons and details of the ancient and modern atomic theories will be carried out in chapter 17. For now it suffices to visualize tiny, indestructible atoms, absolutely solid—and unable to rarefy—of all sorts of shapes moving randomly through the void, colliding with each other, either hooking with one another and clustering (condensation), or unhooking and dispersing (rarefaction). Thus with atoms moving in the void, we understand how condensation and rarefaction are actually carried out—how matter can move, assemble, and stay together, or disassemble. And the consequent rigidity or softness is determined by the density of an object, that is, by how many atoms are cramped together within an object and by how much void is between them through which to move: the less the void, the more rigid the object; the more the void, the softer the object. Could the ancient atomic theory have been discovered through such type of analysis?

I don't know if the atomists, Leucippus and Democritus, discovered atomism by analyzing rarefaction and condensation in the two extreme limits just described, but they were certainly capable of doing so, especially the great geometer Democritus. For such type of thinking, which in mathematics is part of what is known as the theory of limits, had already been invented and applied by him in other cases (e.g., in the calculation of the volume of a cone). In fact Democritus's knowledge on limits was commended by the great astronomer Carl Sagan (1934–1996) in this quote: "Perhaps if Democritus' work had not been almost completely destroyed, there would have been calculus by the time of Christ."[9] By "work," of course, Sagan meant, among other things, Democritus's knowledge of limits, for to invent calculus a prerequisite knowledge is the theory of limits. Calculus was finally invented independently by Isaac Newton and Gottfried Leibniz (1646–1716) in the late seventeenth century.

One other way that the mathematical analysis of rarefaction and condensation might have aided the discovery of atomism is discussed below.

Continuous versus Atomic

The challenge to understand how condensation and rarefaction themselves are carried out was important for the evolution of scientific ideas because it forced Anaximenes's successors to think profoundly about the nature of matter. Consequently, they discovered two antithetical views: the continuous and the atomic (the discontinuous). The complexities associated with the former guided the mathematical genius Democritus, the last of the pre-Socratics, to atomism. Erwin Schrödinger argued that the mathematical challenges of the continuum were related to similar challenges of a continuously distributed model of matter.[10] For example, we cannot tell how many points a purely mathematical line has. Analogously, if we have a material line (or in general an object), we cannot tell how many material points it has and how these points behave during rarefaction and condensation. Namely, how can an unchangeable substance of matter (e.g., Anaximenes's air), distributed *continuously* within an object, rarefy or condense? "What should recede from what [so that an object can rarefy, or what should approach what so that an object can condense]? . . . if it is a *material* line and you begin to stretch it—would not its points recede from each other and leave gaps between them? For the stretching cannot *produce* new points and the same points cannot go to cover a greater interval."[11]

In other words, can matter, modeled as continuously distributed in space, really move through other matter in order to condense or rarefy? Can new matter move into and occupy the space that is already occupied by other matter? When matter moves, where does it move into, and what does it leave behind? How do condensation and rarefaction really work if matter is continuous? They do not! They work only if matter is discontinuous: made of disconnected, indivisible, and incompressible pieces—the atoms of Leucippus and Democritus—

moving in the void. Rarefaction occurs when the atoms in an object recede in the empty space around them, and condensation occurs when they come closer to each other.

Modeling matter as discontinuous (atomic) constituted the very first quantum theory, the precursor of the modern. In modern quantum theory both matter and energy are quantized (discontinuous): matter is composed of disconnected elementary particles, the quarks and leptons, and energy comes in discrete (quantum) bundles (e.g., photons are the particles of the energy of light).

CONCLUSION

All challenges of rarefaction and condensation could be accounted for only through atomism (to be introduced fully in chapter 17), since such a great idea required first the development of all other great ideas conceived by Democritus's predecessors, but also in the light of mathematics (Democritus, the principal contributor of the atomic theory, was a brilliant mathematician). The significance of mathematics, not just as an abstract field of knowledge but also as a practical method to describe nature, had been realized early on, especially by the great Pythagoras as a consequence of his passion for numbers.

CHAPTER 11

PYTHAGORAS AND NUMBERS

INTRODUCTION

Pythagoras (ca. 570–ca. 495 BCE), a pioneer in applying mathematics to the investigation of physical phenomena, consequently initiated the mathematical analysis of nature, a cornerstone practice in modern theoretical physics. "Things are numbers" is the most significant Pythagorean doctrine.[1] While its exact meaning is ambiguous, it probably signifies that the phenomena of nature are describable by equations and numbers. Therefore, nature is quantifiable. That is, properties such as hotness, brightness, loudness, wetness, softness, and in fact all characteristics of nature, are measurable. Based on this, the underlying principle of nature is not material (e.g., water, air) but is rather a mathematical form (an equation). Since the mathematical representation of nature is not readily realized, the doctrine was emphasizing that sense perception was merely revealing an untrue version of nature (reality), a truer version of which could be glimpsed by the intellect through modeling nature mathematically. How one begins to model nature mathematically will be discovered in this chapter via the work of Pythagoras and his students. They quantified pleasing sounds of music, right-angled triangles, even the motion of the heavenly bodies.

THE MAN

Pythagoras founded a school in Croton in southern Italy, open to both men and women, where he and his students pursued various studies, including religion, science, philosophy, mathematics, and music. They practiced a common way of life: asceticism (through body exercise, a vow of silence, a special diet that

135

avoided meat and fish) and secrecy (probably to keep their discoveries exclusive to their students in order to attract more members to their school). Therefore, it has always been difficult to distinguish exactly what philosophical views belong to him or to some other Pythagorean. Aristotle avoids such difficulty by often referring generally to "the Pythagoreans." Plato in *The Republic* writes specifically about Pythagoras and says that he was uniquely respected and loved by his students, not only for his knowledge but also for teaching them "the Pythagorean" way of life, best known for its high ethical standards. Wisdom, justice, and courage were among the sought-after virtues. Friendship was also highly valued. Pythagoras's dogmatic influence on his students was evident by their reference to his opinions as prophesies with the characteristic phrase "He himself said so."[2]

COSMIC HARMONIES

Pythagoras's first application illustrating the role of numbers in nature was the mathematical description of mellifluous sounds of music. First, he discovered that in stringed instruments the sound of a plucked string depends on its length and tension. For example, the sound of a plucked guitar string is of a higher pitch as it is pressed down and made shorter by your finger. Then he observed that the blended sound produced by two plucked strings of the same tension is more pleasing when their lengths are in ratios of small integers—for example, 2:1 is the octave, 3:2 the fifth, 4:3 the fourth, and 5:4 the third—thus numbers forming a discrete, a *quantum* set—which for the example given is the set of 1, 2, 3, 4, and 5 (obtained by arranging in sequence the numbers of the above ratios).

The Pythagorean theory of music was a significant milestone in the evolution of science from two points of view. First, since the phenomenon of sound can be quantified, that is, it is represented by mathematical formulas, why not all phenomena? Second, if all things truly can be represented by numbers, then mathematics is the underlying and unifying principle of every natural phenomenon, even the seemingly dissimilar. With this in mind, everything may somehow be related at least mathematically—in other words, a kind of master mathematical equation may be invented that can describe everything. So phenomena that have no apparent relationship with one another, on a deeper level, mathematically, may prove to obey the same mathematical principle and thus have something subtle in common. We have already spoken about unification efforts undertaken nowadays in search of a theory of everything. But the first step of cosmic unification in search of a universal law was taken with the intellectually bold Pythagoreans when they connected mathematically two seemingly unrelated phenomena: their earthly harmonies with the heavenly motions! How did they do this?

First they supposed that, similar to the way an object on earth moving through air can produce a sound (slow movement making a low pitch, fast movement a high one), the stars (including the sun), moon, and planets (including earth), moving through ether (the purer air believed then to fill the universe) can produce their heavenly sounds. But these sounds must blend into a song harmoniously. They reasoned that the ratios of the length of strings that produce the harmonious sounds in string instruments *must be the same* as the various ratios formed by the speeds of the revolving heavenly bodies. This requirement restricts basically the speeds and orbits of the heavenly objects to certain discrete, quantum numbers, an idea that resonates with the essence of modern quantum theory!

Relative speeds of heavenly bodies could be easily deduced by comparing each body's rising time. For example, the moon rises about fifty minutes after the stars (some reference group of them that on some day rises together with the moon), but the sun rises only about four minutes after the stars, well-known facts in antiquity. With this in mind, the apparent revolution speed of the stars is the fastest, of the sun the second fastest, and of the moon the slowest. Since their speeds were different, the sounds they would produce as they move through ether would also be different, to say the least. But they also had to be harmonious, Pythagoras conjectured, since nature was a "*cosmos*,"[3] a term credited to him, a beautiful and well-ordered universe for which a cacophonous music of heavens was unaesthetic. The music of the heavenly bodies is inaudible, the Pythagoreans explained (as Aristotle tells us), because it is continuously playing and "the sound is in our ears since our birth, thus it is indistinguishable from its opposite silence; sound and silence are distinguishable only via their mutual contrast."[4] In a parallel example, a cook does not smell his own food after a few hours of cooking it. The earthly string harmonies, which could be heard, inspired the Pythagoreans to deduce by analogy the heavenly harmonies, which could not be heard. This type of approach, to come up with a general law by analogy of something specific, is common practice in science.

Such unusual interconnection was celebrated first in 1619 with the harmonic law of Johannes Kepler (1571–1630) when the astronomer discovered that, as planets revolve around the sun in their elliptical orbits, the ratios formed by each planet's fastest speed at perihelion (a planet's closest distance from the sun) over its slowest at aphelion (its greatest distance from the sun), are very close to the Pythagorean ratios of pleasing harmonies in stringed instruments. In his book *The Harmonies of the World*, Kepler wrote, "The heavenly motions are nothing but a continuous song for several voices, to be perceived by the intellect, not by the ear."[5]

Moreover, in the beginning of the twentieth century, the seminal era of quantum theory, physicists Niels Bohr (1885–1962) and Arnold Sommerfeld (1868–1951) conceptualized the atom as a miniature solar system, with the electrons orbiting the nucleus of an atom like the planets are orbiting the sun. In their theory the orbits of the electrons are restricted to certain discrete speeds and sizes (as were the heavenly bodies in the Pythagorean theory) that are expressible in terms of specific integers called quantum numbers that "display a greater harmonic consonance than even the stars in the Pythagorean music of the spheres [heavenly bodies]."[6] Remarkably, unlike the Pythagorean theory of planetary motion, which was quantized, the Newtonian theory was not: planets, according to Newton's theory of gravity, do not have a restriction in their speeds or orbital sizes. But they should, according to quantum theory, although their quantum behavior is negligibly small because of their large mass.

Even more so, according to the latest developments in string theory and in the words of string theorist physicist Brian Greene (1963–), "everything in the universe, from the tiniest particle to the most distant star is made from one kind of ingredient—unimaginably small vibrating strands of energy called strings. Just as the strings of a cello can give rise to a rich variety of musical notes, the tiny strings in string theory vibrate in a multitude of different ways making up all the constituents of nature. In other words, the universe is like a grand cosmic symphony resonating with all the various notes these tiny vibrating strands of energy can play."[7] A subtle cosmic interconnection between all things in nature, describable mathematically, was an idea envisioned by the great Pythagoras and has been consistently reaffirmed by modern physics. Mathematics nonetheless is not always rational.

THE IRRATIONALITY OF A NUMBER

The proof of the Pythagorean theorem—that in a right-angled triangle the square of the hypotenuse is equal to the sum of the squares of the other two sides—was the epitome of the newly born notion of mathematical deductive reasoning, in which general theorems are proven starting from the least number of axioms. It was especially encouraging to the most important Pythagorean doctrine, "things are numbers." But it ended up also being a bad omen. For soon after the theorem's proof, its application on a special kind of right triangle—the isosceles, with its equal sides having a length of one unit—led to the discovery of a new type of numbers, the *irrational* numbers, which perplexed the Pythagoreans and shook the very foundation of their number doctrine. It was found that the length of the hypotenuse of this right triangle is equal to the square root of two, that is,

$$\sqrt{2} = 1.41421356237309504880168872420969807856967187537694807317667973799073247846210703885 04 \ldots,$$

which does not have a precise numerical value; it can only be approximated—shown above it is truncated to 85 decimal places, for there is *literally* not enough paper in the entire universe to write such a number completely! That is, one cannot write down a precise number for the length of such hypotenuse, only an approximate, but we must emphasize that an approximate number is only approximate; it represents not the true length of the hypotenuse but only an approximate length. So how can things be numbers when some things cannot be assigned a precise number? To answer we need to understand irrational numbers a bit more.

In the history of mathematics, integers (. . . , -4, -3, -2, -1, 0, 1, 2, 3, 4, . . .) were supposed to be the only numbers needed, since with their various ratios (fractions) every number that exists (including non-integers) could, so it was thought, be written down. For example, the positive non-integer $1/3$ is expressed as a ratio of two integers, obviously 1 and 3; negative non-integer $-5/4$ is the ratio of the integers -5 and 4; even 0 may be thought of as the ratios $0/2$ or $0/7$, and so on; in fact even integers themselves may be expressed alternatively as a ratio of two integers, for example, $8 = 16/2$. Numbers that can be expressed as ratios of integers are called rational. So, for the Pythagoreans (and in general, up to that point in history) only rational numbers were thought to exist.

Since for the Pythagoreans every number was expressible as a ratio of two integers, so, too, should the length of *every* geometrical line. But they were shocked to discover that the length of the hypotenuse of the aforesaid type of isosceles right triangle could not be expressed as a ratio of two integers! That length was not a rational number. It was equal to the square root of two ($\sqrt{2}$), which turned out to be an irrational number. Irrational *literally* means that there is no ratio, none at all, that can provide an exact numerical value for $\sqrt{2}$. Hence the $\sqrt{2}$ can only be approximated. For example, truncated to one decimal place, the $\sqrt{2}$ is equal to the number 1.4 (which, in this approximate value, can be thought of as the ratios $14/10$ or $7/5$); to two decimal places, the $\sqrt{2}$ is equal to 1.41 (which, in this approximation, can be thought of as the ratio $141/100$). There are infinitely many irrational numbers, all numerically inexpressible by ratios. The famous number π (pi) is irrational.

How, then, can all things be numbers if some things cannot be given an exact numerical value? They cannot if exact numbers is all that we have in mind. But they can in some other broader sense, that of the most advanced theory of matter: the quantum theory. For according to this theory the microscopic particles of matter (quarks and leptons) are regarded as mathematical forms.[8] These forms are really the solutions to the equations of quantum theory and are useful for calculating numerically the average values of various particle properties (e.g., position, velocity, and energy). Since every macroscopic object in nature is composed of these microscopic particles (which are regarded as mathematical forms that generate numbers), then indeed all "things are numbers."

Irrational numbers have been playing a critical role in the advancement of mathematics and physics since the time of Pythagoras. But they still present an epistemological challenge because they provide only a numerically *approximate* knowledge of nature. This "approximate knowledge" is a significant point that will be picked up again in chapter 14, "Zeno and Motion" in order to try to understand the fascinatingly well-reasoned but paradoxical view of Zeno, that apparent motion is not real—that, an apparently flying arrow, for example, is not really moving!

The irrationality of the $\sqrt{2}$ was so shocking to the number doctrine that, according to legend, Pythagoras's student Hippasus of Metapontum (from fifth century BCE), said to have discovered it, was drowned in the deep sea in an act of divine retribution.

Mathematics represents timeless and universal truths, as does nature. And even when the law of a natural phenomenon has not yet been discovered, the law is assumed to exist, as is the mathematical equation that can express it. This in fact is the very premise of science. Without such an attitude science cannot be done and truth cannot be found.

GEOCENTRIC VERSUS HELIOCENTRIC: THE RELATIVE TRUTH

Pythagorean Cosmology

Being a great geometer who understood well the relationships of spheres, flat surfaces, and lines, Pythagoras was probably the first to deduce that the earth is spherical. Several observations might have aided him in reaching such a conclusion. During a lunar eclipse the shadow of the earth on the moon is a circular arc. The masts of receding ships disappear last (and, equivalently, appear first when ships are approaching). Pythagoras himself knew that the evening and morning "stars" are really the planet Venus.

But the most notable achievement of the Pythagoreans in cosmology (often credited to the Pythagorean Philolaus [ca. 470–ca. 385 BCE]) was when they displaced the earth from the center of the universe and imagined it in motion. So, the earth revolves around a center occupied by fire, called Central Fire, and so do the moon, sun, planets (Mercury, Venus, Mars, Jupiter, Saturn—the ones known to antiquity, as only these are visible without a telescope), and the fixed stars—termed so because of their apparently fixed position with respect to one another; on the other hand *planet* means literally "wanderer" because planets were changing their position among the fixed stars.[9] Central Fire is invisible because the inhabitable hemisphere of the earth faces always away from it, whereas the side of earth that always faces it is uninhabitable because it is too hot. Incidentally the moon's synchronous motion—according to which its rotational period around its axis is the same with its revolution period around the earth—produces the same effect: the near side of the moon always faces the earth, whereas its far side always faces away from earth, making it always invisible to an earthbound observer. Revolving around Central Fire is another body, the anti-earth, termed so because of its position.[10] It is imagined to always be in the same direction as the uninhabitable hemisphere of earth, so, like Central Fire, it, too, is invisible. It is not certain why anti-earth was required (some scholars speculate that it was needed to explain eclipses), or even whether anti-earth was really a planet at all—for due to its position anti-earth might have simply been the uninhabitable hemisphere of earth.

In addition to its revolution around Central Fire, earth also rotates around its own axis daily, accounting for the apparent revolution of the sky. This understanding was in audacious opposition to the popular view of an immobile earth at the center of the universe as well as to the evidence of the senses that do not feel earth's motion. In an analogy, to understand the apparent revolution of the sky, pretend to be the earth and stand at the center of a room. Then begin to rotate around the axis of your body, say, counterclockwise. The walls, which you can think of as the sky (with the sun and stars), appear to revolve around you in the reverse direction, clockwise.

Only Central Fire is self-luminous; all other bodies are shining with reflected light from it. In fact this might be the justification of its postulated existence: since the moon is shining with reflected light—an ancient knowledge—and by the cycle of day and night, so, obviously, is the earth. But why not the sun, which in many ways is like the moon—in motion, shape, size, eclipses, color—and all heavenly bodies? If yes, a source of light had to be speculated, hence the Central Fire.

That the Pythagorean system is neither geocentric nor heliocentric is actually a quite justifiable cosmological theory. Since every visible celestial body appears to be moving, perhaps so should the earth, the Pythagoreans might have thought. Now the difficulty with imagining the earth in motion has its origin in our deceptive senses, namely our eyes. Our inability to detect the actual distance of what we see, in particular the stars, is tricking us into thinking that all bodies are the same distance from us. Therefore, being also in different directions, they appear fixed on a hemispherical dome, which is part of what we call the sky. As the sky appears to revolve daily around us the new stars brought into view appear, for the same reason, to also have the same distance from us as all the rest. Thus we imagine every star fixed on a spherical sky—even though at any one time we see only a hemispherical sky—with us on earth at its center. In addition to that, the apparent daily revolution of the dome-like sky around us is easily tricking us into thinking that the earth is absolutely motionless at the sky's center and therefore is at the absolute center of the universe, as if the earth occupies a special position in the universe. This is in fact the geocentric view, which, due to our imperfect senses and our initially uninformed intellect, naturally emerged as the first cosmological model. But the Pythagoreans were well aware of the unreliability of the senses, and they were also accomplished mathematicians with sharp, critical minds. Moreover, being people of virtues, such as humbleness, the Pythagoreans had no difficulty displacing the earth and themselves from the center and purpose of the universe.

No Special Center

Influenced by the Pythagorean cosmology, post-Socratic Greek philosopher Heraclides Ponticus (ca. 390–ca. 310 BCE) proceeded to devise his own. He explained correctly the varying brightness of Mercury and Venus as the result of their varying distances from earth. The additional observations that these planets seem to always follow the sun—when visible, each of them (but independently of the other) either rises just before the sun or sets just after the sun does, especially so Mercury because it is closer to the sun—prompted him to imagine these two planets revolving around the sun—thus justifying their varying distance from earth and consequently their varying brightness—and the sun revolving around an immovable earth at the center of the universe. This partial heliocentric view—with only two planets revolving directly around the sun—became a full-blown heliocentric theory when Aristarchus proposed that all planets including earth revolve around the sun.[11] While this theory was rejected in favor of the geocentric view, it was revived much later by Copernicus.

Traced back to Pythagorean cosmology are the first steps away from the prejudices of the geocentric and anthropocentric worldview and the inspiration for the discovery of the heliocentric worldview. However, perhaps due to scientific misrepresentation of the topic, the popular perception is that the heliocentric model is correct and the geocentric model incorrect. But the profundity of the heliocentric model is really this: (1) it is *another* point of view *as good as the geocentric*—though initially, like the geocentric, it, too, was incorrectly perceived as absolute, as if the sun were the absolute center of the universe (as the *Dialogue Concerning the Two Chief World Systems* of Galileo Galilei [1564–1642] was implying)—and (2) since another center is as good as the previous one, the notion of an absolute center of the universe is abolished. In fact, in modern physics the any-center view is correct. A particular center is chosen merely for its conceptual and mathematical convenience for the understanding of a physical phenomenon and is not to be misinterpreted as absolute or uniquely correct. This view is supported by special relativity (see next subsection). It is also supported by astronomical observations, including the discovery of numerous new galaxies, each with billions of stars revolving relative to the galaxy's center, and generally observations indicating that the universe is isotropic, thus no one location is more special than another. Finally, such view may be accepted based merely on pure humility, that neither the earth nor the sun should occupy a special center, and in general that no point in the universe should be more centered or privileged than another. The universe has neither an edge nor a center, and the laws of physics apply equally the same everywhere. "The merit of the Copernican hypothesis [that (1) annually *earth* revolves around the sun, not the sun around the earth and (2) diurnally *earth* rotates on its axis, not the sky around earth] is not *truth*, but simplicity; in view of the relativity of motion, no question of truth is involved."[12] Equally correct (as will be emphasized a bit more at the end of the next subsection) we could imagine that (1) annually either the earth revolves around the sun or the sun around earth and (2) diurnally either earth rotates on its axis, or the sky revolves around earth. Space and time were absolute in Newtonian physics but became relative in Einstein's theory of special relativity. This means that for relativity an *absolute* frame of reference—a special location of observation that can be used to refer to absolute motion—does not exist. There exist only *relative* frames of reference that can be used to refer to relative motion. Hence we can choose any center *relative* to which something can be at rest or in motion. But a special center for absolute rest or absolute motion is utterly meaningless. Space, time, and motion are all relative. Let us elaborate.

Newton versus Einstein

In Newtonian physics space and time are absolute and thus independent of an observer's relative motion. This means that space distances and time intervals are unchanged by motion. For example, the length and mass of an object are the same for all observers independently of their location or motion relative to the object or relative to one another. The same for them is also the way time passes. Twins, for instance, have always the same age with respect to one another, whether they move or not relative to each other. In general, two events that are simultaneous for one observer are simultaneous for every observer— absolute simultaneity. Space is a kind of preexisting passive (unaffected, in a sense "disconnected" from everything else) playground where objects exist and events occur while time flows steadily in the background the same exact way for everyone (so time, too, is unaffected by everything else). But Einstein's theory of special relativity proved all these to be false, in spite of the fact that all these are how we experience the world daily.

In special relativity the speed of light in a vacuum, designated c, is always 671 million miles per hour—put differently, in one second light travels as far away as is the distance of eight times around the earth. It is the same in all reference frames (for all observers, moving or not relative to the light source). It is also a kind of cosmic speed limit, for although it could be approached, absolutely no material object can travel as fast as or faster than light. This fact has nothing to do with engineering. It is not because we don't have high-powered engines to accelerate an object to the speed of light; rather, this fact is how nature behaves. It is a law of nature that has withstood the scrutiny of experiments since 1905, the year it was postulated by Einstein in his theory of special relativity.

Two of the most dramatic consequences of the constancy of the speed of light concern space and time: they are no longer absolute, they are relative—dependent on an observer's relative motion. They are combined mathematically (by the so-called Lorentz transformation) into a continuum called spacetime. Space distances and time intervals *do* change with respect to an observer's relative motion. Relative space means that a moving object contracts in the direction of motion, as seen by (relative to) a stationary observer—a phenomenon known as length contraction. Relative time means that the passage of time in a moving clock (say, aboard a moving spaceship) is dilated; it is slower relative to the passage of time in a stationary clock on earth—a phenomenon known as time dilation. To function properly the Global Positioning System (GPS) takes time dilation into consideration. If it didn't, the GPS receiver in your car would miss your destination. Parenthetically, the GPS must also take into account another time effect, predicted by general relativity. That clocks in orbit, where gravity is weaker compared to the ground, run faster relative to clocks on earth.

Interestingly, time dilation makes time travel possible because using it we can travel into the future. Suppose Earthly and Heavenly are twins. Heavenly likes to journey in space, while Earthly prefers to stay on earth. If Heavenly travels at a speed close to c upon her return to earth she will realize that she has aged less than Earthly (and all the other people or things on earth). How much less depends on the duration of her trip as well as how close her speed was to c. For example, if her speed was 99.5 percent of the speed of light, then for every one year that Heavenly ages during her trip, Earthly ages ten years. But if her speed was 99.99 percent of the speed of light, then for every one year that Heavenly ages, Earthly ages seventy-one years. So a twin who takes a trip into space will age less with respect to the twin on earth. Here we emphasize that Heavenly feels no different, as regards the passage of time, while traveling. The difference in age is noticed when the twins compare notes, for example, meet again.

In general, if you travel at a speed close to the speed of light, the time elapsed for you will be less compared (relatively) to the time elapsed for those not traveling with you. Hence, because of time dilation, you can then travel into the future of those not taking the trip with you; like the astronauts in the 1968 film *Planet of the Apes*, who aged only eighteen months during their near-light-speed journey and returned to find a post-apocalyptic earth where the elapsed time since they left was 2,006 years. So by taking a trip at high speeds you may return to earth at some future century of your choice. You may enjoy great developments of a more advanced civilization in that future century. The downside, if it is too far into the future, none of your familiar people may be alive to welcome you. Would such a trip be worth taking?

There is yet another fascinating consequence of the constancy of the speed of light. It allows us to see the past. In fact we do it all the time. Looking out in space is looking back in time. And the further out we look, the further into the past we see. This is so because starlight takes time to travel from the distant stars to our eyes. The speed of light is not infinite, so light messages are not transferred instantly. The light from the sun, for example, takes about eight minutes to reach our eyes. This means that observing the sun at, say, 12:00 noon is actually seeing how the sun was at 11:52 a.m. But Polaris, the North Star, is about 434 light-years away from us (i.e., its light takes 434 years to reach us). So looking at Polaris tonight is actually seeing how Polaris looked 434 years ago. Polaris may not even be there now!

There are still other effects of special relativity. A moving object becomes more massive relative to an identical one at rest. Also events that are simultaneous for one observer may not be so for another observer at a different location and/or in motion relative to the first observer. All these so-called relativistic effects become evident only at high speeds though, those comparable to the speed of light. Because the everyday phenomena involve speeds so much smaller than the speed of light, we are tricked into thinking that Newtonian physics is true. It is nevertheless an excellent approximation of truth, for at the limit of low speeds the equations of special relativity reduce to the Newtonian ones. The equivalence of mass and energy (expressed by the equation $E = mc^2$), length contraction, time dilation, and the relativity of simultaneity, are some of the most startling consequences of special relativity.

In light of special relativity, for which space and time are relative, as regards annual motion, according to the geocentric model, relative to the earth (that is, relative to an earthbound observer) the sun appears to revolve around the earth in one year. But equally correct, according to the heliocentric model, relative to the sun (that is, relative to a hypothetical sun-bound observer) it is the earth that appears to revolve around the sun in one year. Likewise, as regards diurnal motion, the sky (with the sun and stars) appears to revolve westward relative to the earth daily (and so the sun appears to rise from the east and set in the west relative to the earth daily)—recall that in our earlier analogy the walls appear to revolve clockwise relative to you. But equally correct, the earth appears to rotate eastwardly on its axis relative to the sky daily—you appear to rotate counterclockwise on your axis relative to the walls. This difference (of what is moving with respect to what) "is purely verbal; it is no more than the difference between 'John is the father of James' and 'James is the son of John.'"[13] The view of the daily eastward rotation of the earth (relative to the sun and all other stars in the sky) is more economical (for only one object, earth, is in relative motion) than the view of the daily westward revolution of the sun and of the myriad stars of the sky (relative to the earth).

CONCLUSION

Within the context of the most advanced theory of matter, the quantum, "things are numbers" indeed. What's more, many apparently unrelated things (phenomena) have already been unified; they are found to obey the same fundamental mathematical equation and thus the same natural law (e.g., the electroweak unification). These findings point clearly to the subtle cosmic interconnection (of mathematical nature) anticipated by the Pythagoreans. But in addition to the aid of mathematics, to find the Logos (reason) of such inconspicuous connections, one needs to be unconventional, to be able to unite diverse fields of knowledge, and to focus a keen eye on the elusive. For only then may one unveil the common characteristics that different phenomena have in all of nature's changes, the perceptible but also the discreet.

CHAPTER 12

HERACLITUS AND CHANGE

INTRODUCTION

Everything is constantly changing, and nothing is ever the same, Heraclitus (ca. 540–ca. 480 BCE) proposed, and in accordance with Logos, the intelligible eternal law of nature. Thus everything is in a state of becoming (in the process of forming into something) instead of being (reaching or already being in an established state beyond which no more change will take place). This means that things, *permanent* things, no longer exist—for they contradict his theory of constant change—only events and processes exist. His doctrine has found strong confirmation in modern physics, for, according to it, absolute restfulness and inactivity are impossibilities; and all happenings, it is speculated, are consistent with a single universal law.

The philosophy of Heraclitus is often stated as aphorisms and has cosmological, ontological, and anthropological significance. The depth of his thoughts as well as the ambiguity of certain concepts he used, such as Logos, fire, and god, earned him the characterization as a dark and enigmatic philosopher. When Socrates was asked to comment about Heraclitus's treatise he replied, "What I understand is excellent; what I don't probably is too, but it would take a Delian [skillful] diver [of the intellect, like the divers from the island of Delos who dive in the deep sea] to recover it."[1]

STRIFE AND HARMONY

For Heraclitus everything in nature is characterized by opposites that are struggling. "We must recognize that war [the competition between opposites] is common, strife is justice, and that all things happen according to strife and necessity."[2] So without strife, as Homer had wished, the universe would be led to its destruction because events and processes could not have existed without some force that promotes change: "Heraclitus criticizes the poet who said, 'would that strife might perish from among gods and men' [Homer *Iliad* 18.107]; for there would not be harmony without high and low notes, not living things without female and male, which are contraries."[3] Hence "strife is justice" because change, for Heraclitus, is caused by the strife of the opposites. Without strife change would not occur.

Now, like Anaximander, Heraclitus, too, requires cosmic justice by such strife. In fact, he argues that not only is absolute dominance not allowed by any of the opposites but quite the reverse, that harmony is born from their strife. "Attunement [harmony] of opposite tensions, like that of the bow and the lyre."[4] This harmony of strife is the result of a subtle underlying unity shared by the opposites; generally, it is the result of the common characteristics that different things have. For example, the property of mass is common to both the different objects the earth and the sun. As a result (according to Newton's third law discussed below) each body attracts the other with the same strength! Discovering and understanding such unity is understanding Logos, but to manage this is difficult because "nature loves to hide."[5] Nonetheless Newton and several scientists thereafter (such as those who created quantum theory) have managed it.

ACTION-REACTION

Newton's action-reaction law, his third law of motion, describes the strife of opposite forces but also their subtle unity and harmony. According to it, for every action force there is an equal reaction force in the opposite direction. For example, the force exerted on a nail by a hammer has the same strength and is in opposite direction to the force exerted on the hammer by the nail. The competing opposites are the competing forces acting in *opposite* directions, but they do so with *equal* strength—so their unity in strength is mathematically expressible, in other words, action force = reaction force. A force (the action) cannot exist by itself; it exists only in relation to its opposite (the reaction)— thus Homer's wish to eliminate strife is unrealizable in Newtonian physics. In fact it is generally shown in physics that "physical action always is *inter*-action, it always *is* mutual."[6]

To appreciate the depth of Newton's third law—in light of Heraclitus's philosophy of the hidden unity of opposites in strife—it suffices to discuss a few more examples: (1) in Newtonian gravity, the earth is attracting you *downwardly* with the *same exact* force as you are attracting the earth *upwardly*! To find out how strong this mutual force is, just jump on a scale and read your weight. Even a freely falling object attracts the earth upwardly *as strongly* as the earth is attracting it downwardly. However, because the mass of the earth is huge, its upward acceleration (toward the object) is negligibly small, but the object's downward acceleration (toward the earth) is noticeable. (2) Analogously, the earth and sun attract each other with forces of *equal strength* (unbelievable but true) and opposite directions, and as a result both celestial bodies move harmoniously through space and time. (3) Both bow and lyre (in the example of Heraclitus) obey Newton's third law, too. In a bow, the cord is pulling each of the two limps (the flexible upper and lower parts of a bow) in one direction— the action force, which is along the cord and toward its midpoint. Whereas the limps respond by pulling the cord in the opposite direction—the reaction force, which is also along the cord but away from its midpoint. Furthermore, action and reaction are forces of equal strength. So the bow's apparent rest is really the result of the constant strife between opposite tensions, the action and reaction. Newton's third law applies even while the cord is being drawn in order to shoot an arrow (or as the cord is being released and shooting the arrow); or as the strings of lyre are at rest, or as they are plucked, producing their sweet notes of music. Even more impressive is that the apparent inactivity, at the macroscopic level, of the bow or lyre at rest (or *any* other object), is, at the microscopic level, really a frantic and endless activity of particle exchange; for force, on the microscopic level, is really an eventful process. And constant change, even the imperceptible, is indeed a fact.

FORCE IN QUANTUM THEORY

According to the standard model of quantum theory the forces of attraction or repulsion between the particles of matter (the quarks and leptons) are caused by the constant exchange of particles of force—called force-carrying particles or messenger particles since they carry the message of the force. The exchange of force particles transfers energy between the particles of matter, causing a change in their own energy, speed, and direction of motion and making them attract or repel.

The massless photons mediate the electromagnetic force; the massless gluons transfer the nuclear strong force (gluons, "glue," bind the quarks to form protons and neutrons for example); the massive W^+, W^-, and Z^0 particles (of positive, negative, and zero electric charge, respectively), the nuclear weak force; and the massless gravitons are speculated to mediate gravity.[7] As regards the gravitons, a theory that describes them has yet to be discovered, and, equally importantly, no experiment so far has confirmed their existence.

The electric repulsive force between two electrons, for instance, is mediated by the continual exchange of photons that, traveling at the speed of light, are emitted and absorbed by the electrons. Namely, one electron rebounds by emitting a messenger photon, and the other electron rebounds by absorbing the photon. Repeated processes of this kind mean that the exchanged photons knock the interacting electrons further and further apart. It is this continual exchange of photons that manifests itself as the electric repulsive force between the two electrons. Similar processes can explain the other forces.

Through the continual exchange of the particles of force, the particles of matter move nonstop and combine with one another to form atomic nuclei, atoms, molecules, and composite objects like bows and lyres. Thus even an apparent static equilibrium of an object at the macroscopic level, down to the microscopic level, is really an eventful, complex, and endless process of particle exchange. Nature *is* constantly changing.

LOGOS

Newton's third law of motion or the more detailed description of a force by the standard model may be viewed as part of Logos. In the third law the underlying unity is the equality of the strength of the opposite forces. In the microscopic interpretation of force, unity is expressed by the conservation laws obeyed by the particles through their interactions (strife); that is, as the particles of matter collide with the particles of force, their net energy (or momentum, to name just two properties that are conserved) before collision *equals* their net energy (or momentum) after collision—again, the Heraclitean unity between competing opposites, is expressible mathematically in physics, that is, energy before = energy after, or, momentum before = momentum after. Of course the actual equations are more descriptive, detailed, and written with mathematical symbols.

Also, matter and antimatter are opposites in strife. Their Logos are the various mathematical laws that they obey, including gravity (the Newtonian concept or Einstein's theory of general relativity), electromagnetism (of Maxwell or quantum theory's), the standard model, string theory, and so on. And the underlying unity consists of the various conservation laws with which each process involving matter and antimatter must comply. The resulting harmony in the strife of matter and antimatter is the general organization of the world (a notion to be revisited at the section "Organization").

In modern physics we are striving to understand various phenomena, first by isolating them and finding which laws they obey. But as in Heraclitean philosophy, according to which true understanding is achieved by identifying common characteristics that different things have, the real picture emerges only when we manage to connect our understanding of isolated and seemingly different phenomena and discover the bigger truth, the Logos they all obey. In modern physics one of the key scientific principles, which is part of Logos, is the Heisenberg uncertainty principle. It will help us understand the doctrine of Heraclitean change from within the context of quantum theory.

THE UNCERTAINTY PRINCIPLE

The most consequential, mind-boggling law of quantum theory—its very heart and soul—is the Heisenberg uncertainty principle. This principle discusses how nature limits our ability to make exact measurements regardless of how smart or patient we are or the sophistication of our experimental apparatus. Namely, as a consequence of the very act of observation, the observer disturbs the object being observed a certain minimum way, causing the result of a measurement to be uncertain. We can measure very accurately the position and velocity of a large-mass object, such as a car or a planet, without significantly disturbing it. We can watch it move and even predict its path of motion. But if instead we had a small-mass object, such as a microscopic particle—an electron, a proton, an atom, even a molecule—we could not measure exactly both its position and its velocity; nor could we observe it in a path of motion or predict its path. Before the uncertainty principle was discovered, absolute accuracy in a measurement, at least in theory, was considered axiomatic, but not anymore.

Suppose we want to observe an electron, hoping to "see" where it is and determine how fast it is moving. To do so we, the observer, must shine light of certain wavelength ("color") upon it—bounce a photon off it. The light (the photon), which is scattered by the electron, will then enter our microscope, be focused, and be seen by our eye. It is the scattered photon that we actually see in an act of observation. Now to illustrate how the observation itself creates the uncertainty in a measurement, we discuss such act in two steps. Step one discusses what happens to the electron when light is shined upon it—when the photon collides with it. Step two discusses how clearly the electron can be seen through the microscope. It is the combination of the effects from these two steps that produces the celebrated uncertainty principle.

Step One: The Collision

As a result of their collision, the bouncing photon transfers some of its energy (and momentum) to the electron and disturbs it (much like when one billiard ball disturbs the motion of another when they collide). But there is no law that can determine the amount of energy imparted on the electron by the photon.

Thus the photon pushes and disturbs and changes the velocity of the electron unpredictably. This means that the electron may have a range of possible recoil velocities, hence its velocity cannot be known precisely: there is an uncertainty in its velocity. On the other hand, the disturbance introduced by a photon bounced off a car or a planet is undetectably small because, compared to an electron, the mass of a car or a planet is huge; just think how much more difficult it is for anyone to push and disturb a real car, which weighs a lot, compared to pushing a toy car, which does not weigh much. The velocity and location of a car or planet can be measured almost with absolute precision. This is in fact another reason that classical physics (e.g., Newtonian physics), which does not include the uncertainty principle, works quite well for macroscopic objects.

Now, concerning the electron, we can reduce the uncertainty in its velocity by using a photon of smaller energy so that its push to the electron is gentler. But a photon's energy is inversely proportional to its wavelength: the smaller the energy, the longer the wavelength (the "redder" the color is), a relationship that brings me to step two. Unfortunately, while a photon with a longer wavelength has less energy, which reduces the uncertainty in the velocity of the electron, it simultaneously increases the uncertainty in the position of the electron—the image of the electron gets fuzzier. Why?

Step Two: The Microscope

Because the determination of the position of the electron depends on the wavelength, too. This dependence, which is known as the resolving power of a microscope, regulates how clearly something can be seen—how well the scattered light can be focused and thus how accurately the electron can be located. The longer the employed wavelength, the fuzzier the image of the electron will be, and the greater the uncertainty in its position. What we see through the microscope is really a fuzzy flash created from the photon scattered by the electron. The electron, which is a point-particle, is somewhere within this flash, but where exactly is indeterminable. Its position cannot be known precisely. The flash may be focused into a region no smaller than the wavelength of light (the law of the resolving power states). Hence the uncertainty in the position of the electron may be equal to or greater than the wavelength of light, but never smaller than it! So at best, the minimum uncertainty is equal to the wavelength. Since the position cannot be known precisely, the electron has a range of possible locations it can occupy, just as it has a range of possible recoil velocities to move with.

Given that light of zero wavelength does not exist—that is, we cannot observe if the light source is turned off—the uncertainty in the position can never be zero—we cannot see with *absolute* precision where the electron is. Nonetheless, we can reduce the uncertainty in the position by using light of a smaller wavelength, though unfortunately this action simultaneously increases the uncertainty in the velocity—for as seen in step one, the smaller the wavelength, the greater both the light energy and the disturbance imparted on the electron (i.e., the greater the range of possible recoil velocities).

Position-Velocity Uncertainty

The wavelength of light used in an observation has conflicting effects; there is a trade-off in the determination of the position and velocity of a particle. The result is the position-velocity uncertainty principle: the more precise the position, the more uncertain the velocity, and vice versa. Heisenberg proved that the product of the two uncertainties can never be less than a certain minimum positive number—which is roughly equal to Planck's constant, a fundamental constant of nature, divided by the mass of the particle.[8] Consequently, *absolutely* precise knowledge of either property is unattainable because if one of the uncertainties were zero, their product would also be zero, a result that would be in clear violation of the principle. In classical physics, on the other hand, of which the uncertainty principle is not part, these uncertainties could each be zero—thus a particle's position and velocity could, at least in principle, be determined exactly—leading to what is known as classical determinism, which is the opposite of quantum indeterminism (that is, quantum probability), the consequence of the uncertainty principle.

CLASSICAL DETERMINISM VERSUS QUANTUM PROBABILITY

In the macroscopic world of classical physics, by knowing the forces that act on an object as well as the object's exact position and velocity at some initial time, we can determine its exact position and velocity (its trajectory) for all future time. So its motion is precisely determinable: a path can be plotted, even watched live, point by point continuously from an initial instant to any future one. Because of this capability classical physics is said to be deterministic. We can plot the precise orbit of a space shuttle, for example, just by knowing the forces acting on it and the initial conditions (its position and velocity at some initial instant), and we can watch it fly through space and time as predicted by our equations. It is therefore easy to predict a solar eclipse—when the earth, the new moon, and the sun will align—but absolutely impossible to measure or predict where an electron in an atom is or will be. Why?

Because, according to quantum theory, the subatomic world of particles is profoundly different than everyday experience; it cannot be described by classical physics. Inherent in the uncertainty principle, which limits the accuracy of a measurement, particle properties (such as position, velocity, momentum, and energy) cannot be assigned an exact value, neither initially nor at any time later. Thus they must unavoidably be expressed only as probabilities, which then lead to quantum indeterminism. Solutions of the so called Schrödinger equation can be used to calculate such quantum probabilities—for example, the probability of finding a particle at a certain location at a certain instant of time. A probability is a number that represents the tendency of an event to take place, not its actual occurrence. Hence, the best we can do is theoretically to predict only *probable* outcomes, and experimentally measure *also* only probable outcomes. Consequently, a particle's path of motion can neither be predicted (nor plotted), nor can it be observed; it is indeterminable: establishing a definite, traceable, point-by-point orbit is an impossibility. In fact, the very notion of an orbit is inadmissible in quantum theory. The determinism of classical physics is therefore replaced by the probability of quantum theory. And the consequences of this fact are staggering. The true nature of nature, for example, is different than the way it appears to be through mere observation.

OBSERVATIONS ARE DISCONNECTED EVENTS

During the act of observation all we see through the microscope is a flash of light somewhere within which the particle exists. But where exactly it is within this flash at each instant, and what it does when we are observing it, whether at rest or in motion, are all indeterminable. Even worse, nature does not allow us to know what happens between consecutive observations. Consecutive observations have time and space gaps; flashes are seen one at a time and spatially separated. Hence, inherent in the uncertainty principle observations, *any* observations of both the microscopic and macroscopic world, are always disconnected events! Roughly speaking, it is as if we are observing nature by continuously blinking.

This is a profound result and in direct contradiction with apparent reality according to which the changes in the daily phenomena are observed to occur continuously. The act of watching an arrow in flight, for example (an interesting thought experiment to be revisited in the section "The Arrow Paradox" in chapter 14) is really a series of disconnected observations, which to our imperfect eyes appear to occur continuously only because the time and space gaps between subsequent observations are undetectably short. The arrow's apparent continuity of motion is therefore an illusion. The shortness in these gaps is, incidentally, a consequence of the fact that in an observation, the disturbance introduced by a photon bounced off a macroscopic object such as car or an arrow is undetectably small because these objects have comparatively more mass than the mass of microscopic objects such as electrons and protons. This is, recall, also the reason that macroscopic objects have (actually, appear to have) definite orbits while microscopic objects do not.

So observing anything, anything at all, can happen only discontinuously. It is, roughly speaking, like cinematography (motion pictures), where a series of separate drawings, each, say, of a ball at a different position, is flashed before us rapidly (with short time gaps). Now, (1) if the position of the ball is changed *gradually* in each subsequent drawing, that is, the distances (the space gaps) between each new position of the ball and the previous one are sufficiently short, then, when the drawings are flashed before us, the ball is observed to move continuously (thus with a definite orbit). But this continuity in observation is really an illusion of the deceptive senses that cannot notice the short gaps. Case (1) corresponds more to how we observe macroscopic objects. On the other hand, (2) if the space gaps are sufficiently long, then, when the drawings are flashed before us, the ball is observed to move discontinuously. Case (2) will correspond more to how we observe microscopic particles, but only after the following two modifications: first, do not think of the ball to be the actual particle; rather, it roughly corresponds to the flash of light somewhere within which the observed particle exists; and second, as we will argue in "Nature as a Process" below, even if we do observe a similar *type* particle (say, an electron) at consecutive observations, it is indeterminable if it is the same *particle* (electron), even when the time and space gaps between observations are short. These two modifications, which capture more accurately how we observe microscopic particles, make it impossible to plot a definite path of motion for any microscopic particle.

Now, the reason that the phenomena are observed to occur discontinuously might be that the very phenomena *themselves* occur discontinuously (even when we are not observing); they may not just be *observed* to occur discontinuously. In any case, the discontinuity in observations has astounding consequences: in the section titled "Nature as a Process" we will use it to question the very identity of a particle, and in chapter 14, "Zeno and Motion," we will use it to question the reality of motion *itself*.

CHANGE

The Heraclitean doctrine that everything is constantly changing and nothing is ever the same has three implications. First, there is a change; second, the change is constant; and third, because nothing is ever the same the constant change is unidirectional. Modern physics agrees with all three: first, change occurs in two different ways: (1) through motion and (2) through the transformations of matter and energy; second, the uncertainty principle of quantum theory as well as general relativity affirm that change is constant, and in addition, as seen just above, quantum theory ascertains it is also discontinuous; and third, the second law of thermodynamics discusses how change is unidirectional—the universe becomes increasingly disordered.

(1) Motion Causes Change

Change caused by motion is discussed in the following three cases.

A. The motion of the particles of matter causes their rearrangement in an object and consequently causes change in its various qualities (e.g., density and temperature). For example, atoms are more compressed in denser objects and jiggle faster in hotter objects.

B. Space itself is changing because matter distorts it, a phenomenon that can be understood when we describe how gravity works within the context of the theory of general relativity.

Gravity, in general relativity, is explained by giving space properties, namely, by regarding it as a flexible medium distorted by matter—like a trampoline surface that is stretched and warped by a bowling ball resting or moving on it. In the case of the earth and the sun, for example, the distortion of space caused by one body influences the motion and is felt as gravity by the other.

In a simplified analogy, the flexible trampoline fabric (which plays the role of space) is curved when a bowling ball (which plays the role of the sun) rests on it. The geometry (shape) of the fabric depends on (1) the mass of the bowling ball and (2) the distance from it: (1) the more the mass, the more curved the fabric (space) becomes; (2) the closer to the bowling ball, the greater is the curvature of the fabric. The distorted fabric in turn influences the motion of a small marble (which plays the role of the earth) rolling on it. Depending on how we start the marble moving (i.e., with what initial speed and direction, and from what location), it will move on the distorted fabric around the bowling ball by following a particular path (circle, ellipse, parabola, spiral, straight toward the bowling ball, etc.), and thus will appear to be attracted by the bowling ball. For example, a marble released from rest moves on the distorted fabric caused by the bowling ball and plunges onto it, like an apple falls from its tree onto the ground by moving through the distorted space caused by the earth. In the trampoline analogy the distorted fabric *is* gravity; and the greater the bowling ball mass, or the smaller the distance from it, the stronger gravity (the distorted spacetime) becomes.

In the earth-sun case the sun distorts the spatial fabric around it (and time, too—it passes more slowly as one gets closer to the sun). Traveling at the speed of light, this distortion reaches and affects the motion of the earth—analogously, a water disturbance, a water wave, travels in the sea but with a much smaller speed. In turn the earth traverses the distorted space as if space pushes the earth through it. Of course the earth distorts the space around itself, too (although its smaller mass produces a much smaller distortion than that of the sun); and so do the moon, planets, stars, and galaxies. Gravity is really the twists, curves, ripples, bumps, depressions, and in general all these distortions (the changing geometry) of spacetime. And each body's motion is actually a response to the space distortions from all other bodies around it. Because these bodies are in constant motion in the universe, the pattern (the geometry) of space distortions that they create is in a state of constant flux—and so the motion of matter causes change in the geometry of space. In our analogy, as it rolls, the bowling ball transfers the warping of the trampoline surface to different locations. Because both space and time are distorted by matter, spacetime in general relativity becomes a four-dimensional malleable (distortable) continuum. In turn, these spacetime distortions, which we usually call gravity, influence the motion of matter.

C. In addition to its constant warping, space as a whole is also expanding, thereby carrying all the galaxies with it and causing them to move away from each other. Here again, motion, which in this case is a consequence of the expansion of space, produces change. Known as the expansion of the universe this was first predicted theoretically by the solutions of the equations of general relativity, shortly after their publication in 1916. It was later confirmed experimentally by astronomer Edwin Hubble (1889–1953) in 1929 when he observed a redshift in the light emitted by the distant galaxies. The redshift is a measure of the relative velocity between a galaxy and the earth. Specifically, it means that distant galaxies are rapidly receding from us. The greater the distance, Hubble discovered, the faster the recession speed, a result known as Hubble's law. This law is included in the big bang model.

Galaxies are not moving out into preexisting space, a common misinterpretation of the phenomenon of expansion, but they are moving away relative to each other, and what carries them is space itself as it is expanding (stretching); the result is that the size of the universe is increasing with time. Furthermore, the recession of galaxies does not make the earth the center of the universe or in any way a more special place than any other. Quite the opposite, because the universe is isotropic the expansion would look the same from any location in the universe. In a classic analogy, imagine how any dot (i.e., galaxy) on an inflating balloon is seen receding from the perspective of any other dot as the expanding membrane (i.e., space) itself carries all dots with it. Since galaxies are observed as receding from each other and the universe as expanding, in the past they must have been closer to each other, and the universe must have been much smaller—imagine our balloon to be deflating. In the extreme case, the whole universe (all the galaxies, all matter, energy, and space, even time) is imagined to have been a mere point, its explosion of which is the premise of the big bang model. Whether such a point-like universe actually existed is still only a hypothesis, and why it exploded is still puzzling. Nonetheless, we do know that the universe must have been very, very small (if not point-like) when it exploded, as described by the big bang theory, and that it has been constantly expanding (stretching), thus changing ever since.

(2) Transformations Cause Change

Transformations of matter and energy also cause change in the various qualities of objects (or the universe in general) via two types of processes: first, when the various particles of energy convert into particles of matter, back and forth, materializing and dematerializing; and second, when one type of material particle converts into another. An example of the former is the materialization of the energy of light (or invisible gamma rays) into an electron-positron pair, or the dematerialization of such pair into energy (light or gamma rays); and an example of the latter is the conversion of two protons into a proton-neutron nucleus (a heavy form of hydrogen called deuterium), a positron, and an elusive neutrino (all common reactions in the stars energizing them with their light). Of course more everyday-type transformations of matter and energy, such as from solid to a liquid (as the melting of ice) or liquid to a gas (as the evaporation of water) cause change, too.

But things are not merely changing, they are *constantly* changing, a conclusion required by the uncertainty principle.

Change Is Constant

To avoid violating the uncertainty principle, motion in nature must be perpetual. If a particle could sit still, it would mean that its velocity would be exactly zero and so then would be the uncertainty in its velocity. Consequently, the product of the position and velocity uncertainties would also be zero, a result in violation of the Heisenberg uncertainty principle. The principle holds only if motion is perpetual. A particle cannot sit still, ever. This result is also supported by the third law of thermodynamics, which states that the absolute zero—the lowest possible temperature for which every particle in a substance would be motionless—is unattainable (temperature is a measure of particles' average energy of motion). Now, since the motion of particles is perpetual, so is change. Incidentally, since motion is constant, then motion is also involved even when change is caused by the transformations between the particles of matter and the particles of energy.

Support of the "constant change" view comes also from the expansion of the universe that has been happening ever since the big bang. What's more, the spacetime continuum fluctuates constantly at microscopic scales, like a turbulent sea—the distortions of space are changing violently—a phenomenon resulting from efforts to reconcile the theory of general relativity with the uncertainty principle of quantum theory.

Change Is Unidirectional

But change is not merely constant; it is also unidirectional, meaning nothing is ever the same; "you could not step twice into the same river."[9] Heraclitus parallels the constant unidirectional change in nature to the ever-changing waters in a river. If the state of a river at one moment were ever the same as the state at another moment, it would have been possible for one to step twice into the same river. Since one cannot do that, nothing ever is the same, and so change is not only constant; it is also unidirectional. In fact, one "could not step twice into the same river" not only because a river's waters are ever-changing, but also because one's own body is also ever-changing. *Everything* is constantly changing, and *nothing* is ever the same.

Now, according to the second law of thermodynamics, net entropy—the degree of disorder (randomness) in the universe—is always increasing. Thus, nature is in a state of becoming, but it is a disorderly becoming.

Indeed, then, everything is constantly changing, and nothing is ever the same! Heraclitus's doctrine of change includes everything, even the seemingly unchanging, such as a rock in its apparent state of rest or even a human body. In fact, even the gradual biological evolution by descent and variation—that the more complex life-forms do not arise spontaneously but evolve from simpler ones through modifications—is a principle to be expected as a consequence of the Heraclitean theory of constant change.

NATURE AS A PROCESS

Heraclitean View

A profound consequence of the Heraclitean theory of universal constant change is the view of nature as a process made up of events. For the notion of "a thing" is inconsistent with a theory of constant change. To be able to be spoken of and defined, the thing must remain absolutely the same for at least a period of time; it must have some permanence and must be identifiable. But the notions of sameness and changelessness are contradictory to a theory of *constant* change. Consequently, it is more appropriate to consider a thing as an event (something happening somewhere at some instant of time) and not as something permanent. Thus, what changes is not something material or initially permanent; what changes are the events. Groupings of events constitute processes, which in turn make nature the ultimate process.

Quantum View

This notion is supported by quantum theory. We will argue that microscopic particles are better understood to be events rather than permanent things.

We learned earlier that because of the uncertainty principle observations are disconnected events. Now, without continuity in observation, without the ability to keep a particle under continuous observation (even for the smallest time duration), how can we establish its identity or permanence? With time (and space) gaps between observations during (and within) which we cannot see what a particle is doing, how can we be sure whether, say, an electron observed at location A has moved there from location B, or whether it is really one and the same electron as that observed at location B, regardless of their proximity? We cannot! Since observations are disconnected events, consecutive observations of identical particles—such as electrons, all of which have the same intrinsic properties, for example, charge, spin, rest mass—might in fact be observations of two different particles belonging in the same family (e.g., two different electrons), and not observations of one and the same particle (e.g., the same electron) that has endured for a certain period of time. It is therefore impossible to ever determine whether the observations of two identical particles could actually be observations of one and the same particle, and consequently whether a particle endures for a period of time.

So without the ability to keep a particle under continuous observation, it is impossible to establish experimentally its identity or permanence. Because of this, the notion that a particle is an identifiable individual and a permanent *thing* breaks down (or, it is an ambiguous notion, to say the least). The alternative is to consider a particle to be an event.

General Relativity View

Particles, in the view of general relativity, can endure up until they convert to energy. Until then they are identifiable permanent entities because general relativity has not yet been reconciled with the uncertainty principle. Still, particles are events in general relativity, too. This is so because matter is intricately connected with the fabric of spacetime (they are continuously affecting each other). So as time is constantly changing, so are, in general, the properties of space and matter. Hence a point in the continuum of spacetime is regarded as an event. And so a particle occupying a space location at an instant of time is treated also as an event. Two events are separated by their spacetime interval, which involves a spatial distance and a time interval.

In conclusion, matter, energy, space, and time are all intimately linked, interacting with one another constantly, causing changing events and processes. Nature is the perfect process and therefore is in a state of becoming; nothing ever *is*. But what causes change in the theory of Heraclitus, and what causes change in modern physics?

FIRE AND ENERGY

A permanent primary substance of matter is contradictory to a theory of constant change. The only element of permanence in such theory is change itself. What really causes change then? For Heraclitus it was the "everlasting fire,"[10] and for modern physics the eternal energy. The strife of the opposites or the interactions of matter are fueled by fire and energy, respectively. Matter, the events as has been argued, is the result of the transformations of the fire or of the energy. Fire and energy also represent particular processes. They cause cooling and condensing or heating and rarefying or forming and dissolving. "The transformations of fire [energy] are, first of all sea [liquids]; and half of the sea is earth [solids], half whirlwind [gases]."[11] And the transformations of fire, as it is with energy, occur with measure—by obeying conservation laws—since in a metaphor Heraclitus argued, "All things [matter] are an exchange [is a transformation] for fire [of energy] and fire [and energy is a transformation] for all things [of matter], as goods [as one type of matter] for gold [transforms into another type of matter or energy] and gold for goods [back and forth]."[12] In another statement he writes, "This world, which is the same for all, none of the gods nor of the humans has created, but it was ever and is and will be, an everlasting fire [energy], flaring up with measure [conservation] and going out with measure [transforming back and forth between its various forms by obeying the law of conservation of energy]."[13] Change, which in each respective theory is caused by fire or energy, is guaranteed to always occur since fire and energy are conserved while they are also changeable. From all these qualities fire and energy seem to qualify as the primary substance of the universe in each theory. As argued by Karl Popper, if the world is our home, then for Heraclitus fire is not *in* the house, "the house [the *house*] is on fire," a "somewhat more urgent message."[14] Equivalently, we can argue, the world is energy.

Fascinated by the similarities between the Heraclitean fire and energy Heisenberg wrote: "Modern physics is in some way extremely near to the doctrines of Heraclitus. If we replace the word 'fire' by the word 'energy' we can almost repeat his statements word for word from our modern point of view."[15] And also, "Energy is in fact that which moves; it may be called the primary cause of all change, and energy can be transformed into matter or heat or light. The strife between opposites in the philosophy of Heraclitus can be found in the strife between two different forms of energy."[16]

ORGANIZATION

But while everything is constantly changing, and defining a permanent fundamental particle of matter is impossible, something must endure, at least for some time. For, in all this constant change, still, definable "things" such as rivers exist and so do we, and rivers are distinct from us, and one river (or one person) distinct from another. So both we and the rivers are constantly changing, but simultaneously both we and the rivers are recognizable. "We step and do not step into the same river; we are and are not,"[17] said Heraclitus. How can this be?

In a constantly changing nature, which is best regarded in terms of events, each event is different. But *collections* of events can endure, by creating a certain macroscopic average, which, for a period of time, is a recognizable "organization";[18] the identifiable plethora that we call things may then be viewed as different organizations.

From the above Heraclitean quote, in the part "we step . . . into the same river; we are . . ." Heraclitus treats the "we" and "river" as two identifiable organizations, thus as two different and permanent things—at least for some period of time. Whereas in the part ". . . and do not step . . . and are not," he seems to imply that while the "we" and the "river" appear as permanent things these are so only on the average. In reality neither "we" nor the "river" are ever the same.

Heraclitus realized that while a thing (an organization) can on average be identifiable for some period of time, strictly speaking the thing is uniquely different at each instant of time. And so while there may be a collection of events that can produce an identifiable organization called river, which on average persists unchanged for a certain time duration, this river is never ever really exactly the same. Logos (the cause of change, the law of nature), in the view of Heraclitus, is the only thing truly eternal. Although through a truly strict interpretation of a continuously changing nature without anything ever the same, even Logos should be changing. The modern physics equivalent of this is the hypothesis that the fundamental constants of nature, numbers that describe the various laws we have and are the reason the universe is what it is (such as the speed of light, Planck's constant, or the gravitational constant), might after all be functions of time. If this is discovered to be true, then the order and organization of tomorrow's nature (especially in the long run) will be so unknowingly different from today's, a real intellectual treat for those inquiring minds who like the constant search and discovery, for in such a case, the mysteries concerning the nature of nature will be forever changing, and so will our very knowledge about them.

CONCLUSION

Heraclitus declares the being (that which exists, nature) but identifies it with becoming. All follows from that: everything is constantly changing, material sameness is impossible, there is a plethora of different events that make nature a process, and described by warring opposites that nonetheless obey Logos. But Parmenides declares just the *Being*; only what is, is, what is not, is not. All follows from that: change, he argues, is logically impossible and so what is, is one and unchangeable! This dazzling and absolute monism is in daring disagreement with sense perception.

The Heraclitean and Parmenidean worldviews are therefore antinomies (contradictions), for starting from a being the two philosophers developed a unique series of logical arguments and arrived at opposite results: for the Heraclitean, being is becoming, but for the Parmenidean, Being just is. It is Heraclitean change and plurality versus Parmenidean constancy and oneness. But it is a controversial oneness, for Being's exact nature is uncertain.

CHAPTER 13

PARMENIDES AND ONENESS

INTRODUCTION

Philosophy was seriously shocked by the logic of Parmenides (ca. 515–ca. 445 BCE). Being the first philosopher of ontology, Parmenides questioned the nature of existence itself and created his monistic philosophy by contemplating the most fundamental of questions: How can something exist? And, what are the properties of that which does exist? And through purely rational arguments he marvelously reasoned out an answer that overturned completely the common perception of the world around us! In particular he asked, How could there be something instead of nothing? What does it mean to say that something exists? Can existence (nature) come to be from nothingness? Is there such a thing as nothingness? Has nature been caused by a primary cause—that is, by an absolutely first cause that permits no cause (no explanation) of its own? Does nature have an ultimate purpose that permits no purpose of its own? What is the nature of nature? Remnants from his profoundly abstract thought are present in modern cosmological models describing one indivisible and whole universe, unborn, eternal, and imperishable.

IRONCLAD LOGIC

First he argued that we can think only about that which exists, the *Being*, "for the same is the thinking and the Being."[1] On the contrary, he thought, we can neither speak about nor think about something that does not exist, *Not-Being*. For if we could it would mean that Not-Being had properties (those mentioned speaking or thinking about it). But true nothingness is property-less. Therefore, the notion of nothingness (Not-Being) is impossible! This is in fact the critical premise of Parmenides's theory. And to understand his arguments we must always remember that for him what does not exist, does not exist, neither now nor before or after, neither here nor there; that is, we cannot assume what does not exist now (or here), could exist later (or there), or could have existed before somewhere. No! Only what is, is—only Being exists. What is not, is not—Not-Being does not exist.

With this premise in mind he proceeded to figure out if change is logically possible. Change, he maintained, requires that the notion of nothingness exists. But since such a notion is an impossibility, so then is change; Being is unchangeable—for if it could change, it would change into something that Being is not already, into something new that does not yet exist, thus into Not-Being, but this is an impossibility, for Not-Being does not exist ever anywhere. Analogously, if it could change, it would cease to be what it once was, thus what once existed (Being) would no longer exist; it would become Not-Being, but this is again an impossibility for Not-Being does not exist. In other words, that which exists (Being) cannot change because change requires that the notion of nothingness (Not-Being) exist. Because only then could Being have been it (Not-Being) and could have again become it. Equivalently, we can say that change is impossible because it requires that something is either created from nothing or destroyed into nothing, but since the notion of nothingness does not exist, change does not exist either.

Being is also unborn—it is uncaused, that is, it has not been caused by anything, thus has no beginning—and it is imperishable—it has no end, no ultimate purpose. It just is! It could not be born, Parmenides thought, for if it could, it would be born from either (a) Not-Being—but this is an impossibility, for Not-Being does not exist—or (b) Being—also an impossibility, for something cannot be born if it already exists, that is, something cannot be born from itself. Analogously, Being cannot perish; for nothingness, which Being must become in order to perish, does not exist. Hence, what is (Being) just is; it neither comes to be from nothing nor perishes into nothing. This remarkable Parmenidean thesis was embraced by the pluralists Empedocles, Anaxagoras, and Democritus, and as we will see, it was applied in their own theories. Well, then, since Being is unborn, unchangeable, and imperishable, there is neither becoming nor passing away—nature just is!

Being is also always everywhere: there is neither a place nor a time where or when what is, is not already complete (e.g., of the same amount, appearance, and generally of the same properties). For if somewhere or sometime Being were less than complete, if it lacked something, it would mean that somewhere or sometime that something that Being would lack would not exist; it would be Not-Being, but since Not-Being does not exist, there is never any expectation for Being to be it. So Being is always complete everywhere. Hence, diversity and plurality are illusions of the senses. It is also motionless—for being always everywhere, there is never anywhere to go where it is not already. Similar-type arguments lead to the various properties of Being.

The nature of nature (of Being) is of the purest oneness: there is only one thing, Being. It is an indivisible eternal whole, unborn, unchangeable, imperishable, continuous, indestructible, finite, and uniform (always the same everywhere). This is a dazzling but provocative oneness, for it is (or so it seems, anyway) logically sound, yet it is also daringly in stark contradiction to apparent reality. And what its exact meaning really is depends on how these properties of Being are regarded, literally or metaphorically, of material or immaterial nature. Being is that which *is*. All other characteristics beyond that are quite uncertain and debatable, for what *is* (what exists) is *really* the question for Parmenides. But irrespective of its nature, Being captures a highly valued place in pre-Socratic philosophy and modern physics alike—namely, oneness!

THE NATURE OF BEING

Modern physics embraces a kind of monism and wholeness, too, for it tries to ultimately unify all four forces (gravity, electromagnetism, the nuclear strong, and the nuclear weak) and all particles under a single overarching principle in which there will be only one unified force or, equivalently, one type of fundamental particle, suggesting a subtle interconnection and oneness in all apparent plurality. Hence, should the properties of Being be interpreted metaphorically, Being might be a metaphor of the one, unchangeable, universal, eternal, objective truth of nature (a unified force of a theory of everything), which can be discovered through a combination of observation and rational contemplation. If so, Parmenides's philosophy is not necessarily against the notion of change but rather is in support of the view that the true way things change is not at all as perceived by the senses. I believe this view is reasonable since in his philosophy "the same is the thinking and the Being," that is, we can think only about something that exists; then since (1) we think about the observed changes, they must exist, though they must occur much more complexly than they appear (and in fact they do) because (2) we *also think about the complex and subtle ways that these changes occur* (through our scientific models, e.g., general relativity, quantum theory, biological evolution, etc.).

On the other hand, should the properties of Being be interpreted literally, then nature is one, uninterrupted, indestructible, indivisible, eternal, and material whole; it's a kind of full and solid block of matter without parts (uniform). Such a type of full nature implies that there is no void (empty space). The void is Not-Being for Parmenides, true nothingness—it does not exist. But interestingly without the void it is difficult, if not impossible, to accept that motion and change are real. For the easiest way to understand the occurrence of motion and thus change, too, is to imagine the existence of empty space within which things could move. Assuming there is no empty space, motion is an illusion and so is change. The atomists Leucippus and Democritus found this conclusion utterly absurd but inspiring as well. For to create their atomic theory and explain motion and change rationally they had to employ both: Parmenides's Being—its material nature in particular—and his Not-Being.

ATOM AND BEING, VOID AND NOT-BEING

So by giving Parmenides's ideas a straightforward, literal, and material meaning, the antithetical notions of his Being (existence) and Not-Being (nonexistence) evolved in the minds of the atomists into the antithetical notions of "the full"[2] (the atom), and "the empty"[3] (the void, empty space), respectively, and became the essence of their atomic theory. Incidentally, the intellectual continuity in our efforts to know nature is unquestionable in this case. Now, there were many atoms (Beings) with key properties of Being (i.e., whole, indivisible, indestructible, solid, with no parts) and lots of empty space (Not-Beings) within which atoms can move. Interestingly, although the atomists could not counter Parmenides's arguments against the existence of Not-Being, they still identified it with a certain kind of nothingness that *existed*, the empty space. But empty space's perception has been controversial ever since its conception.

NOTHING COMES FROM NOTHING

For Parmenides, there is no empty space, for empty space is nothingness, Not-Being, and Not-Being does not exist. How can something, which is nothing, really exist? Parmenides thought. How can something be defined and assigned properties when it is supposed to be property-less? It cannot, he argued. We are unable to even think of nothingness, he reasoned. Nothingness is a meaningless concept, for if nothingness existed, it would not really be nothingness; it would be something-ness. If we could refer to something and give it properties, that something could not be nothing; it would be something real and would exist.

Parmenides wanted to understand change, motion, and the empty space via purely logical arguments. For the empty space especially, he thought there was no good logical argument in support of its existence. Whether the empty space is a true nothing or not is a notion to be revisited in chapter 17, "Democritus and Atoms." Nonetheless, Parmenides thought that empty space was a true nothing, and as seen above, he also argued that *nothing* comes from nothing—Being neither comes to be from nothing (it is unborn) nor passes away into nothing (it is imperishable).

This principle has in fact a counterpart in modern physics in the notion of energy, which includes mass, since for relativity mass and energy are basically the same, as implied by $E = mc^2$. Like Being, energy can neither be created—it does not come to be from nothing—nor destroyed—it does not pass away into nothing. It just is and its total amount is unchangeable and enduring. Particles of matter and antimatter are constantly being created and annihilated, for example, but not out of nothing and into nothing. To occur these processes require something to already exist—namely, energy. They are created from energy and annihilated into energy. There is no mechanism in modern physics that violates the basic Parmenidean idea that something can neither be created from nothing nor pass away into nothing. All interactions require something, energy (or matter, since they are equivalent), but also space and time. Indeed, nonexistence is an impossibility even in modern physics, and the uncertainty principles of quantum theory (see next paragraph) may be considered as statements in support of that. Why use these principles? Because these principles are relationships between space, time, matter, and energy, concepts that constitute the essence of nature (of something-ness). And if we hope to prove that the notion of nothingness is an impossibility—that nothingness is not derivable from something-ness—well, we had better begin from an analysis of principles that describe the essence of something-ness.

So to argue for this (that nonexistence/nothingness is an impossibility) we first recall the position-velocity uncertainty principle: the product of the uncertainties in the position and in the velocity of a particle must be greater than Planck's constant divided by the particle mass—that is, such product is greater than zero. Analogously there exists the time-energy uncertainty principle. It states that the product of the fluctuations in the energy of a particle and the time interval that the particle endures must be greater than Planck's constant—again such product is greater than zero. For these uncertainty principles to hold spatial distances, time intervals, velocities, and energies are forbidden from ever being absolutely zero—that is, their nonexistence is forbidden. For example, the smaller the confining space of a particle is (or the briefer the time interval the particle endures in such confinement), the more frantic its motion and energy are. But neither the confining space nor the time interval can ever be exactly zero, for if they were, the uncertainty in position and the uncertainty in time would have been zero, too, and consequently both uncertainty principles would have been in violation—the product of the uncertainties in position and velocity, and in time and energy, would have been zero, too (instead of greater than zero). Similarly, both uncertainty principles would have been in violation if a particle had zero velocity or energy. The principles hold only if spatial distances, time intervals, velocities, and energies are nonzero; they must exist; they cannot be nothing! (In a sense such result is expected because our physics relationships, equations or inequalities, are in the first place conceived to describe *something*, not nothing; the notion of nothingness is indescribable.) Hence, as per Parmenides's reasoning and as per the uncertainty principles, nothingness is not only not allowed to exist—for *nothing* comes from it (e.g., the uncertainty relationships are violated and thus cannot be used to account for what exists)—but, equally profoundly, existence is required, that is, spatial distances, time intervals, velocities, and energies must be nonzero (for only then do the uncertainty relationships, which describe something-ness, hold).

In fact, one of the fundamental tenets of quantum theory is that information cannot be lost from the universe (recall the section "Black Holes: Challenges in the Quest for Sameness" in chapter 8). In Parmenidean terms this means that, what is—Being, information already present—cannot become Not-Being—information cannot be lost. So Stephen Hawking might be onto something with regard to his new analysis of black holes.

Quantum theory (the essence of which is the uncertainty principles) is then in accordance with Parmenidean philosophy, "for the same is the thinking and the Being": for we can think only about that which exists, in other words, the uncertainty principles describe only something-ness and forbid nothingness. With this in mind Parmenides's main question, How can something exist? may now be answered: within the context of modern physics, something (Being) must exist because nonexistence (Not-Being) is impossible. Now, what is the nature of that which exists?

AN INDIVISIBLE WHOLE

Relativity

The view of Being as an indivisible whole is supported by Einstein's theory of general relativity: for *everything that there is*, space, time, matter, and energy are no longer independent of each other (that is, they are not absolute), as was the case with Newtonian physics, but are intimately interwoven, affecting one another constantly. "Time and space and gravitation have no separate existence from matter."[4] Spacetime is a malleable continuum distorted by matter.

Yes, it is true that for the sake of practical calculations in physics we often isolate, in our mind, a phenomenon of interest by assuming that it is disconnected from the rest of nature (disconnected from the whole). For example, we study the gravitational interaction between the sun and the earth by neglecting the gravitational effects of the rest of the heavenly bodies. But as in the philosophy of Parmenides modern physics is about oneness, not isolation. And in reality all things in nature are part of the whole and are entangled far more intricately than the theory of relativity alone could discover.

Quantum Entanglement

One of the most fascinating consequences of quantum theory is the phenomenon of quantum entanglement. According to it, there are no perfectly isolated particles (or systems). The notion of an individual particle disconnected from the rest of the universe is inaccurate. Rather, all particles in the universe are part of a unified whole. They are in constant and *instant* interaction, affecting and determining the behavior of each other regardless of how far apart they are. Quantum theory suggests that everything that happens in the universe influences instantly everything else. In this sense the universe is indeed a Parmenidean indivisible whole. To explain this concept further we use the following thought experiment.

Suppose, for simplicity, that a mother particle could initially be at rest and with zero spin, and that later it decays into two daughter particles, A and B. To conserve momentum (linear and rotational) the daughter particles must take off away from each other as well as spin in opposite directions. In 1935, Einstein, with Boris Podolsky (1896–1966) and Nathan Rosen (1909–1995), argued through this thought experiment (which is known as the EPR, from the initials of their last names), that the daughter particles must have a fixed spin *since the moment of their creation*. To conserve rotational momentum one must spin clockwise, the other counterclockwise. Which particle spins in what direction is determined with a measurement. So if Alice measures that particle A spins clockwise, she is also certain that particle B must spin counterclockwise, as it is so confirmed when Bob measures it. Einstein's view is really the deterministic view of classical physics: that a particle has a fixed property *even before we measure it*.

But according to quantum theory, the spins of the particles A and B become fixed *only when an observation (a measurement) is made*. Until then, not only do we not know how the particles spin, but even worse and unlike Einstein's view, the particles *do not have a fixed spin*; each particle is assumed to spin simultaneously in both directions until a measurement is performed that will force them to take on a fixed spin—a peculiar concept, which is known as the Copenhagen interpretation of quantum theory.[5] It is this interpretation that Einstein found illogical and aimed to refute. And so did Erwin Schrödinger: to capture the peculiarity of the indeterminate spin state that particles A and B were assumed to be in before the act of measurement, he used a metaphor, the famous Schrödinger's cat. Briefly, he argued that according to the Copenhagen interpretation, until an actual observation is performed, a cat in a sealed opaque box, which also contains radioactive atoms with a chance to decay and spread poison, is both dead and alive at the same time. Namely, the state of existence of the cat before the observation is a mix of two possible outcomes because the status of the cat depends on the status of the radioactive atoms, which, per the Copenhagen interpretation, themselves are in a mixed state of two quantum probabilities, that of the decay outcome, which will kill the cat, plus that of the non-decay, which will preserve the cat. Only after opening the box and observing can the observer actually determine whether the cat is definitely either dead or alive. Before the observation, the cat is both dead and alive, in the Copenhagen interpretation. But according to classical physics even before opening the box, the cat is in a definite state of existence: it is either only dead or only alive.

So, according to the quantum view, the spin direction of each particle is fixed by the very act of measurement. For example, if Alice measures that particle A spins clockwise, *then and only then the spin of particle A becomes fixed* (contrary to Einstein's view, for which A would have been spinning clockwise since its creation); and, equally importantly, *then and only then the spin of particle B becomes fixed, too*, and it is counterclockwise (also contrary to Einstein's view for which B would have been spinning counterclockwise also since its creation). In general, measuring a property of particle A instantly forces a certain fixed property for particle B, even though particle B is not measured directly. This view, which is really the phenomenon of quantum entanglement, appeared absurd to Einstein because it meant, he argued, that the measurement of the spin of particle A affects and fixes *instantaneously* the spin of particle B, even when such measurement is performed while the particles are light-years apart and across the universe. This instantaneous "spooky action at a distance," as nicknamed by

Einstein, was not required by his analysis since according to it the particles had presumably fixed spin since their creation. How can such instantaneous influence exist, Einstein thought. How is it that measuring a property of one particle instantly affects and fixes an earlier indeterminate property of another particle? How is it that the very moment that the spin of particle A is measured, A communicates instantly how it spins to particle B so that particle B can spin opposite (to conserve momentum)? It is a strange type of communication that occurs faster than the speed of light, in fact instantaneously, and appears to violate one of the main principles of relativity, that information cannot be transferred with a speed faster than that of light because it would violate causality. This bizarreness caused Einstein to believe that quantum theory was not a complete theory of nature. Observing solely particle A would not in any way influence particle B, which is spatially separated from A, he thought.

But he was wrong. These opposite views appeared for a while to be part of the unverifiable realm of metaphysics. But in 1964 physicist John Bell (1928–1990) found a way to convert each point of view into an experimentally testable calculation, which is known as Bell's inequality.[6] The experimental verdict found Einstein's view false and favored the spooky action at a distance of quantum entanglement! Indeed, by measuring the properties of particle A we instantly affect the properties of particle B regardless of how far apart they are. And so, generally speaking, by measuring a property of one particle in a system, what we actually measure is a property of the whole system—which includes us, the observer, too—or, more precisely, of the entire universe. The universe is indeed an indivisible entangled whole. In the Copenhagen interpretation the observer is really part of what he observes—there is a mutual influence between observer and what is being observed. Whereas in classical physics the observer is thought of as an outsider separated from what he observes—there is no influence at all between observer and what is being observed. The classical-physics view of an observer is therefore like someone watching a movie—if the movie is nature, then an observer eating popcorn and drinking soda while watching has no influence on the movie plot (on nature). Whereas the quantum view of an observer is like someone being in the movie—his actions are *part* of the plot.

Of course this constant and instant interconnectivity between things in the universe, this quantum entanglement, exists not just when we curious observers of nature exercise our free will and decide to make a measurement but is an intrinsic property of nature. For just as in an act of measurement, for which we observers cause willingly particles to interact in order to satisfy our inquisitive mind—for example, photons are shined upon electrons to see where they are and how they spin—particles in nature are in constant interaction anyway (without us having to cause it at will), as if nature itself is constantly self-measuring (self-observing). Now, with self-measuring in mind, we have an additional reason to reinforce a previous conclusion, that, not only are the phenomena *observed* to occur discontinuously (as a result of the very act of observation, as argued in chapter 12), but the phenomena *themselves* might occur discontinuously even when *we* are not observing, for *nature* is—*self-observing*.

The whole universe experiences the phenomenon of quantum entanglement. If two particles have a chance to interact initially (that is, to become entangled like particles A and B that were created from the decay of the same mother particle), they continue to interact (they remain entangled) even when they are later separated. With this in mind, the entire universe may be considered an entangled whole (where everything in it is in constant and instant interaction with everything else, a perfect Parmenidean whole), for initially, according to the big bang, the entire universe was a mere microscopic size, possibly just a single point, where certainly everything was in close interaction and thus entangled with everything else, and so must then continue to be so today even with everything so far apart. The cosmic interconnectivity of mathematical nature anticipated by the Pythagoreans is now taking a concrete form through quantum entanglement.

In concluding this section, I would like to emphasize that information that travels faster than the speed of light is still impossible as stated in the theory of relativity. That is, while Alice's measurements of the properties of particle A influence instantaneously the properties of particle B, still information, that is, what Alice knows concerning the properties of either particle A or B, cannot be communicated to Bob faster than the speed of light; each person's knowledge can be communicated to one another at best at the speed of light, say, by a radio signal. Only then can Alice and Bob verify the remarkable correlation between the properties of particles A and B due to the phenomenon of quantum entanglement. Before such communication, the outcome of Bob's measurement concerning the spin of particle B would appear to him as random, as dictated by the laws of quantum probability, even when Bob does his measurements after Alice has done hers.

So as attested indirectly by the motion, change, and plurality of everyday experience—when these of course are investigated rationally by the human mind—the universe is indeed an indivisible whole. But there is a hypothesis that such universal oneness was once truly absolute.

THE ABSOLUTE ONENESS

The ultimate example of Parmenidean oneness, wholeness, and completeness, as properties of the universe, comes perhaps from the cosmological model of the big bang. It speculates that all matter and energy, all of space and time, the absolute wholeness of the universe of today, was once, about 13.8 billion years ago, contained in just a singular point. This primordial point, we must emphasize, was not within the universe; this one point *was* the universe, the *whole* universe; infinitely small, hot, and dense, containing a single type of particle and obeying one grand law—the absoluteness of oneness.

Our current big bang model begins its speculations on the universe's properties as early as the unimaginably small period of about 10^{-43} seconds after the initial bang—then the universe was extremely small but not of size zero. As the universe expanded and cooled down, its absolute oneness—the single primordial point that is hypothesized to have once been—and intrinsic simplicity manifested themselves as plurality and complexity, as particles of matter, the quarks and leptons, and as four groups of force-carrying particles, the photons, the gluons, the W's and Z's, and the gravitons. Quite possibly we will discover other types of exotic particles. The particles of force coalesced the particles of matter into nuclei, atoms, molecules, planets, stars, galaxies, books, and readers. But if the universe was once characterized by an absolute oneness, should it not continue to be characterized as such even today?

Unfortunately, the properties of the universe at this hypothetical primordial singular state cannot be described even in principle by our current physical theories. At this singularity—when the universe's size and age are both identically zero—all our physics equations break down; they are meaningless. Could this breakdown be an indication that such a singular state of the universe is really an impossibility, a Not-Being? If so, the universe might have been very small but not point-like. But was it born?

UNBORN AND IMPERISHABLE

Parmenides's philosophical worldview is, so he says, presented to him as a revelation by a goddess and is described in his poem *On Nature*. The main parts of the poem are the "Way of Truth"[7] (which discusses his philosophy) and the "Way of Opinion"[8] (which, among other things, discusses the philosophies of other philosophers). His primary goal was not so much to create a specific physical theory that would explain particular phenomena of nature but rather a theory attempting a logical explanation of existence itself: how can something be? It just is, he reasoned, for there is no such thing as nothing. Nature is unborn and imperishable. That which exists can neither be created from nonexistence nor obliterated into nonexistence. If the universe had a beginning, it would mean that it once did not exist—for if it existed it could not begin. But if the universe did not exist, it would have been Not-Being, and so again it could not begin (for Being cannot come from Not-Being). So the only way to explain why the universe is, is to assume that what is, has always been, unborn, without a beginning.

Now, on the one hand, his view of an unborn nature means that nature has not been caused; it does not have a primary cause. On the other hand, the opposite idea is that nature has been caused by a primary cause. This latter view is in a sense antiscientific since the premise of science is comprehensibility. But a primary cause cannot be understood—if it could, we would know what caused the "primary" cause, hence the "primary" cause would not have been really primary. Conversely, an unborn nature seems, at least at a first glance, to be more in accordance with the scientific premise, because something unborn/uncreated does not require a primary cause (an explanation) of why it exists—for it has always been.

That said, the notion of an unborn (uncreated, uncaused) nature (or, analogously, an imperishable nature having no ultimate purpose) must be examined with more caution. For it does not exclude the possibility of a god coexisting in the whole—as is in fact the case of the Parmenidean "Way of Truth," according to which the apocalyptic goddess Parmenides, and all the rest of nature, all just are. Moreover, an omnipotent and omniscient god could have made nature appear uncreated to us mere mortals. The point is that science cannot prove or disprove the existence of a god, and therefore such notion, as Parmenides might have put it in his "Way of Opinion," will always remain a matter of subjective belief. In science we must always begin with an assumed something (a Being)—and if we happen to finally explain such assumed something, we explain it with a new assumed something. Science cannot begin from Not-Being: *there is no scientific explanation of a universe coming to be from nothing!* Why there is something instead of nothing is scientifically unanswerable. Causality in our theories explains only later effects by earlier causes, but it cannot explain the primary cause (the beginning). And as a consequence, there is no way to ever know if there is an ultimate purpose. Even if the truth of the universe is revealed to us, the only way we can know that such truth is *absolute* is if we ourselves have absolute abilities—so that we can comprehend the absoluteness in the revealed truth. But we do not. And so again, the true nature of such hypothetical revelation is subjective.

Among our best cosmological models in science, the big bang does not and cannot answer why there is a universe; it only assumes that there is one (that might have begun or might have always been) and then continues to describe it. But it cannot answer why there is what there is. The prediction of the big bang model, that the age of the universe is 13.8 billion years, is only a *relative* age, namely, that our scientific theories can begin describing the properties of the universe roughly since 13.8 billion years ago. But we emphasize that with regard to what the universe might have been doing before that, we are clueless.

Interestingly, in an effort to avoid the breakdown of the equations at the hypothetical big bang singularity, some cosmological models attempt to model mathematically a self-reliant universe with neither space nor time boundaries—that is, an unborn nature having neither beginning nor end. A geometrical analogy of such type of universe is the surface of a sphere: no place on it can be considered more of a beginning than an end, more of a center than an edge, or more special in any way. Of course once more we emphasize that a mathematically/scientifically unborn universe does not say much about the true origin of the universe—that remains a subjective matter.

Lastly, the Parmenidean view is a hopeful philosophy because within its context consciousness is part of Being—I think, therefore I have consciousness. Hence, consciousness can never become Not-Being, even with the body's apparent death.

CONCLUSION

After Parmenides, any new natural philosophy would be considered incomplete unless it could address successfully his various conclusions, which, though unconventional, were logical. And as if that by itself was not a formidable task, Parmenides's best student, Zeno, assertively supports his teacher's views by adding to the complexity with his famous paradoxes that question the very nature of plurality, space, time, and the reality of apparent motion.

CHAPTER 14

ZENO AND MOTION

INTRODUCTION

Through a series of so-called paradoxes, Zeno of Elea (ca. 490–ca. 430 BCE) tried to argue for the astonishing conclusion that motion is impossible and plurality is an illusion. Could he be right? We present four of his most daring paradoxes: the dichotomy, the Achilles, the arrow, and the space, which challenge various views on space, time, and motion, and examine them within the context of modern physics. We also refer briefly to the conclusion of his paradoxes on plurality, which deal with whether there are many things or just one.

There is still no commonly accepted resolution for any of Zeno's paradoxes, a fact that preserves their legacy as the most difficult and long-standing puzzles. Part of the reason for this is the involvement of key notions such as space, time, and matter, of which their true nature is far from known even by the standards of modern physics. The real resolution of the paradoxes might require an even more radical understanding of these notions than the one presently provided by general relativity and quantum theory. Proposed solutions have often aimed to prove that motion is real. We will argue in favor of Zeno that at best, whether motion occurs or not is not experimentally provable.

THE DICHOTOMY PARADOX

According to Aristotle's account, Zeno said "Nothing moves because what is traveling must first reach the half-way point before it reaches the end."[1] In order to interpret this quote we must suppose that space is either (1) infinitely divisible (where space is imagined to be divided to ever smaller fractions) or (2) finitely divisible (where space cannot be divided beyond a fundamental length).

(1) Infinitely Divisible Space

The paradox can be interpreted two different ways, both of which are essentially the same. In the first interpretation, the question is this: can a traveler ever start a trip? To begin a trip of a certain distance a traveler must travel the first half of it, but before he does that he must travel half of the first half, and in fact half of that, ad infinitum. Since there will always exist a smaller first half to be traveled first, Zeno questions whether a traveler can ever even start a trip.

In mathematical language, the traveler will be able to start his trip only if he can first find the smallest fraction (the "last" term) from the following infinite sequence of fractions of the total distance: $1/2, 1/4, 1/8, 1/16, \ldots$. But such smallest fraction does not exist; it is indeterminable (in fact this is what is meant by calling such a sequence of fractions infinite). So the paradox is this: while on the one hand Zeno's argument, which questions the very ability to even start a trip, is logical; on the other hand, all around us we see things moving. Hence either Zeno's reasoning is wrong or what we see is false.

In the second interpretation, the paradox can be reformulated in a sort of reverse manner. In such case the question will be: assuming a traveler can somehow start a trip, can he ever finish it? To finish a trip of a certain distance a traveler must first travel half of it, then half of the remaining distance, then half of the new remaining distance, ad infinitum. Since there will always exist a smaller last half to be traveled last, Zeno questions whether a traveler can ever finish a trip.

"Answers"

First note that by getting up and walking, as Antisthenes the Cynic[2] did after listening to Zeno's presentation and thinking that a practical demonstration is stronger than any verbal argument, is not at all a refutation of Zeno's paradoxes of motion, because Zeno does not deny *apparent* motion; he questions its truth. The pre-Socratics were well aware of the deceptiveness in apparent reality; what we see happening is not necessarily happening the way we see it.

An "Answer" Based on Simple Mathematics

With the first interpretation in mind, to start the trip the traveler must first figure out the smallest fraction of the total distance, that is, the "last" term of the infinite sequence of numbers 1/2, 1/4, 1/8, 1/16, ... Only then he will know where to step first and begin the trip. But such term is indeterminable. After infinite subdivisions of the total distance, the "last" term of the sequence is indeed infinitesimally small and *approaches* zero, though is not *exactly* zero: there will always exist a smaller first half to be traveled first. Now since such term approaches zero we might want to approximate it to exactly zero. But with such approximation the traveler will step first where he is already at, the beginning. This might be interpreted to mean that the trip cannot start, thus motion is impossible. Nonetheless, this is not necessarily the best conclusion since it is reached only after our convenient approximation of the "last" term with the number zero. Since the actual value of the "last" term is indeterminable, a better conclusion would be that, indeterminable must also be the status of the trip (whether the trip can ever begin). Thus the notion of motion is, to say the least, ambiguous. The same result is obtained through similar arguments applied to the second interpretation of the paradox.

An "Answer" Based on Modern Mathematics

Often an answer of the paradox is sought through calculus. Suppose the trip distance is 1 meter. Then, as per interpretation two, a traveler first travels half of the trip distance, that is 1/2 of a meter, then half of the remaining distance, that is, an additional 1/4 of a meter, then half of the new remaining distance, that is, an additional 1/8 of a meter, ad infinitum. To find out if the traveler covers the trip distance of the one meter, we must add all the segments traveled by him, that is, 1/2 + 1/4 + 1/8 + 1/16 +. . . . Because the sum of this infinite geometric series converges on 1, some argue that the distance traveled by the traveler after infinite steps is 1 meter, thus he has moved and the paradox is resolved.

But this argument has a flaw hidden in the details of calculus. To be able to do calculus (i.e., calculate series sums like the one in hand) irrational numbers must be approximated with rational. Recall from chapter 11 that there exist infinitely many irrational numbers along any space distance. For example, between the point zero (the beginning of the trip distance) and the point of 1 meter (the end of the trip distance), there are infinitely many irrational numbers—such as $\sqrt{2} - 1 = 0.414213562\ldots$, or half of it, or one third of it, and so on—that must be approximated with rational numbers before any sum is calculated. For example, approximated to four decimal places, the rounded-off value of $\sqrt{2} - 1 = 0.4142$. Zeno, however, seems to tacitly question these very approximations that are required in mathematics to make the series convergent to a practical and calculable answer. Because nature, he would claim, does not have to behave according to the result of such convenient and ambitious human approximations.

Furthermore, some argue that the convergence method does not address the paradox because it does not explain how an infinite number of tasks (going from the first half of the distance to half of the remaining, etc., ad infinitum) can be carried out in finite time. But can a finitely divisible space solve the paradox?

(2) Finitely Divisible Space

In a finitely divisible space it is assumed that there exists a fundamental length that cannot be divided further. Therefore, there exists only a finite number of fractions of the total distance, and the paradox appears resolved, or, more precisely, the question of the paradox, as we will see, is revived in a new form.

For example, say space is finitely divisible and composed of fundamental lengths equal to 1/4 of a meter (which means that space cannot be divided in smaller lengths than 1/4 of a meter). How could a traveler complete a trip of distance 1 meter? Having in mind the first interpretation, the traveler steps first at the 1/4-meter location, then at the 1/2-meter, and finally at the 1-meter, the final destination. Analogously, by interpretation two, the traveler steps first at the 1/2-meter, then at the 3/4-meter, and finally at the 1-meter, the final destination. But a finitely divisible space generates a series of new unresolved questions. For example, how can the traveler move to the 1/4-meter location without passing first through all other locations (such as the 1/8-meter or the 1/16-meter, etc.)? Also, what determines the fundamental length of a finitely divisible space—whether 1/4 of a meter or some other number?

THE ACHILLES AND THE TORTOISE PARADOX

"In a race the faster runner can never overtake the slower. Since the faster runner must first reach the point from which the slower runner departed, the slower runner must always hold a lead" (Aristotle's account of Zeno).[3]

The paradox says that in a race between, say, fast Achilles and a slow tortoise, initially separated by some distance, Achilles can never overtake the tortoise because before he achieves that he must first reach the starting location of the tortoise. But by the time he arrives there the tortoise will have had the chance to move to a new location forward; and by the time he arrives at the tortoise's new location, the tortoise will move farther forward to another new location, ad infinitum. Therefore, despite that faster Achilles will be constantly approaching the slower tortoise, still there will always exist some small and ever-decreasing distance separating them (though not necessarily in fractions of half, as in the dichotomy paradox). This is a paradox because, despite Zeno's argument that a faster runner cannot overtake a slower one is logical, fast runners apparently do overtake slower ones. Is Zeno's reasoning flawed, or are our senses false?

This paradox is basically the same as that of dichotomy, so everything mentioned earlier applies here. One important difference is that the Achilles paradox is complicated further by contemplating the relative motion between two things (of Achilles and the tortoise), and between those things and their potential meeting point (destination); whereas the dichotomy paradox contemplates the relative motion of just one thing (of a traveler) with respect to a destination. Furthermore, the Achilles paradox contemplates the nature of time more directly (i.e., whether infinitely divisible, that is, continuous, or finitely divisible, that is, discrete), since it involves the time required by Achilles to reach the tortoise and how during such time the tortoise has the chance to move forward.

In the dichotomy paradox, the first interpretation (for which a traveler cannot start his trip) seems to deny motion more directly than the second interpretation (for which a traveler is assumed to move, although he cannot ever finish his trip). In the Achilles paradox, Achilles and the tortoise are assumed to be moving, but motion seems to not work in the conventional way, for the faster Achilles cannot overtake the slower tortoise. In the arrow paradox Zeno is even more audacious, for he directly denies motion by any interpretation. Reconstructing it reads as follows.[4]

THE ARROW PARADOX

An arrow is at rest whenever it is in a space equal to itself. A launched arrow goes through its flight one instant at a time. Since the arrow is in a space equal to itself each instant of the flight (just as it is when it is at rest), then the arrow must be at rest at each such instant as well. Since it is at rest at any one instant, it must be at rest for the entire duration of the flight. Hence the flight is apparent, not real; the arrow does not move.

This is a paradox since its conclusion is based on a logical argument that contradicts the apparent reality of sense perception according to which a flying arrow changes positions each moment of its flight and thus apparently moves. Again, is Zeno's logic flawed or are our senses?

I believe the formulation of the arrow paradox must have been triggered by a simple observation, that an object at rest occupies a space equal to its own size. A book, for example, resting on a desk occupies a space exactly equal to its own size. That said, I am not implying that such observation validates Zeno's conclusion that an arrow in apparent flight does not move. But could he be right? Could it be true that an apparently flying arrow is really motionless? Using quantum theory I will argue that at best, it is not provable whether the arrow is moving or not. Motion, in general, is an ambiguous concept.

MOTION IS AMBIGUOUS

While motion is part of apparent reality and is also the very premise of important theories of physics, on a fundamental level (i.e., concerning the motion of microscopic particles, to say the least) motion has not yet been experimentally proven, and in fact never can be! Therefore, motion is essentially a postulate inferred from sense-perceived experiences, but its truth is actually ambiguous. This is so because inherent in the Heisenberg uncertainty principle observations are disconnected, discrete events; consecutive observations have time and space gaps—we can observe only discontinuously (as seen in chapter 12). The concept of continuity in observation must be dismissed. It is a false habit of the mind created by the observations of daily phenomena—as of an arrow in flight (although, as explained in the section titled "Observations Are Disconnected Events" in chapter 12 and as will be reemphasized below, the arrow's apparent continuity of motion is an illusion due to its large mass that makes the time and space gaps between consecutive observations undetectably small). Now without the ability to observe continuously, motion not only *is observed* to be discontinuous but the very notion of motion *itself* becomes ambiguous. How so?

Motion occurs when during a time interval a particle (e.g., an electron) changes positions; a particle should be now here and later there in order to say that it moved. But since nature does not allow us to keep a particle under continuous observation and follow it in a path, and also since a particle is identical to all other particles of the same family (for example, all electrons are identical), it is impossible to determine whether, say, an electron observed in one position has moved there from another, or whether it is really one and the same electron with that observed in the previous position, regardless of their proximity. Since observations are disconnected and discrete events—with time (and space) gaps in between, during (and within) which we don't know what a particle might be doing—subsequent observations of identical particles might in fact be observations of two different particles belonging in the same family and not observations of one and the same particle that might have moved from one position (that of the first observation) to the next (that of a subsequent observation). Without the ability to determine experimentally whether a particle has changed position, its motion—and motion in general—is a questionable concept.

In summary, (1) without the ability to keep a particle under continuous observation (2) it is impossible to establish experimentally its identity, and therefore (3) it is also impossible to prove experimentally that it has moved.

Reinforcing this conclusion is the fact that, when we observe a microscopic particle all we see through a microscope is just a flash of light, and somewhere within it is the particle. But where exactly within it the particle is each moment of time, and whether it is at rest or in motion, are all indeterminable; while we do detect a particle, we detect it neither at rest nor in motion. Hence indeed, neither immobility nor motion are experimentally provable. Motion is ambiguous.

Trying to capture the peculiar consequences of the Heisenberg uncertainty principle concerning motion, physicist J. Robert Oppenheimer (1904–1967) wrote: "If we ask, for instance, whether the position of the electron remains the same, we must say 'no'; if we ask whether the electron's position changes with time, we must say 'no'; if we ask whether the electron is at rest, we must say 'no'; if we ask whether it is in motion, we must say 'no.'"[5]

Now, since motion is ambiguous for microscopic particles, then in a stringent sense it must be ambiguous for arrows, too, for arrows are made of microscopic particles. Observing an arrow in flight moving continuously does not prove (in the strictest sense of the word) that one and the same arrow has endured and moved, simply because there is no proof that any of its component microscopic particles have endured and moved. Besides, quantum theory (hence the uncertainty principle) is true for both the world of the large and the small. It is only for practical purposes that the world of the large is assumed to behave according to classical physics—for which objects appear to endure and move in definite traceable paths—because the consequences of the uncertainty principle for large objects are undetectably small, although not zero. Well, then, how do we explain everyday apparent motion and in fact the apparent continuity of apparent motion for any object, such as an arrow in flight, Achilles chasing a tortoise, or anyone taking a trip?

CINEMATOGRAPHY AND APPARENT MOTION

We can explain them with cinematography. In an analogy, consider a series of identical and disconnected red lightbulbs, closely spaced along the arched outline of George Washington Bridge. Now imagine it is nighttime and that the first lightbulb in the series is turned on briefly for a few seconds then off forever; after a brief time gap, lasting a minuscule fraction of a second, the second lightbulb is turned on and off the same way, then the third, and so on, until each lightbulb is turned on and off in this sequential manner. The events, the on-off turnings of each lightbulb, are (1) identical (in the sense that an observer sees the same red light) and (2) disconnected: the space gaps are the distances between the bulbs; the time gaps are the minuscule fractions of a second. Furthermore, (3) assuming that the space and time gaps of these events are small enough, to a distant observer, this phenomenon will appear as if *one red object* (the first lightbulb) has *moved* and has moved *continuously* along the outline of the bridge, when in fact no object has. Motion in this case is an illusion of the senses created by observing a series of identical and appropriately disconnected events. In particular, the first two facts, (1) and (2), create the illusion of motion, and requirement (3) creates the illusion of continuity of motion.

The way the red light appears to be moving is similar to the way an arrow in flight appears to be moving. In each case apparent motion and apparent continuity of motion are in reality the result of (1) a chain of identical observations (of an apparently identical red light or arrow), which (2) are also disconnected (for the arrow, this is due to the uncertainty principle) with (3) undetectably small space and time gaps in between (for the arrow, this is due to its large mass). Specifically, (1) and (2) create the illusion of motion, of one and the same object, the red light or the arrow, and (3) creates the illusion of *continuity* of motion.

But to refine the analogy, we must add this: unlike the case of the bridge for which several identical lightbulbs are assumed to preexist along its outline, for an arrow in flight we cannot assume several identical arrows to preexist along its apparent path; only that, each one of our *observations* is of an apparently identical arrow; though it is uncertain whether our observations are absolutely of one and the same arrow, for, as we learned in chapter 12, it is impossible to establish experimentally the identity of microscopic particles, and since arrows are made up of such particles, it is also impossible to prove unambiguously whether our observations are actually of one and the same arrow (that is, of an arrow composed of the *same* microscopic particles at each location of its apparent flight)—a fact that makes motion an ambiguous notion for any object, microscopic or macroscopic. The most we can say is that, at subsequent observations the observed arrows have similar bulk properties, similar general *form* or *organization* (just as the Heraclitean river); it is this general organization that seems to endure, at least during some interval of time and within some region of space, creating the illusion of a permanent thing (e.g., an arrow) in motion.

While on the one hand all around us certain things appear to endure (appear as *permanent things* at least for some time and within some region of space), and whenever they appear to move there appears to be continuity in their motion, on the other hand, neither permanency in things nor motion are experimentally provable ideas. Thus motion is, to say the least, an ambiguous concept, a result to be expected because of the very definition of motion, which requires that permanent things exist so they can move: motion occurs when during a time interval a *particle* (a *thing* in general) changes positions; but to refer to a particle and define its motion, the particle must *remain the same for the duration of its motion*; motion cannot be defined if a particle does not remain the same for a period of time. Now for Heraclitus and modern physics there are no identifiable particles, no permanent things, only events. And without an enduring thing, without the ability to establish the sameness of a thing at two different moments, motion remains an ambiguous concept. While ambiguous, can it nonetheless be a practically useful way to explain the phenomena?

ADEQUACY VERSUS TRUTH

The answer to the previous questions is sometimes. Causality in classical physics is deterministic: a single cause produces a single effect, and both cause and effect are precisely determinable at least in principle. In quantum theory causality is probabilistic: causes and effects are expressed in terms of a probability; this is actually the reason it is impossible to determine whether the observations of two identical particles might be observations of one and the same particle, for we cannot causally connect these observations with deterministic (absolute) accuracy. Still, to make sense of the phenomena from the point of view of quantum theory, we often assume certain causal chains of events. For example, an electron here collides with a photon and recoils there (as if the electron endures). Thus while neither a cause nor an effect are certain, and motions are untraceable, still the assumption of a certain chain of events and of motion is often an adequate way to model a practical explanation. Supposing that particles endure, we have previously argued that they are constantly moving.

But while motion may be an *adequate* and useful concept in devising a certain practical explanation of nature—especially so for macroscopic objects such as arrows, cars, and planes—as a *true* property of nature it is, to say the least, an ambiguous concept, for it lacks the support of experimental confirmation from the microscopic constituents of matter that make up all macroscopic objects. Therefore, the merit of modeling an object (an electron or an arrow) as moving is a practical necessity of everyday life, not a confirmable truth. It should also be pointed out that even practicality cannot be applied consistently (especially so in the microscopic world).

Before quantum theory, and within the context of classical physics, concepts such as position, velocity, and motion (in general), were intuitive, self-evident, and could be used in a definitive way to characterize an object; an object moves with a specific velocity, it is now passing from here and will later go there. However, after quantum theory all these concepts became counterintuitive and could not be used in the same definitive way to describe the behavior of microscopic particles; such particles have neither definite position nor velocity nor a path of motion—as seen, this ambiguity of motion first came up in the natural philosophy of Zeno. A better way to describe the behavior of a particle (and in general the phenomena), then, is not through motion but through the probabilities of quantum theory. It is the concept of probability that is the fundamental (intrinsic) property of matter and not properties such as position, velocity, and motion. Within this context, as initially argued in chapter 11, a particle is truly a mathematical form. Could this mean something concerning change and motion in nature?

A COMPROMISE: YES TO CHANGE, NO TO MOTION

In the view of Zeno the arrow itself exists but does not move, and there is no change. On the other hand, in the view of Heraclitus, the arrow as a permanent thing does not exist (only events exist), but constant change and constant motion do; only the *organization* of the arrow exists and endures at least for some time. Could these antithetical views be reconciled by modern physics? Well, we can observe an electron (or an arrow) here now and an electron (or an arrow) there later. So obviously we can experimentally confirm that there is a certain change of events (at least in what we observe and where and when we observe it, that is, our observations are of different phenomena, electrons or arrows, at different places and times); but we cannot experimentally confirm that anything has moved. So the compromise might just be this: that constant changes do exist in nature (as Heraclitus posited), but motion does not (as Zeno theorized). But do these changes occur in a passive, playground-like space, or are space, time, and matter somehow related?

THE SPACE PARADOX

"If everything that exists is in some space, then that space, too, will exist in some other space, ad infinitum" (Aristotle's account of Zeno).[6] We may reconstruct this quote as follows.

(1) Things that exist do so in some space.
(2) Space exists (for if it didn't [1] wouldn't hold and things could not exist).
(3) Since space exists and everything that exists is in some space, then space, too, must exist in some other space, ad infinitum.

With this paradox Zeno seems to be arguing that requiring space, that is, void (as the atomists do, by treating it as a sort of a playground to put things in), is as problematic as denying space (as Parmenides does). And that, if we shouldn't completely deny space, we also shouldn't treat space as a playground—as if space is supposed to exist independently of the objects merely for the objects' sake, namely, for them to exist in it.

As discussed in chapter 11, the theory of special relativity replaced the playground-like space of Newtonian physics (for which space and time are absolute) with the spacetime continuum (for which space and time are relative). For relativity things do not just exist in a passive space with time flowing steadily in the background. Instead, space, time, and matter are complexly intertwined—with astonishing effects such as length contraction, time dilation, relativity of simultaneity, and space distortions (the latter being true only in the theory of general relativity).

So Zeno's space paradox is a paradox because while, on the one hand, his argument against a passive playground-like space is logical, on the other hand, it contradicts sense perception of exactly that kind of space. With the theory of relativity in mind, the space paradox may be considered resolved.

The peculiarity of a playground-like space implied by the space paradox is appreciated further when we try to construct a similar-type time paradox (though this is not one of Zeno's paradoxes). For example, if everything that exists does so for some time, then that time, too, will exist for some other time, ad infinitum. This time paradox argues against an absolute (Newtonian) time, flowing the same way for everyone while things happen.

Last, in his effort to show that a nature made up by many things is as problematic and contradictory as the Parmenidean oneness, Zeno devised several other paradoxes. Based on them he concluded that if in nature there are many things, they must simultaneously be (a) infinitely small and (b) infinitely huge and (c) finitely many and (d) infinitely many.[7] We will not cover these paradoxes here.

CONCLUSION

Zeno's paradoxes challenge our views on the very nature of space, time, and matter. Are these notions somehow connected? Is there just one primary substance of matter, or are there many? Is the nature of matter continuous—spread everywhere and also infinitely divisible for which matter can be cut to ever smaller pieces? Or is the nature of matter atomic—and so finitely divisible, for which matter cannot be cut beyond some fundamental pieces that are spread unconnectedly because they are surrounded by void? In 1916 Einstein addressed successfully the first question with his theory of general relativity, in which spacetime is a continuum in constant and intricate interaction with matter. Empedocles, Anaxagoras, and Democritus took up the other three. Matter is atomic for Democritus but continuous for Empedocles and Anaxagoras. And while plurality in the number of primary materials is possibly speculated first by Empedocles, all three philosophers had a unique take on it.

CHAPTER 15

EMPEDOCLES AND ELEMENTS

INTRODUCTION

Empedocles (ca. 495–ca. 435 BCE) managed to reconcile the antinomies between the Heraclitean becoming (the constant change) and the Parmenidean Being (the constancy) by introducing four unchangeable primary substances of matter: earth, water, air, and fire, later called elements, and two types of forces, love and strife. Change was produced when the opposite action of the forces mixed and separated the unchangeable elements in many different ways, an idea in basic agreement with modern chemistry or, more fundamentally, with the standard model of particle physics.

ELEMENTS AND FORCES

Unlike Thales, who taught that the primary material can transform and change its nature (for example, water can become ice), Empedocles held (as did Anaximenes) that the nature of a primary material must always remain the same, like the Parmenidean Being. But with a single primary material of unchangeable nature, he could not account for the observed material diversity of the world. Thus he postulated four such materials, the elements, which were uncreated and imperishable—neither born out of nothing nor perishable into nothing. His choice for these elements was wise because with them he could explain the three phases of matter: the element earth could account for the solid phase, water for the liquid, air for the gaseous. Furthermore, through fire he could account for light. Now, not only do the elements not change into one another; they do not change at all. But that did not matter. Because Empedocles explained nature's enormous diversity by imagining love (the force) mixing the elements with one another and strife (the other force) separating them from each other, in infinitely numerous proportions and combinations, forming composite objects or dismantling them. For example, love can mix earth and water to produce mud, but strife can separate the earth and water from mud. Hence love causes attraction of unlike elements (thus, in a sense, per Aristotle, indirectly it also causes repulsion of things that are alike). And strife causes repulsion of unlike elements (thus, in a sense, indirectly it also causes attraction of those that are alike).

Empedocles explained the unique properties of objects in terms of the proportions of the elements they contain. A hot object, for example, contained more fire than a cold object. And a wet object contained more water than a dry one. Thus, the quantitative difference of the various materials present in an object determines the qualitative difference between objects.

Birth and growth occur while the elements mix, as in a blooming flower, and decay and death occur while the elements separate, as in a shriveling flower. Coming to be (the birth, the generation of something) occurs simply from a mixture of things (the elements) that already exist, not from Not-Being (nothingness)—that is, there is no absolute birth. And perishing (the death of something) occurs simply from a separation into things that also already exist, not into Not-Being—that is, there is no absolute death. That there isn't absolute birth or absolute death is of course part of the Parmenidean thesis and is also accepted by Anaxagoras and Democritus.

Like the elements, the forces were corporeal, uncreated, unchangeable, and imperishable. But it was *their* motion through the elements that caused the elements to move, too—either pushing them together to mix or pulling them apart to separate. Hence, forces were the source of motion and consequently of change.

Force, in natural philosophy, appears for the first with Empedocles, who interprets nature in terms of matter and forces. Matter and force, however, became popular with Newton's work: first, with his three laws of motion, and second, with his law of the universal force of gravity. According to his second law of motion, for example, the cause of motion is a force: you pull an object and the object moves. Also matter can produce a force: the sun produces gravity, or an electron the electric force. Nonetheless, while the matter-force interpretation of nature is still immensely practical, it began fading away in twentieth-century physics: forces, in modern physics, gradually became no longer essential. This is a topic to be revisited in chapter 17. Force, in particular an action-at-a-distance type (as is Newton's force of gravity), we will see there, remarkably was never required in the atomic theory of Democritus.

EMPEDOCLES AND THE STANDARD MODEL

Empedocles's idea of forces mixing and separating a fixed number of primary materials is in fundamental agreement with the standard model of particle physics. Whereas Empedocles proposed two forces and four primary elements (renamed "particles" by physicist Leon Lederman),[1] the standard model considers three fundamental forces—the electromagnetic, the nuclear weak, and nuclear strong (recall gravity is not part of the standard model)—and twelve types of particles of matter—the six quarks and six leptons (even though various other considerations can increase the number to forty-eight: each quark comes in one of three "colors" [these are variations, not real color], and each matter particle has an antiparticle). Of course, unlike Empedocles's elements, in modern physics quarks and leptons are changeable—they transform to energy or from one material particle into another. Still quarks and leptons are brought together by the forces in a multitude of combinations and proportions to form atomic nuclei, atoms, molecules, flowers, and in general all the plethora of small and large objects, animate and inanimate, similar and dissimilar; but the forces can also break down larger objects into smaller ones.

In Empedocles's chemistry every object is made by a unique mixing proportion of the elements—for example, a bone, he says, is two parts earth, two parts water, and four parts fire (though the sources do not explain how he derived that). Analogously, in modern chemistry every chemical compound is made by a fixed proportion of the chemical atoms—for example, a water molecule, H_2O, is always made of two atoms of hydrogen and one of oxygen. Of course modern chemistry can be analyzed even more fundamentally within the context of the quarks and leptons of particle physics, and still preserve Empedocles's notion of fixed proportions. That is, H_2O, for example, is really a fixed mixture of two protons (one from each hydrogen nucleus) plus eight more protons as well as eight neutrons (from the oxygen nucleus) plus two electrons (one from each hydrogen) plus eight more electrons (from the oxygen). Now, electrons belong in the lepton family of particles, thus they are fundamental (they are not made of other types of particles), but protons and neutrons are not fundamental: a proton is made from three quarks, two up and one down ("up" and "down" are quark names); a neutron is made from one up and two down quarks. In addition, quarks and leptons are kept together or pushed apart via the continual exchange of force particles, the photons and gluons, in our example. Analogously, in Empedocles's theory the fixed proportions of the elements are achieved via the constant competition of love and strife.

Empedocles was interested not only in the composition and changes of individual objects but also of the world as a whole.

THE CYCLES OF THE WORLD

The structure of the cosmos is spherical for Empedocles, and the changes in it occur without an ultimate purpose or divine intervention (the latter is also the view of the atomists). Instead, nature is ruled by necessity and chance: namely, only some outcomes are possible (this is what is meant by necessity), but which ones actually occur is completely the result of chance. Interestingly, this is the meaning of probability in quantum theory.[2] Which outcomes (necessities) are possible and what the probability (chance) of their occurrence is are calculable by the mathematical laws of quantum theory. While these laws restrict the outcome of an experiment to any one of a group of possible ones (this is the element of necessity), the laws do not specify which one should occur (this is the element of chance). A hydrogen atom, for example, cannot really have any energy. When observed it has only one energy from a group of allowed ones (necessities). But while in one of the allowed energy states, a transition into another of the allowed energy states occurs by chance.

Nature in the cosmology of Empedocles goes through everlasting cycles of growth and decay, gradually and continuously, through four basic periods.[3] In the first period of the cycle love dominates totally but temporarily, mixing the elements completely. In the second period strife begins its influence, and so there is a gradual transition to partial mixing and separating. In the third period strife dominates totally but also temporarily, separating the elements from each other completely, so each, in its pure form, occupies a different region of space: one region of the universe is occupied only by earth, another only by water, another only by air, and the last one only by fire. In the fourth period love makes its gradual comeback, and so again there is a partial mixing and separating of the elements. Life (the evolution of plants and animals) and nature in general as we know it (with the sun, planets, stars) are happening during the second and fourth periods. The state of our cosmos is temporary for Empedocles, and it is gradually being succeeded by another. Interestingly, if we are not myopic in our comparisons, these four periods have several similarities with modern cosmological models of the universe.

CYCLES IN MODERN COSMOLOGY

According to the big bang model, initially everything was completely mixed together, space, time, matter, and energy (like Empedocles's first period). Life as we know it was then impossible because the universe was tiny and superhot, without stars or planets, just a super-dense mixture of tiny particles. The universe has since then been evolving, reaching its astronomical size and diverse state of today, with galaxies, stars, planets, and life (as in Empedocles's second period). Now, if, as speculated by various big bang models, the universe is "open," it will continue to expand forever, increasing its size so much that ultimately everything in it will be completely separated (as in Empedocles's third period). It will then be a cold, lifeless universe without planets or stars, only isolated tiny particles. But if, as also speculated by other models, the universe is "closed," then after it goes through a third period (a state of maximum, though not necessarily complete, separation of everything in it, during which stars might fade out and die), it will stop expanding and will begin contracting, resulting again in life-bearing partial mixing and separation (as in his fourth period). The fourth period is much like the second, for as matter is brought together in a shrinking universe, the particles coalesce again to form countless light-giving stars and life-sustaining planets to orbit them. But in a "closed" universe the contraction will continue until the crushing force of gravity ultimately collapses the universe in on itself, and brings once more everything completely together (the first period all over), causing a "big crunch" (the opposite of the big bang). If things are so, we live in an ever-changing universe going through endless cycles of big bangs, expansions, contractions, and big crunches. But we are not sure. Still, we could describe rather accurately the main events in the universe and when they occurred by starting from the first moment of the big bang until now.

COSMIC CALENDAR

In modern cosmology all events in the universe span 13.8 billion years in time. To gain a perspective of such time vastness we often employ a cosmic calendar. It is a metaphor by which 13.8 billion years, the estimated age of the universe since the big bang, are compressed into just one calendar year. The initial bang, the big bang, is supposed to have happened at precisely midnight, 00:00:00 (which, in the twenty-four-hour time notation, is the 0th hour, 0th minute, 0th second) on January 1, causing the expansion of the universe. What caused the bang is still unknown, although it is speculated to have been a kind of repulsive gravity that is predicted from the equations of general relativity. But what is known is that this expansion has been happening ever since and up until now, the last moment of December 31 at 24:00:00, increasing the size of the universe from an unimaginably small size initially, possibly point-like, to today's immensity. What banged (expanded, stretched)? Spacetime did and still does. Within a minuscule period of time after the initial bang, possibly by a mere 10^{-36} second, the universe underwent an immense faster-than-light expansion, a *big* bang, an idea known as cosmic inflation. In the blink of an eye it expanded by a factor of 10^{30}![4]

By about fourteen minutes (380,000 years) after the big bang, at 00:14:00 on January 1, the universe expanded, became less dense, and cooled significantly and as a result became transparent to light (as a clear-air day is to visible light), allowing for the first time the "afterglow" of the big bang, formally known as the cosmic microwave background, to travel freely through space and time, from there and then to here and now, and to be seen today (by radio telescopes) coming from every direction in the universe—a triumphal proof of the big bang theory. Earlier than the first fourteen minutes, the young universe was very dense and hot and thus opaque to light—as a foggy day is to visible light—thus light could not travel far. January 1, at 00:14:00, is also the instant that the simplest and lightest of the chemical atoms, hydrogen, first formed when a relatively cooler universe allowed electrons and protons to capture each other via the electric force.

Stars and galaxies began to form by around February 1 (about a billion years after the big bang) from matter pulled together by gravity. Stars shine because of nuclear fusion, the process via which light nuclei combine to form heavier ones, converting mass into energy and releasing light. Nuclei heavier than iron, including silver and gold, are synthesized via fusion when a super-giant star (more massive than the sun) becomes a supernova—dies violently in a cosmic explosion, producing as much light as a galaxy of stars!

Its death is also life's birth! For gradually after millions or billions of years, a supernova's scattered debris, an interstellar cloud of gas and dust, collapse again under the crushing force of gravity and grow into a new star with its orbiting planets that may also develop life. A perfect example is our own solar system. It was born much, much later, around September 3 (about 4.5 billion years ago) from the gravitational collapse of a massive interstellar cloud that was composed from the atoms that were synthesized earlier in the universe, including the heavy atoms made in the stars. Thus earth and everything on it, including us, are all made of these ancient atoms—if you are wearing a gold ring, you are in a sense actually wearing a portion of a star, for your jewelry's atoms were once manufactured in a supernova-destined star! Even more impressive, in the words of the great Carl Sagan, we are all "star stuff"! In other words, most of the atoms we are made of were once made inside stars that lived and died millions or billions of years before we or our own solar system were even born.

Primitive microscopic life-forms were thriving on earth by September 29 (3.5 billion years ago), so the first type of life must have evolved much earlier than that. On December 30 (sixty-five million years ago) an asteroid collided with the earth and caused the extinction of many species, including the dinosaurs. But that was a good day for primates because that's when they started to evolve. *Homo sapiens*, which are primates, evolved on the last hour of the last day of the cosmic calendar, December 31 at 23:52 (only eight minutes ago, two hundred thousand years ago). And at different moments during the last minute of the last day various other significant events occurred. Humans painted fine cave art one minute ago at 23:59 (thirty thousand years ago). They domesticated plants and other animals and gave birth to civilization twenty-three seconds ago at 23:59:37 (about ten thousand years ago).

Recorded history, which preceded the construction of the pyramids by a few centuries, began only eleven seconds ago at 23:59:49 (about five thousand years ago), and the birth of science occurred just six seconds ago, at 23:59:54 (2,600 years ago). Our innovative Internet was implemented about 0.08 seconds ago at 23:59:59:92 (in the 1980s), and a twenty-year-old reader of this book was born only 0.05 seconds ago at 23:59:59:95. If wisdom is, as the wise say, acquired with time, then human wisdom is only infinitesimal, not at all like that of the cosmos, infinitely universal.

What will a second such cosmic calendar be like for the universe? Will the universe continue to expand? Will it stop and begin to contract? We are not sure. While the cosmic microwave background together with Hubble's law constitute two of the most significant experimental proofs of the universe's expansion, an experimental proof concerning the universe's fate (if a particular one does exist) is yet to be found. Experiments are important because they verify or falsify a scientific hypothesis. Empedocles is known to have done an experiment, possibly the first in the history of science.

IT'S EXPERIMENT TIME

While air was the primary substance of matter in the philosophy of Anaximenes, still it was not accepted as a real corporeal substance for two reasons: (1) it is invisible and (2) because other objects appear that could be placed in air or move through it. So, within the context of these reasons, air was thought, at least by the Pythagoreans, to be really the void. But using a clepsydra (a device to lift and transfer liquids) Empedocles overturned such belief by experimentally proving that air is indeed a material substance.[5]

Submerge a straw (which is much like a clepsydra) in a glass of water. Water flows into the straw through its bottom opening and fills it as high as is the water level in the glass. But if before you submerge the straw you first cover its top opening with your finger, no water (or, actually, very little) will flow into the straw. This happens, Empedocles argued, because some invisible material, which is already trapped in the straw, presses on the water (through the bottom opening of the straw) and keeps it out; water, in this case, cannot move through this material. This material is air. Only when you uncover the top opening can water once again flow in the straw. For in this case the air in the straw escapes through the top opening, and so an equal volume of water flows in to take its place. (Incidentally, why water or air or any object can move will occupy the mind of the atomists, as will be seen in chapter 17). In conclusion, since it is not always true that an object can move through air or be placed in it, air must be a material substance, regardless of its invisibility. Empedocles's reasoning is correct.

THE ORIGIN AND EVOLUTION OF THE SPECIES

In his effort to understand the origin of the species and their adaptation to their environment Empedocles, like Anaximander, conceived of an evolutionary theory by natural selection. In the beginning a chancy mix of the "immortal"[6] (permanent, unchangeable) elements created all imaginable "mortal"[7] (temporary) organic "forms, a wonder to behold."[8] These, though, were just parts, from humans, animals, and plants. And so "many heads sprouted without necks, and arms wandered bare and bereft of shoulders, and eyes strayed up and down in need of foreheads."[9] That is, until love mixed them in countless ways more so that the species of plants and animals formed. But only the fit survived; the unfit died. When a human head, for example, combined with a human body, the creature acquired a fitting form and survived, Empedocles thought; but when a human head combined with an ox body, he continued, the creature was unfit and died. Chancy material combinations and natural selection (that is, survival of the fittest and adaptation) are important aspects in both Empedocles's and modern theories of biological evolution.

CONCLUSION

Empedocles's pluralistic philosophy was a crucial turn away from the monistic philosophies we have discussed so far (i.e., those that considered water, the infinite, or air as the only primary substance of matter, or the philosophy of Parmenides about oneness), for it paved the way for the most successful ancient pluralistic philosophy, the atomic theory of Leucippus and Democritus. Their theory required myriad particles: the atoms. But before the theory of atoms, pre-Socratic philosophy had to go through yet another theory of remarkable originality; four primary substances of matter for Empedocles, but infinitely many for the nous of Anaxagoras.

CHAPTER 16

ANAXAGORAS AND NOUS

INTRODUCTION

"Nous [the mind] set everything in order,"[1] thus it has the ability to understand nature rationally, Anaxagoras (ca. 500–ca. 428 BCE) proposed. Order though, according to him, is not achieved through the consideration of just one primary substance or even four but through a countless number of them, including things such as gold, copper, water, air, fire, wheat, hair, blood, bones, and in general all other existing substances. However, unlike Empedocles's four elements, which are pure, Anaxagoras's substances are not; "in everything there is a portion of everything,"[2] a notion as bizarre as two of the most popular interpretations of quantum theory, the Copenhagen and the many-worlds.

IN EVERYTHING THERE IS A PORTION OF EVERYTHING

All Materials Simultaneously

Every piece of substance, however large or small, contains some portion of everything—portions can be large but infinitesimally small, too, because for Anaxagoras matter is infinitely cuttable. Hence, no one substance is more fundamental (that is, smaller, simpler, purer) than any other. But "each thing is most manifestly those things of which it has the most."[3] A piece of gold, for example, contains gold as well as everything else—copper, wheat, hair—but appears as a distinct golden object because its gold content is the greatest. This does not mean, however, that this golden object contains the substances in pure form, side by side, separated, and identifiable, and the amount of pure gold in it happens to be more. No! To the contrary, no matter how small a bit we may cut from such golden object, it will still contain a portion of everything—it will never be pure gold. Therefore, despite that this is a golden object, *every part* of the object is also *simultaneously* watery, woody, milky, bloody, bony, hairy, and every other material, but not just that—it gets stranger.

All Qualities Simultaneously

A strict interpretation of "in everything there is a portion of everything" means that an object has not only a portion of each type of substance but also a portion of all opposite qualities. In fact, according to some scholars, it is not really necessary to speak of the substances separately from the qualities because the qualities determine the type of a substance anyway.[4] Now, as with the substances, these qualities are not to be assumed to be side by side in an object or separated somehow, as if, say, an object has its right side wet and its left dry. Rather, "Things in this one cosmos are not separated from one another, nor are they split apart with an axe, neither the hot from the cold nor the cold from the hot."[5] So every part of an object is all the qualities simultaneously. For example, something hot is to some degree also cold. Or white snow, Anaxagoras argued, is to some degree simultaneously black, too—a statement of the same unusual meaning as Schrödinger's cat being simultaneously both dead and alive.

ANAXAGORAS AND THE COPENHAGEN INTERPRETATION

So for Anaxagoras an object is simultaneously hot, cold, wet, dry, hard, soft, sweet, sour, black, white, bright, dark, dense, rare, dead, alive, spinning clockwise, spinning counterclockwise, and all other opposite qualities. This is a peculiar interpretation of nature, for before we observe an object, the most we can say about the state of its existence is that it is a mix of all possible outcomes—of all opposite qualities though each with a different degree (portion) of contribution. Only after we observe the object can we describe it in a specific way, in terms of "those things of which it has the most," say, as golden, yellow, dry, cold, and heavy.

Remarkably, such interpretation is similar to the most popular interpretation of quantum theory, the Copenhagen view. According to it, before we observe something, the state of its existence is a mix of all possible outcomes (qualities), each of which has its own quantum probability to actually occur. If the idea of portion in Anaxagoras's theory is roughly associated with the idea of probability in quantum theory, then indeed, "in everything [a system of interest] there is a portion [is described by the quantum probabilities] of everything [of every possible outcome]." Recall how before an observation Schrödinger's cat is simultaneously both dead and alive (or how an electron spins simultaneously both clockwise and counterclockwise). And each of these potential outcomes has its own probability to actually happen. Only after we observe, the Copenhagen interpretation states, can we determine whether the cat is definitely either dead or alive (or whether the electron spins definitely in the one or the other direction), and in general, whether an object is, as Anaxagoras states, definitely golden, yellow, dry, cold, and heavy.

Now the reason Anaxagoras required that various portions of all qualities had to coexist simultaneously everywhere within an object and at all times is that he wanted to remain in accordance with the Parmenidean thesis, that Not-Being does not generate Being, and that Being does not become Not-Being. Something must always exist if it is to be observed, the thesis says. That is, if a quality were not already present everywhere in an object always, it could not have come to be later; because if it did come to be later, it would mean that Being could be generated from Not-Being, but that's impossible. Hence a hot object, for example, has to contain simultaneously both hotness *and* coldness everywhere within it and always, though in different portions. For if a hot object did not contain coldness, too, coldness would have been Not-Being (at least for that object), and therefore it could have never come to be (coldness could have never become a reality, a part of Being)—it would then be impossible for the hot object to be cooled down.

This concept has a certain similarity with the Copenhagen interpretation but also a certain difference. Concerning the similarity, the reason we may observe the cat to be alive (or the electron to spin clockwise) is that the cat's (or the electron's) state of existence before the observation is a mix of all possible outcomes, that is, a mix that includes a portion (the quantum probability of occurrence) of the alive quality (or the clockwise spin) *together* with a portion of the dead quality (or the counterclockwise spin). In quantum theory this mix state is expressed mathematically. And the outcome with the highest probability (portion, in the language of Anaxagoras) is the one most likely to be observed.

Analogously, the reason we may observe, say, hotness, in Anaxagoras's view of our previous example, is that the object's state of existence before the observation is a mix that includes a portion of hotness and a portion of coldness, but with the hotness portion being the highest.

But Anaxagoras is even bolder than the Copenhagen interpretation, a fact that brings me to their difference. He insists that the notion of the simultaneous existence of all qualities is true all the time, even after an observation. Hence in Anaxagoras's view, the cat is still both dead and alive (or the electron still spins in both directions, or the object is both hot and cold) even after we observe the cat to be only alive (or the electron to spin only clockwise, or the object to be only hot). But in the Copenhagen view, after an observation the cat is only alive (or the electron spins only clockwise, or the object is only hot). Thus, although we observe only one quality, and so the cat appears only alive (or the electron is detected to spin only clockwise, or the object to be only hot), for Anaxagoras the other qualities never cease to exist; he insists on this because he does not want to violate the Parmenidean thesis that if a quality ceased to exist, it would mean that that part of Being became Not-Being. Now, can Anaxagoras be somehow right on this, too? Can the cat somehow be both dead and alive even after we observe it to be only alive?

ANAXAGORAS AND THE MANY-WORLDS INTERPRETATION

Fascinatingly yes! According to the second-most popular interpretation of quantum theory, the many-worlds view, even though *we* observe the cat to be alive (or the electron to spin clockwise), in *another universe* (world, reality) the cat is dead (or the electron spins counterclockwise)! That is, an outcome that is possible but does not occur in our universe still occurs in another universe. In general, every outcome that could have occurred in our present reality (universe) but did not, branches off as an alternative reality (it gets realized) in a parallel (i.e., separate) universe; each parallel universe thus has its own unique reality that consists of events that could have happened in our universe but did not.

Hence, the many-worlds view is in closer agreement than the Copenhagen view with both Anaxagoras's theory as well as the Parmenidean thesis. For, Parmenidean Being (being everything that there is) can easily be interpreted to include every possible outcome of an observation. Then, in the view of many-worlds, it is not only before an observation that all possible outcomes coexist in a mix and are part of Being (as is also required by the view of Anaxagoras), but all such outcomes, in a sense, continue to coexist and are thus still part of Being even after an observation (also as required by the view of Anaxagoras); for each possible outcome occurs in its own parallel universe even when such outcome is not observed to occur in our own universe. Whereas on the other hand, in the view of the Copenhagen approach, though before an observation all possible outcomes coexist in a mix and are part of Being (as also required by the view of Anaxagoras), after an observation only what is observed to occur continues to exist (to be part of Being), and what is not observed no longer exists, as if part of what once existed, part of Being, became Not-Being (a situation in clear violation of both the view of Anaxagoras and the thesis of Parmenides). With the Parmenidean thesis in mind one might then say that the many-worlds interpretation of quantum theory is more accurate than the Copenhagen.

THE PERFECT "IN EVERYTHING . . . IS . . . EVERYTHING"

The ultimate example of "in everything there is a portion of everything" is the big bang singularity, the hypothesis that the primordial state of the universe was once, 13.8 billion years ago, a mere point. "In everything"—in the singularity, which *itself* was everything that existed, the whole universe—"there" was "a portion of everything"—matter, energy, space, time, *and* the laws (or ultimate law) they obey. And the reason the universe is diverse, with planets, stars, people, plants, is that, as Anaxagoras might have explained, there is only a *portion* of everything in everything and "each thing is most manifestly those things of which it has the most."

Now, what remains enigmatic is this: (1) if indeed the notion of "in everything . . . is . . . everything" was true at the singularity, why would it not be true always? And (2) can plurality, all the beautiful and diverse nature of today, unfold from an absolute oneness, the singularity? Both are open questions.

On the first question, Anaxagoras would have answered "in everything . . . is . . . everything" always and everywhere. On the second question, all three pluralists, Anaxagoras, Empedocles, and Democritus, believed that plurality must be absolute; that is, neither could plurality (the many) have come to be from an initial singularity (from what is initially one), nor a singularity (the one) from an initial plurality (from what is initially many). Nevertheless, whether nature is truly monistic or pluralistic is a question that has yet to be answered. The singularity hypothesis is problematic (the physics equations are meaningless at such state of existence), and, as we saw in the section titled "Unborn and Imperishable" (in chapter 13), some cosmological models try to avoid it. Will our nous ever know the nature of nature?

NOUS

Nous for Anaxagoras is the only thing that is truly pure, containing nothing else except itself. Only living things have nous. Nous is infinite, timeless, has all knowledge about everything, and is the primary cause of motion and thus of all unfolding physical phenomena.

I believe that Anaxagoras did not mean literally and in a mechanical way that nous can cause motion, change, and "set all things in order." Rather, he meant that our nous is capable of a scientifically logical explanation of nature, such as one that assumes motion. And in general that any model of nature our nous conceives is adequate, provided that with it the phenomena are set in order and explained *scientifically* (rationally). This might be what matters after all, especially if an absolute knowledge about nature, a much-pondered topic since antiquity, is an impossibility, for example, because the truly first and last causes of the universe are indeterminable. Of course if reality is one (that is, objective), so ultimately should the scientific model be that our nous will conceive to explain it. And since for Anaxagoras "the phenomena are a sight of the unseen,"[6] that is, what we see contains subtle information about things we cannot directly see, along with reason (nous), Anaxagoras says the senses also have a catalytic role in our attempts to set things in order. However, what is the origin of our ability to reason? Why do humans have an intelligent nous (mind), the most intelligent, in fact, of all the species that we know?

FROM WALKING TO THINKING

Anaxagoras believed that the cause of human superiority over animals is the hand. Although other primates walk upright occasionally, only humans walk upright habitually, and as a result only humans have freed their hands permanently and have been using them consistently. This unique trait of ours has been significant for both our biological and intellectual evolution because when Lucy, a species of the genus *Australopithecus*, about 3.2 million years ago "decided" to walk upright more habitually, it meant that two of her four legs were starting to evolve into hands, increasing her potential to use and make tools. Toolmaking stimulates thinking (silent and out loud, thus speech, too), which in turn refines toolmaking, which stimulates further thinking, in a continuous cycle, ultimately advancing both technology and the intellect, and making Lucy's distant relative, the *Homo sapiens* (us), indeed *sapiens* intellectually superior to any other animal (at least on earth). And thus the origin of such superiority might truly be the hands. What actually ignited this development was a purely chance mutation in the spine that allowed our hominid ancestors to stand upright and evolve their forelegs into hands, become environmentally fitter, and get naturally selected further.

Upright posture (and consequently free hands), speech, and a complex brain (nous) are among the most unique traits of the human species and the source of our resourcefulness. The brain is the center of operations. Speech is controlled by the left side of the brain, but the coordination of the movements of our hands comes from the back part of the organ. Such coordination took literally millions of years of evolution to be mastered, during which the brain was driving the hands, which in turn were driving the brain, causing the enhancement of both organs, developing each to its present advanced stage of evolution compared to the hands and brains of other species. Without this type of evolution we would not have been able to make our first tools; stack up stones; build homes, the pyramids, the Parthenon, the Empire State Building; or create cell phones, computers, spaceships, MRIs, or the LASER, but also we would not have been able to pursue other more abstract endeavors, such as religion, philosophy, science, and the arts.

At the same time, however, I wonder if there is a limit to such brain-hands interactive enhancement. Even worse, I wonder, what the risks today are from the plentiful technology made by our hands as a result of the ingenious science conceived by our brains (nous). Do we get to depend more and more on machines to think for us and on pills to save us, risking the weakening of both the mind and the body? If yes, our natural abilities might atrophy and our evolution might be stalled. We might even devolve, for habits have, as seen, a say on whether we get to evolve.

The name of Galileo is often associated with the first major conflict between religion and science. He was tried by the Inquisition of the Roman Catholic Church for his support of the heliocentric model, which the Church considered in contradiction of biblical accounts of an immobile earth in the center of the universe (i.e., the geocentric model). But it was really the science of Anaxagoras that caused such conflict first. He was charged with blasphemy by the Athenian democracy for thinking that "the sun is a fiery stone"[7] and not a god. He was tried and found guilty. Although he was defended by his student Pericles, the famous statesman, still by one account he was exiled, by another he was sentenced to death. Anaxagoras was an original thinker. He is credited with explaining eclipses correctly and for introducing philosophy to the Athenians. When asked why he was born he replied, "To theorize about the sun, moon, and heaven."[8] When told "You are deprived of the Athenians," he replied, "No, they are deprived of me."[9]

CONCLUSION

Whether the nature of matter is infinitely cuttable (without smallest pieces) or finitely cuttable (with ultimate smallest pieces that make up everything) is still an open question. Anaxagoras held the former, but Leucippus and Democritus held the latter: namely, matter is atomic and thus made up of disconnected, indivisible pieces known as the atoms and surrounded by empty space. What revolutionized science was the atomic theory of matter, an idea that is two and a half millennia old.

CHAPTER 17

DEMOCRITUS AND ATOMS

INTRODUCTION

Perhaps the greatest scientific achievement of antiquity, possibly of all time, was the realization of the atomic nature of matter. "There are but atoms and the void" Democritus (ca. 460–ca. 370 BCE) proposed.[1] And he understood the great diversity of material objects as complex aggregations of uncuttable atoms, the building blocks of matter, moving in the void, the empty space between them. Leucippus, who flourished between 440 and 430 BCE, invented the atomic theory, and Democritus, a true polymath and a prolific philosopher, developed it extensively. Uncuttable (the actual meaning of *atom* in Greek) are also the modern elementary particles of matter, the quarks and leptons, and although void is a controversial concept still, a kind of void is required to explain nature.

ATOMS

Ancient Atoms

Atoms, in the ancient atomic theory, are the uncuttable smallest pieces of matter, disconnected from each other because they are surrounded by void, space devoid of matter that was required to enable the atoms to move. Atoms are invisible, impenetrable, solid (absolutely rigid), indestructible, eternal, unchangeable, unborn (not generated by something else more fundamental), and imperishable (they do not transform into something else more fundamental). Atoms are therefore like many Parmenidean Beings. Unlike the elements of Empedocles, which represented four different types of known materials, or those of Anaxagoras, which represented infinitely many types, all atoms are made from one and the same type of material (although not from any particular one of the everyday, such as water or air). Atoms have no internal structure (they are homogeneous) but differ from each other only as regards their size and shape. Their only behavior is motion. Roaming around in the void of an infinite space there exist infinitely many atoms of various shapes: angular, concave, convex, smooth, rough, round, sharp, and so on.

Their motion is perpetual (so, then, is change in accordance to the Heraclitean worldview). Motion continues by itself without requiring a force (or a causal agent in general). The fact that constant motion continues by itself was discovered via experimentation first by Galileo and was later restated by Newton through his first law of motion, also known as the law of inertia (a body will remain at rest or in uniform motion until affected by a force). Atomic motion is also thought random because space was correctly assumed to be isotropic (having no special location or direction, i.e., space has no absolute up, down, right, left, in, out, center, or edge). Hence, when left to themselves (undisturbed), atoms had no reason to move more one way than another—all directions of motion were equally probable. Motion was not explained by Leucippus and Democritus; it was simply postulated to have always been, without a beginning. In fact atoms and the void were also postulated. Not accounting for the cause of motion received Aristotle's intense criticism, even though it is actually a normal scientific procedure. For we need to remember that postulating the truth of a certain beginning and proceeding from there to understand nature in a causal and rational way is the only way to do science. Science must begin from something (a postulate, an axiom, a primary cause); it cannot begin from nothing (Not-Being). For Democritus the atoms, the void, and motion were part of a primary cause, which by definition requires no cause of its own.

As they move, atoms collide with each other, bounce, rotate, some hook together (whenever their shapes are complementary) and assemble in a multitude of arrangements, forming all kinds of macroscopic (compound) objects that appear "as water or fire, plant or man,"[2] or unhook and disassemble, deforming (destroying) the objects. As atoms aggregate objects form and increase in size, and as atoms segregate objects change form and decrease in size (i.e., they perish). Just as the words "tragedy" and "comedy" are formed when letters (which can be thought of as atoms) from the same alphabet are combined differently, Aristotle had explained, the immense plethora of diverse objects is formed when atoms are arranged in space differently through their motion.[3] In fact, atoms were required by Leucippus and Democritus to explain exactly this diversity in nature: "For from what is truly one a plurality could not come to be, nor from what is truly many a unity, but this is impossible."[4]

Atoms have none of the conventional properties of matter such as color, taste, smell, sound, temperature, or even weight. The proof of this, Democritus thought, is in the fact that various objects are perceived differently by different people. Something sweet to me might be spicy to you. "To some honey tastes sweet to others bitter but it is neither," said Democritus.[5] So he explained the conventional properties in terms of the shape of the atoms and their motion in the void; shape and motion cause unique macroscopic arrangements (in objects, in people) and thus unique conventional properties. For example, the atoms of a hard object are more closely packed with less empty space (void) between them than the atoms of a soft object. Now, since the atoms of a soft object have more void to roam around they can be pushed there more easily. Hence, such objects feel squeezable and soft. Metals are made of atoms with hooks that hold them firmly interlocked, but liquids are made of round atoms so they can flow by each other easily. Sweet objects are composed of round good-sized atoms; bitter of round, smooth, crooked, and small; acid of sharp (so they can sting the tongue) and angular in body, bent, fine; oily of fine, round, and small. Even light was made of atoms (particles)—incidentally, Einstein won the Nobel Prize in Physics by interpreting light in terms of discrete particles: the photons. Black, white, red, and yellow were considered primary colors and associated with different shapes and arrangements of atoms.[6] Combinations of these four colors were in turn used to account for all color variations. Democritus worked out a detail theory on sensation. In general he argued that the constant motion of the atoms, which persists even when a composite object is seemingly at rest, causes some of them to be emitted by the object. Flying through the void, these atoms in turn ultimately collide with the atoms of a sense organ and create a unique sensation (a flavor, smell, color). By the way, a collision (recall section "The Uncertainty Principle" in chapter 12) is the first out of two steps/events that generate the famous Heisenberg uncertainty principle. Because Democritus realized the uncertain nature of atomic collisions (due to the random atomic motion), he argued that the knowledge we acquire by sense perception (e.g., observation) is "bastard,"[7] uncertain—recall that the Heisenberg uncertainty principle also states that the act of observation causes uncertainty.

All changes of the apparent world of sense perception, animate and inanimate, were reduced to the irreducible atoms and their motion in the void. This scientific reductionism is ambitious. In principle, it is also the goal of the modern theory of the elementary particles of matter, the quarks and leptons. But while there are striking similarities between these modern particles of matter and the ancient atoms, there are also serious differences. Only for purposes of comparison let us call the ancient atoms Democritean or D-atoms, and today's building blocks of matter, the quarks and leptons, QL-atoms.

D-atoms and QL-atoms: Similarities and Differences

Before we proceed with this comparison let us first briefly summarize the historic developments in search of the D-atom, the ultimate uncuttable piece of matter. Until about the end of the nineteenth century chemical atoms, such as hydrogen, carbon, oxygen, and so on, of the periodic table of chemistry, were thought to be the fundamental particles of matter, the D-atoms. But this idea turned out to be incorrect when the structure of the chemical atom was probed further and it was found that it was made up of electrons and a nucleus. In 1897 physicist J. J. Thomson (1856–1940) discovered the electron, and in 1908 his student physicist Ernest Rutherford (1871–1937) discovered the nucleus. Electrons (one of the six types of leptons) are still thought indivisible, but atomic nuclei not so; the latter are made from divisible protons and neutrons, though they are themselves composed of indivisible quarks. Six types of quarks were postulated to exist as elementary particles in the 1960s, and all six types had been discovered by the end of the twentieth century. Hence, chemical atoms are not fundamental; they have substructure and in fact are made of the QL-atoms, which are among the particles of the standard model of physics. Like the shadows, which *were* real but were not the real *objects* in Plato's parable of the cave, chemical atoms *are* real but are not the real *fundamental particles* (the smallest cuts of matter as envisioned by Leucippus and Democritus)—although the name "atom" has been undeservingly stuck on them.

On the other hand, both D- and QL-atoms are fundamental because they are not made from other particles; they are disconnected pieces of matter, indivisible (uncuttable), invisible, and the smallest, and their various combinations make up all material things in the universe. Neither the D-atoms nor the QL-atoms have any of the conventional properties of composite objects. These properties are really a consequence of the collective behavior of the D- and QL-atoms that make up these objects.

D-atoms are unchangeable, they do not transform, but QL-atoms do; they transform from one type of material particle to another and also into and from energy (although they do not transform into something more fundamental). But like matter, energy also comes as discrete bundles, as particles (e.g., photons), and so Leucippus's and Democritus's notion of discreteness as a property of nature is preserved. Furthermore, like D-atoms, which are made of the same substance, QL-atoms are made of the same substance, too, mass and energy (which are equivalent as per special relativity). And since D-atoms are indestructible, so is their substance, but so is the substance of QL-atoms, for the total amount of mass-energy in the universe is constant (as per the law of conservation of mass-energy). So the substance of both, the ancient and modern atoms, endures, while nature is constantly changing.

D-atoms have shapes and thus have nonzero size, QL-atoms are considered point-like, thus shapeless and size-less. There is both a challenge and a simplicity associated with each view. In the D-atoms we need to imagine all sorts of complex atomic shapes, but we need not worry about forces. Democritus did not introduce any (a topic to be revisited below). D-atoms, he explained, coalesce into composite objects as a result of their perpetual motion and complementary shapes. Also, composite objects have size since they are made of D-atoms, which themselves have size. On the other hand, QL-atoms lack shape and size, but they still combine. They do so via the exchange of the particles of force (the photons, the W's, the Z's, the gluons, also the gravitons if we find them). Being point-like thus shapeless is in a sense a simplicity, for it means that QL-atoms are internally structureless like the D-atoms. But it is also a complexity, for how can something of zero size, of zero extension in space, have properties such as mass, electric charge, energy, spin, and so on, and, even worse, how can composite objects have size and extension in space when their constituents do not? Within the context of the Parmenidean theory the "size" question may be restated as follows: how can size come from not-size? That is, how can Being (the nonzero size of macroscopic objects) come to be from Not-Being (from zero-size constituents)? This was impossible for Parmenides and Anaxagoras—that's why the latter, recall, posited "in everything . . . is . . . everything." Democritus solved the size challenge by postulating that matter cannot be divisible (cuttable) at infinitum; it must be finitely divisible with the smallest cuts to be the indivisible, the uncuttable D-atoms of nonzero size. For only then, he thought, could composite objects have size, if they are composed of things that themselves have size. Interestingly, the latest hypotheses for fundamental particles of nature include particles that *do* have size: as seen in chapter 8, these are one-dimensional strings or two-dimensional membranes. If these turn out to exist, we may not need to worry about size (Being) coming to be from not-size (Not-Being), an idea that would be pleasing to all three philosophers, Parmenides, Anaxagoras, and Democritus. The size challenge is revisited in the section on "Void or Not?"

D-atoms are postulated to move in order to comply with the apparent world and explain change. On the other hand, since motion is an ambiguous concept from the modern point of view (as seen in chapter 14), QL-atoms are postulated to move only as an *adequate* way of understanding the phenomena. Furthermore, QL-atoms are best regarded as events (as seen in chapter 12) and not as permanent Parmenidean Being-like entities (as are the D-atoms). QL-atoms' existence (their properties and behavior) is best described in terms of the quantum probability, a number that expresses only a potential event: for example, which type of QL-atom might be observed, where, and with which properties and behavior. Within the context of quantum theory, material particles (the QL-atoms) are less materialistic in the sense that they no longer have key properties that material particles were once thought to have in order to be called material: they are neither permanent, nor indestructible, nor unchangeable, nor deterministic, nor have they well-defined shapes or trajectories through space and time, and as a consequence nor have they identity and individuality. They are represented by quantum probabilities, thus, as has also been argued in chapter 11, they are mathematical forms. These mathematical forms have often been correlated with the Platonic forms, from Plato's theory of forms, according to which the physical objects of sense perception are but mere copies, shadows, of a greater truth (form). Hence, having quantum probability in mind, QL-atoms are described by both an element of chance but also of necessity (as these ideas were discussed in the section titled "The Cycles of the World" in chapter15). But so are the D-atoms; their motion is purely random and chancy, although at the same time what drives them to combine or separate is necessity.

Are the QL-atoms the smallest cuts of matter and, within this context, thus the ultimate D-atoms? It is generally not thought so. Are the QL-atoms different forms of the same type of universal substance, the same type of particle, and what that might be? While the Higgs boson particles have some qualities required of a universal substance, the standard model that predicts them does not include the most puzzling force in the universe, namely, gravity. Therefore, although useful, any model of nature that does not incorporate gravity is incomplete.

The basic concept of the ancient atomic theory was highly valued by Nobel laureate Richard Feynman. He said: "If, in some cataclysm, all scientific knowledge were to be destroyed, and only one sentence passed on to the next generations of creatures, what statement would contain the most [scientific] information in the fewest words? I believe it is the *atomic hypothesis* (or the atomic *fact* . . .) that *all things are made of atoms—little particles that move around in perpetual motion, attracting each other when they are a little distance apart, but repelling upon being squeezed into one another.* In that one sentence, you will see, there is an *enormous* amount of information about the world, if just a little imagination and thinking are applied."[8] On a related note, in his book *The God Particle* Nobel laureate Leon Lederman graded thousands of scientists (including himself) for their efforts in their quest for a primary substance of the universe. He started with Thales all the way to 1993, the completion date of his book. Democritus received the only A in the class![9]

So as a general *idea*, the D-atoms, the uncuttable discrete and fundamental pieces of matter that everything is made of, are still part of our most advanced theories of nature, for these basic but important properties are properties of the QL-atoms, too—in fact also of the strings of string theory. But whether the QL-atoms (together with the force-carrying particles of the standard model as well as the Higgs boson, a total of sixty-one particles all confirmed to exist), a zoo of other unconfirmed particles (including the graviton, predicted by various other scientific models), or some new particles of *one and the same type* (a much-desired scientific simplicity of Democritean grandeur) that manifest themselves by way of the familiar particles are/will be the truly uncuttable discrete and fundamental pieces of matter, remains to be seen. How about the void? Does it exist or not? Is it needed, or can it be avoided?

VOID OR NOT?

The atomists Leucippus and Democritus called an atom a *thing*, Being (what-is), and the void *nothing* (*not thing*), Not-Being (what-is-not).[10] And they agreed with the theory of Parmenides (with one interpretation of it, anyway, for which the properties of Being are understood literally) that motion is impossible without the void. But whereas Parmenides denied the existence of the void by considering it Not-Being, the atomists postulated the opposite: Not-Being, the void, exists, for only then, they thought, can motion and change be accounted for. It is the place to put the atoms and enable them to move. For the atomists the void is empty space, and so *in* it there is nothing. But for Parmenides *it, the void, empty space itself*, is nothing, Not-Being; not *in* it there is nothing. The nature of the void has created mind-boggling debates since the time of Parmenides. For if something, for example, the void, is really nothing, how can it exist? How does one define "nothingness"? The answer is not easy.

But first let's summarize Democritus's arguments favoring the void. By accepting the phenomena of motion, change, and diversity to be real, he deduced the void to be real as well, for without the void his impenetrable, indivisible atoms could not move and consequently the phenomena of change and diversity would not occur; but they do occur, so the void must be real. Similarly, by accepting also multiplicity and division of composite objects to be real, he again deduced void to be real, for without it, composite objects could not be divided (cut) into smaller pieces: "division resulted from the presence of void in bodies."[11] As explained further by philosopher and mathematician Bertrand Russell (by paraphrasing Democritus), "When you use a knife to cut an apple, the knife has to find empty places where it can penetrate; if the apple contained no void, it would be infinitely hard and therefore physically indivisible."[12] Here we recall of course that for Democritus divisibility does not continue ad infinitum; it applies only to composite objects and stops at his physically indivisible atoms. So for Democritus both the atoms and the void are real: "thing [atoms] is [exist] no more than not-thing [void]."[13]

Now what does modern physics think of the void? Does it exist or not? Is it a true nothing, the Parmenidean Not-Being, or something else? While "nature abhors a vacuum,"[14] a popular phrase since the Renaissance, yet "nothing works without, well, nothing."[15]

Void?

On the one hand, void is still a useful concept for the understanding of many phenomena. According to quantum theory, electrons in a chemical atom, for example, "move" around their nucleus by keeping their distance from one another as if space between them is empty, devoid of matter—"a regulation against overcrowding"[16] formally known as the Pauli exclusion principle. As a consequence of this principle, the electrons of chemical atoms keep their distance from each other; they do not like to be squeezed together into a small region, so they act as if they were rigid: the closer they get, the faster they move apart—a statement in agreement with the Heisenberg uncertainty principle, for, according to it, the uncertainties in the position and velocity are inversely proportional, so the smaller a particle's region of confinement (the smaller its position uncertainty), the faster its motion to escape such region (the greater its velocity uncertainty), "almost as if it [the particle] were overcome with claustrophobia," Brian Greene wrote.[17] The exclusion principle explains why chemical atoms are mostly empty space and why macroscopic objects (which are made of chemical atoms) have a degree of rigidity, size, and shape. D-atoms are rigid, consequently, in a sense they, too, obey the regulation against overcrowding, for one D-atom cannot occupy the same region of space as another D-atom. Had the exclusion principle not been true, the QL-atoms, which obey it, would not endure as disconnected pieces of matter, thus nuclei would not form, nor would chemical atoms or the molecules of organic chemistry, and consequently nor would the matter that living things are made of; generally, *all* matter in such a scenario would collapse into a uniform, undifferentiated, and lifeless state. The diversity in nature is in a sense a consequence of the exclusion principle: diversity is a law of nature!

Or Not?

On the other hand (that is, to be able to explain other phenomena), in the quantum realm the void is not really devoid of matter but a very busy place, seething with all-pervasive fields of energy (e.g., light and gravity waves, even the much-required Higgs boson field that explains mass—see section "Worlds without Forces" below), known as vacuum energy. These fields cannot be zero, even in seemingly empty space, because the time-energy uncertainty principle would be in violation (recall the section titled "*Nothing* Comes from Nothing" in chapter 13). And they are actually fluctuating constantly, creating and annihilating pairs of particles with their corresponding antiparticles. These particles, which are called virtual, are not created out of nothing or annihilated into nothing but are made out of energy and return to be energy (the vacuum energy). Unlike real particles, which can be directly observed, virtual particles cannot, even though they can still cause measurable effects on real particles, a proof that "empty" space is really not empty.

Moreover, according to the theory of general relativity, the whole of space is filled by a gravitational field with properties (such as strength) that vary from place to place and from one moment to the next. Einstein explained gravity by assigning *properties* to "empty" space; empty space (and time) is a flexible medium that gets distorted by a mass, and gravity is space's (and time's) distortions. These *properties* are the void, and so in the theory of general relativity the void is not the Parmenidean Not-Being, for Not-Being, a true nothing, is property-less.

Furthermore, an astronomical observation completed in 1998 with the aid of the Hubble Space Telescope found that the expansion of the universe is accelerating—so a galaxy's recession speed measured today is faster than its speed measured yesterday. This accelerated expansion is attributed (although reluctantly) to the existence of dark energy, which is hypothesized to permeate the universe and act as a kind of antigravity by stretching space and causing it to expand at continuously faster speeds. Dark energy, which has not yet been detected, is one of the most puzzling mysteries of the universe. Dark matter is yet another great puzzle: though invisible, for it does not emit light, its existence is inferred indirectly by its gravitational pull on neighboring stars. What makes dark matter invisible, no one knows.

Ordinary matter, matter we can see, which makes up flowers, people, planets, stars, and galaxies, is only about 5 percent of the total stuff in the universe. The other 95 percent, which includes dark energy and dark matter, is stuff that we neither see nor know much about, although their subtle presence is deduced in some way.[18] Space, even "empty" space, is a place of constant, frantic activity of virtual particles, light, gravity, dark energy, dark matter, ordinary matter, of Higgs boson particles, possibly strings, membranes, and who knows what else. It is certainly not the Parmenidean Not-Being (the *absolute* nothing). Therefore, once more we emphasize that no scientific theory can base its beginning, its first cause/s, on absolute nothing, on Not-Being; science must begin from something-ness, and what that might be becomes increasingly more complex. In fact even Leucippus's and Democritus's nothing (their notion of void) is really not nothing, since from the point of view of modern physics "it [their void] was the carrier for geometry and kinematics, making possible the various arrangements and movements of atoms."[19]

Fascinatingly these atomic arrangements and movements were imagined by Democritus to be carried out without the requirement of a force of, say, gravity, electricity, or magnetism; other than their direct collisions during which there was a physical contact, D-atoms experience no other force! D-atoms have no weight, they produce no force of gravity.[20] How can this be? How can there be a world without gravity, without forces in general?

WORLDS WITHOUT GRAVITY

Weight or gravity (that is, the tendency of objects to fall or the property of heaviness) was not one of the primary characteristics of atoms but a property that was accounted for by Democritus ingeniously through motion, in particular rotational motion.[21]

Though motion is chaotic, Democritus argued, in an infinite space with infinite atoms there is always a chance that the bulk of the atoms of a certain region move collectively in a preferred direction of motion, rotational in particular, and produce a vortex. The rotational motion of such a vortex, Democritus thought, ultimately causes the bigger atoms (the more massive, the heavier, as we would say today having gravity in mind) to move toward its center, ultimately forming the earth and the water on it, and the smaller atoms (the lighter) to move toward its outskirts, ultimately forming the air, the sky, and the stars. Because the dynamics of our world system is still rotational (e.g., the sky rotates, relative to us, and so in a sense do the moving clouds), objects made of the bigger atoms, like a rock, still fall, and objects made of the smaller atoms, like steam, smoke, or fire, still rise, Democritus argued. Air, on the other hand, generally does not fall, he thought, because of its rapid revolution, just as water does not spill from a cup when it is rapidly spun around. His analysis was logical as regards observation because the earth, which (for him) is made of the bigger atoms, formed in the center of his vortex, water, made of smaller atoms, is on earth, and air, made of even smaller atoms, is above water and earth.

Now, concerning the dynamics of a vortex, in reality it is the reverse that happens: massive objects tend toward the outskirts of a vortex, and lighter ones toward the center (this, for example, happens in a centrifuge, a device employed to separate different substances). Nonetheless this error in Democritus's explanation is really a minor point compared to the fact that he managed a reasonably clever explanation of the world only in terms of a basic property that atoms have, namely, their motion in the void. Thus he saw no need of a force of a weight, of gravity, despite that apples fall as if a force is pulling them through space. The latter, legend says, inspired Newton to conceive his theory of universal gravitation for which gravity *was* a force, only to be abolished as a force by Einstein's theory of general relativity. How so, and what does quantum theory say about forces in general?

WORLDS WITHOUT FORCES

An apple and the earth, or the sun and the earth, feel a force of attraction from each other, Newton thought, as a consequence of a mysterious action at a distance (not to be confused with the action at a distance of quantum entanglement) that he himself admitted he did not understand. How is gravity transmitted if the interacting objects do not touch each other? How does one body feel the other, how do they communicate, if nothing but empty space exists between them and if nothing specific is really exchanged by them?

Einstein provided the answer through his theory of general relativity. He eliminated the need for an action-at-a-distance-type force by recognizing that the agent that transmits gravity is space itself (in fact, time, too, but let's keep things simple) when distorted by matter. Space is no longer the Newtonian passive playground where events unfold but a flexible medium the geometrical shape of which *changes* (gets warped) by matter. As discussed in chapter 12, the distortion of space (the changing geometry of space) in turn influences an object's motion and feels like gravity. With such geometrical representation of spacetime the notion that gravity is a force is abandoned in general relativity. And in the study of the phenomena of gravity, an object's motion through space and time may no longer be regarded as a response to an action-at-a-distance force of gravity acting on it, as in Newtonian physics, but as a response to the warping spacetime caused by the distribution of all other objects around it.

Moreover, as already discussed in chapter 12, according to the standard model of quantum theory, the particles of matter, the QL-atoms, combine with one another via the continual exchange of the particles of force. Recall, for example, that the attractive and repulsive electric force is really a manifestation of intricate particle collisions; even gravity is hypothesized to work via the exchange of gravitons. Matter and force are no longer distinct notions. Instead, forces are really expressions of complicated particle collisions.

And so, as is the view of Democritus, nature can be understood in terms of just particles and their complex collisions—forces were never required in the theory of Democritus and are no longer required in modern physics! Incidentally, although Empedocles's two forces, love and strife, were separate entities from his four elements, still they were not action-at-a-distance-type of force:

via their direct contact, they either pushed the elements together to mix or pushed them apart to separate, so they, too, in a way, acted as colliding particles.

Even mass, and consequently weight, is thought to not be a fundamental property of the QL-atoms, rather, a property the QL-atoms acquire through their interactions with the Higgs boson field. The mass of an object is a measure of its resistance to motion. The smaller the resistance, the less the mass is. Throwing a baseball is easier than a bowling ball: the baseball has less mass than a bowling ball, or, equivalently, it produces less resistance to our attempt to throw it. Now, the Higgs boson field pulls on the other particles (e.g., the QL-atoms) as they traverse through it and impedes their motion. It is this resistance that we interpret as mass. In an analogy, stirring a cup of coffee with a spoon is easy, but a cup of honey is not. Honey is a more viscous fluid, and the spoon feels heavier, more massive, as it moves through it. In this analogy, the spoon is a QL-atom and the fluid the Higgs boson field. Just as fluids with different viscosities cause the spoon that moves through them to feel lighter or heavier by impeding its motion, the standard model imagines that the all-pervasive Higgs boson field creates an analogous effect on the initially massless QL-atoms as they traverse it, endowing each with a unique mass and slowing them down. It is as if the Higgs field had different viscosity for different-type QL-atoms. Similarly, the force-carrying particles W's and Z's acquire their mass, but photons, feeling no resistance by the Higgs field, remain massless and thus can move with the speed of light. The analogy describes what is known as the Higgs mechanism, which explains why some particles have mass and some not (though it does not explain why they have the actual mass they do). What particular agent gives Higgs bosons their mass is nevertheless still an unknown. The idea that mass may not be a fundamental property was prompted by a few interesting and unresolved questions. Why, for example, is there no pattern in the mass values of the particles, a fact in contrast to other particle properties, such as spin or electric charge? For instance, in some units of measurement, the spin of all QL-atoms is $1/2$ and the spin of all the force-carrying particles is 1.

Amazingly, mass is not a fundamental particle property, neither in the standard model of modern physics nor in Democritus's atomic theory! Equally amazing is that in both the modern and the Democritean physics, the non-fundamental property of mass (and the consequent heaviness, weight) is caused (is acquired) by atomic motion—the motion of the QL-atoms through the Higgs field, and the motion of the D-atoms in the vortex!

Particles are by definition discrete entities, thus their existence implies a certain discontinuity in nature. But is the nature of nature truly discontinuous?

CONTINUITY VERSUS DISCONTINUITY

If indeed something does exist always everywhere (including apparently empty space), then the essence of existence is continuous. At the same time, to make sense of the diversity in nature, the continuity of that which exists must vary, from place to place and from time to time. These variations are interpreted as discontinuities in matter and energy and are called particles. But what separates these discontinuities cannot be absolute nothingness, for energy is ever-present and everywhere. If the sea is the energy, the sea waves are the fluctuations of the energy, that is, the discrete particles of matter and the discrete particles of force. But even between the sea waves there exists water, the sea, energy, not nothing. So the view of modern physics is some kind of combination of the Parmenidean Being (of an indivisible, continuous whole obeying one eternal truth), the Heraclitean constant change (of everything in the sensible world), and the Democritean discreteness (of a whole, which while in essence continuous is also inhomogeneous and discrete, for it fluctuates). Now what exists must, we believe, be describable by a single idea or equation, a single type of particle. Can the human intellect ever conceive it? What is the role of the senses in conceiving it?

INTELLECT VERSUS SENSES

For Democritus reality is objective and much deeper than what's revealed by sense perception alone. Trying to capture both the unreliability and the significance of sense perception in our attempts to understand nature rationally, Democritus imagined a hypothetical dialogue between the intellect and the senses.

Intellect: "Sweet is by convention, bitter by convention, hot by convention, cold by convention, color by convention, in reality however there are but atoms and the void."[22]

Senses: "Troubled Intellect! From us you take the evidence and you want to overthrow us? Our fall will be your fall."[23]

The *Intellect* says that what's perceived by the *Senses* is radically different from the way nature really is. Knowledge derived by the *Senses* is "bastard" but by the *Intellect* "legitimate"[24] (Democritus). The *Senses* perceive sight, hearing, smell, taste, and touch, but these sensations are not objective properties of nature. They are only perceptions by convention (in relation to us), only a consequence of the atoms and their motion in the void—the objective truth, in other words, the true nature of things, is only atoms and the void the *Intellect* claims.

This might be true, the *Senses* respond, but though unreliable ("bastard"), the quest for knowledge always begins with sense perception. For the evidence of the atoms and the void is obtained through observation of colors, tastes, and so on, thus the participation of the *Senses*. It is what we see that we use in order to explore what we cannot see, the *Senses* emphasize. After all, "the phenomena [what is seen, occurrences] are a sight of the unseen."[25] At the end, neither the intellect alone nor the senses alone can lead to the truth, but their combination might.

Remarkably, for Democritus the aggregations and segregations of unseen atoms in the void produced not only our own world (with the earth, moon, sun, planets, stars, plants, fish, animals, and including humans) but also countless others.

WORLDS OTHERWORLDLY

The process that creates a vortex from which a world like ours was formed is not unique in the universe, Democritus thought, so he posited the existence of otherworldly worlds. (Democritus therefore, like the Pythagoreans and possibly Anaxagoras, was not fixated on a geocentric model.) In modern terminology such worlds are really the galaxies; the stars we see at night all belong to the Milky Way galaxy, which is one of numerous others in the universe. The idea of multiple worlds does not violate Democritus's isotropic universe, where there is no preferred direction of motion, hence all directions are equally probable. The clockwise rotation of a vortex that creates one world is canceled out by the counterclockwise rotation of a neighboring world, a phenomenon that modern physics calls the conservation of angular momentum.

Today we know that our universe consists of many galaxies, each of which contains billions of stars, each of which may have its own group of planets. In fact earthlike (thus habitable) planets are common! According to the latest research one in five sun-like stars in our Milky Way galaxy has an earthlike planet orbiting it—the nearest case is perhaps of a star only twelve light-years away and visible to the naked eye.[26] Consequently the existence of life elsewhere in the universe besides earth is increasingly more probable. While scientists have not yet discovered any, still one may ponder such a question optimistically. First, one may answer favorably on the grounds of modesty. Our part of the universe is no more special than another (the laws of nature are everywhere the same), hence it is not unreasonable to expect life to evolve elsewhere as it did on earth. This is especially so on earthlike extrasolar planets (planets orbiting other stars). Or, one may answer mathematically by using the famous Drake equation to calculate the odds for intelligent life elsewhere in the universe.[27]

The equation gives a rough estimate of N, the number of advanced civilizations present in the Milky Way galaxy with which we might be able to establish radio communication. The important factors that N depends on are:

$N_{\text{-habitable planets}}$ = the number of habitable planets in the galaxy—those capable of developing and sustaining life;

$f_{\text{-life}}$ = the fraction of habitable planets where life actually evolves;

$f_{\text{-civilization}}$ = the fraction of planets with life where also an advanced civilization (with capabilities of interstellar communication) has developed at some point in the past;

$f_{\text{-now}}$ = the fraction of planets that still have an advanced civilization, now.

N is equal to the product of the above factors, that is:

$$N = N_{\text{-habitable planets}} \times f_{\text{-life}} \times f_{\text{-civilization}} \times f_{\text{-now}}$$

None of these numbers is known with certainty. We can only make rough estimates. Suppose, in a sample exercise, that there exist 1,000 habitable planets in the Milky Way, in other words, $N_{\text{-habitable planets}} = 1,000$. And life evolves on 1 in 10 of them, in other words, $f_{\text{-life}} = 1/10 = 0.1$. Now, 1 in 4 of these planets with life, also develops, sometime, an advanced civilization, in other words, $f_{\text{-civilization}} = 1/4 = 0.25$. Finally only 1 in 5 of such advanced civilizations are still around today, in other words, $f_{\text{-now}} = 1/5 = 0.2$. What is N?

$$N = 1,000 \times 0.1 \times 0.25 \times 0.2 = 5$$

How do you think the human race will react to a discovery of an advanced extraterrestrial civilization? What would it be like to be exposed to another intelligent species' worldview? We have no experience with that since on earth, out of the myriad species, only humans have developed advanced thought. Why?

CONCLUSION

In an attempt to understand nature in a logical and causal manner, the search for the primary substance of the universe in pre-Socratic philosophy comes to an end with the atomic theory of Leucippus and Democritus. Their notion of indivisible, discrete particles without substructure has endured and, according to modern physics, is still one of the most remarkable properties of nature. Could spacetime form a type of discreteness, too? Namely, is there a fundamental irreducible (smallest, without substructure, and finitely divisible) space length and time interval, or is there always a smaller portion of them (i.e., are they infinitely divisible)? Although no experiment has confirmed spacetime discreteness, some modern theoretical models speculate that it might be true. Recall Democritus's notion of atoms was confirmed two and a half millennia after it was first proposed. Interestingly, the hypothesis of space discreteness has been an innovation of Epicurus (341–270 BCE), a post-Socratic, who continued the remarkable work of Democritus.

EPILOGUE

Our ancient quest for knowledge began with our evolution two hundred thousand years ago, and with everything we experienced through our struggles to survive and our efforts to thrive and live fully. We hunted and gathered, painted on caves, told stories, domesticated animals and plants, built homes, wondered about nature, gave birth to civilization and religion, picked up writing, philosophized, and engaged in science. In fact, we keep on doing all these wonderful things, but, amazingly, we have been doing them ever more in the light of science.

Since its birth 2,600 years ago, science has evolved significantly. Nonetheless its ultimate goal still remains essentially the same: to understand nature rationally and to reduce the explanations of all natural phenomena to the least possible number of basic assumptions (first causes, axioms)—ideally to just one. Now, say that has been achieved, will the human intellect be satisfied?

We like the Homer's *Odyssey* so much because it is a story of a journey (in fact a long one) not of a destination. Shortly after Odysseus returns to Ithaca the story ends and we all get melancholy—we like the journey better than the destination, for although, with Odysseus's return, the events in Ithaca were breathtakingly exciting and awaited eagerly from the start of the story, their completion also brought the absolute end of the epic adventure.

The beauty of nature is in her secrets, the magic is in our discoveries. I never want to know everything—to have the journey of knowledge, of search and discovery, ever end. What would be next if I did? What would happen to the magic? Space, time, matter, energy, the human senses to observe, and the intellect to contemplate—it is all nature, and her nature is her many secrets (her shadows). They are many but also intelligible (steal-able, like Promethean fire)! I hope you have a magical, endless journey searching, in the light of science, for the nature of nature!

NOTES

PROLOGUE

1. The Higgs boson, nicknamed "God Particle" by Leon Lederman for its elusiveness as well as significance for our understanding of the structure of matter. Leon Lederman and Dick Teresi, *The God Particle: If the Universe Is the Answer, What Is the Question?* (Boston: Houghton Mifflin, 1993).

CHAPTER 2. WHAT IS SCIENCE?

1. Dionysius, bishop of Alexandria, from Eusebius, *Preparation for the Gospel* 14.7.4, trans. Daniel W. Graham, *The Texts of Early Greek Philosophy: The Complete Fragments and Selected Testimonies of the Major Presocratics* (Cambridge: Cambridge University Press, 2010), p. 521 (text 5). Graham's sourcebook contains fragments and testimonies of the major pre-Socratic philosophers. It is a Greek-English edition that also cites the original ancient source of each text.

CHAPTER 3. URBANIZATION

1. Pliny *Natural History* 2.53. See Daniel W. Graham, *The Texts of Early Greek Philosophy: The Complete Fragments and Selected Testimonies of the Major Presocratics* (Cambridge: Cambridge University Press, 2010), p. 25 (text 5).

2. Smithsonian National Museum of Natural History, http://humanorigins.si.edu/ (accessed March 10, 2014).

3. Peter Tyson, "Who's Who In Human Evolution," http://www.pbs.org/wgbh/nova/evolution/whos-who-human-evolution.html (accessed March 10, 2014).

4. John Savino and Marie D. Jones, *Supervolcano: The Catastrophic Event That Changed the Course of Human History* (NJ: Career Press, 2008); Smithsonian National Museum of Natural History, http://humanorigins.si.edu/evidence/human-evolution-timeline-interactive (accessed June 6, 2014); Toba: The Toba Super-Eruption, http://toba.arch .ox.ac.uk/index.htm (accessed June 6, 2014).

5. Ian Tattersall, *The World from Beginnings to 4000 BCE* (Oxford: Oxford University Press, 2008), p. 89; Savino and Jones, *Supervolcano*, p. 126; Smithsonian National Museum

of Natural History, http://humanorigins.si.edu/human-characteristics/change (accessed March 10, 2014).

6. Richard Wrangham, *Catching Fire: How Cooking Made Us Human* (New York: Basic Books, 2009).

7. Sister Wendy Beckett, *The Story of Painting* (New York: Dorling Kindersley, 2000), p. 10.

8. Ibid., p. 11.

9. Tattersall, *World from Beginnings*, p. 107; Tyson, "Who's Who In Human Evolution," http://www.pbs.org/wgbh/nova/evolution/whos-who-human-evolution.html (accessed June 6, 2014); Smithsonian National Museum of Natural History, http://humanorigins.si.edu/evidence/human-fossils/species/homo-floresiensis (accessed June 6, 2014).

10. Walter Burkert, *Greek Religion: Archaic and Classical*, trans. John Raffan (Cambridge, MA: Blackwell Publishing Ltd. and Harvard University Press, 1985), p. 47.

11. Ibid., p. 248.

12. Isaac Asimov, *Asimov's Chronology of Science and Discovery* (New York: HarperCollins, 1989), p. 12.

13. Hesiod *Works and Days* 120, trans. Hugh G. Evelyn, *Mythology Ultimate Collection* (Houston: Everlasting Flames Publishing, 2010).

14. Ibid., pp. 127–28.

15. Ibid., pp. 133–37.

16. Gen. 4:1–26.

17. Hesiod *Works and Days* 110–20.

18. Ibid., 127–28.

CHAPTER 4. THE MYTHOLOGICAL ERA

1. Bertrand Russell, *The History of Western Philosophy* (New York: Simon & Schuster, 1945), p. 11.

2. Plato *Republic* 3.390E, trans. Allan Menzies, *History of Religion: A Sketch of Primitive Religious Beliefs and Practices, and of the Origin and Character of the Great Systems* (Memphis, TN: General Books, 2010), p. 40.

3. Hesiod *Theogony* 542, trans. Hugh G. Evelyn, *Mythology Ultimate Collection* (Houston: Everlasting Flames Publishing, 2010).

4. Hesiod *The Astronomy* fragment 4. See Evelyn, *Mythology Ultimate Collection*.

5. Apollodorus (pseudo-Apollodorus) Book 1 of Bibliotheca. See Darryl Marks, *Mythology Ultimate Collection* (Houston: Everlasting Flames Publishing, 2010).

6. Gen. 1:27.

7. Apollodorus (pseudo-Apollodorus) Book 1 of Bibliotheca.

8. Sextus Empiricus *Against the Professors* 9.19. See Daniel W. Graham, *The Texts of Early Greek Philosophy: The Complete Fragments and Selected Testimonies of the Major Presocratics* (Cambridge: Cambridge University Press, 2010), p. 613 (text 187).

9. Gen. 14–15.

10. Quoted in Erwin Schrödinger, *Nature and the Greeks and Science and Humanism* (Cambridge: Cambridge University Press, 1996), p. 69.

11. Kathleen Freeman, *Ancilla to the Pre-Socratic Philosophers* (Cambridge, MA: Harvard University Press, 1996), p. 23.

12. Ibid., p. 22.

13. Clement of Alexandria 7.22, trans. Graham, *Texts of Early Greek Philosophy*, p. 109 (texts 31, 32), p. 111 (text 33).

14. Martin West, "Early Greek Philosophy," in *The Oxford Illustrated History of Greece and the Hellenistic World* 1986, ed. John Boardman, Jasper Griffin, and Oswyn Murray (Oxford: Oxford University Press, 1986), p. 110.

CHAPTER 5. RELIGION AND SCIENCE

1. Karl R. Popper, *Conjectures and Refutations: The Growth of Scientific Knowledge* (London and New York: Routledge, 1989), p. 50.

2. Robert Parker, "Greek Religion," in *The Oxford Illustrated History of Greece and the Hellenistic World 1986*, ed. John Boardman, Jasper Griffin, and Oswyn Murray (Oxford: Oxford University Press, 1986), p. 248.

3. Walter Burkert, *Greek Religion: Archaic and Classical*, trans. John Raffan (MA: Blackwell Publishing Ltd. and Harvard University Press, 1985), pp. 276–304.

4. Bertrand Russell, *The History of Western Philosophy* (New York: Simon & Schuster, 1945), p. xiii.

CHAPTER 6. THE BIRTH OF SCIENCE

1. Andrew Gregory, *Eureka! The Birth of Science* (Cambridge: Icon Books, 2001); Bertrand Russell, *The History of Western Philosophy* (New York: Simon & Schuster, 1945); Carl Sagan, *Cosmos* (New York: Random House, 1980); Erwin Schrödinger, *Nature and the Greeks and Science and Humanism* (Cambridge: Cambridge University Press, 1996); G. E. R. Lloyd, *Early Greek Science: Thales to Aristotle* (New York: W. W. Norton & Company, 1970); G. S. Kirk, J. E. Raven, and M. Schofield, *The Presocratic Philosophers* (Cambridge: Cambridge University Press, 1983); Isaac Asimov, *The Greeks; A Great Adventure* (Boston: Houghton Mifflin, 1965); John Burnet, *Early Greek Philosophy* (London: A & C Black, 1920).

2. Russell, *History of Western Philosophy*, p. 10.

3. Smithsonian National Museum of Natural History, http://humanorigins.si.edu/evidence/human-family-tree (accessed March 10, 2014); Peter Tyson, "Who's Who In Human Evolution," http://www.pbs.org/wgbh/nova/evolution/whos-who-human-evolution.html (accessed March 10, 2014).

4. Smithsonian National Museum of Natural History, http://humanorigins.si.edu/evidence/human-fossils/species (accessed March 10, 2014).

5. Jared Diamond, *The Third Chimpanzee* (New York: HarperCollins, 1992); Jennie Cohen, "Did Neanderthals Create World's Oldest Cave Painting?" http://www.history.com/news/did-neanderthals-create-worlds-oldest-cave-paintings (accessed June 7, 2014).

6. Discovery Channel, "Neanderthal," http://www.youtube.com/watch?v=W7UFbxsF3p0&feature=related (accessed June 7, 2014); History Channel, "Clash of the Cavemen DVD," http://www.youtube.com/watch?v=gUifwntZBZw&feature=related (accessed June 7, 2014).

7. Diamond, *Third Chimpanzee*; Discovery Channel, "Neanderthal"; History Channel, "Clash of the Cavemen DVD."

8. Ibid.

9. Isaac Asimov, *Asimov's Chronology of the World* (New York: HarperCollins, 1991).

10. Daniel R. Altschuler and Christopher J. Salter, "The Arecibo Observatory: Fifty Astronomical Years," *Physics Today* 66, no. 11 (November 2013): 45; Sagan, *Cosmos*, p. 290.

11. Oswyn Murray, "Life and Society in Classical Greece," in *The Oxford Illustrated History of Greece and the Hellenistic World 1986*, ed. John Boardman, Jasper Griffin, and Oswyn Murray (Oxford: Oxford University Press, 1986), p. 221.

12. Walter Burkert, *Greek Religion: Archaic and Classical*, trans. John Raffan (MA: Blackwell Publishing Ltd. and Harvard University Press, 1985), p. 4.

13. Russell, *History of Western Philosophy*, p. 208.

14. Schrödinger, *Nature and the Greeks*, p. 84.

15. Erwin Schrödinger, *What Is Life? & Mind and Matter* (Cambridge: Cambridge University Press, 1967).

16. Plato *Phaedrus* 246A–254E. See trans. Benjamin Jowett, *The Complete Works of Plato* (The Complete Works Collection, 2011).

17. Russell, *History of Western Philosophy*, p. 16.

18. Plato *Phaedrus* 245A, trans. Jowett, *The Complete Works of Plato*, Kindle Locations 16329–30.

19. Russell, *History of Western Philosophy*, p. 21.

20. Edith Hamilton, *The Greek Way* (New York and London: W. W. Norton & Company, 1930), p. 224.

21. Bruce Thornton, *Greek Ways: How the Greeks Created Western Civilization* (San Francisco: Encounter Books, 2002), p. 4.

22. Stephen Bertman, *The Genesis of Science: The Story of Greek Imagination* (Amherst, NY: Prometheus Books, 2010), Kindle Locations 52–53.

CHAPTER 7. CLOSE ENCOUNTER OF THE TENTH KIND

1. This chapter was inspired by chapter 2 of *The God Particle*, in which Leon Lederman imagines conversing with Democritus: Leon Lederman and Dick Teresi, *The God Particle: If the Universe Is the Answer, What Is the Question?* (Boston: Houghton Mifflin, 1993).

CHAPTER 8. THALES AND SAMENESS

1. Aristotle *Metaphysics* 983b6–13, 17–27. See Daniel W. Graham, *The Texts of Early Greek Philosophy: The Complete Fragments and Selected Testimonies of the Major Presocratics* (Cambridge: Cambridge University Press, 2010), p. 29 (text 15); Aëtius 1.31, 1.10.12. See Graham, *Texts of Early Greek Philosophy*, p. 29 (text 16); Simplicius *Physics* 23.21–29. See Graham, *Texts of Early Greek Philosophy*, p. 29 (text 17).

2. Aristotle *Metaphysics* 983b6–13, 17–27. See Graham, *Texts of Early Greek Philosophy*, p. 29 (text 15).

3. Aëtius 1.31, 1.10.12. See Graham, *Texts of Early Greek Philosophy*, p. 29 (text 16).

4. Ibid.

5. Ibid; Aristotle *Metaphysics* 983b6–13, 17–27. See Graham, *Texts of Early Greek Philosophy*, p. 29 (text 15).

6. Stephen Hawking, *A Brief History of Time: From the Big Bang to Black Holes* (New York: Bantam Books, 1988), chap. 5.

7. Brian Greene, *The Elegant Universe: Superstrings, Hidden Dimensions, and the Quest for the Ultimate Theory* (New York: W. W. Norton & Company, 1999), p. 144.

8. Gareth Morgan, "Stephen Hawking Says There Is No Such Thing as Black Holes, Einstein Spinning in His Grave," Express, January 24, 2014, http://www.express.co.uk/news/science-technology/455880/Stephen-Hawking-says-there-is-no-such-thing-as-black-holes-Einstein-spinning-in-his-grave (accessed March 12, 2014); Zeeya Merali, "Stephen Hawking: 'There Are No Black-Holes,'" *Nature*, January 24, 2014, http://www.nature.com/news/stephen-hawking-there-are-no-black-holes-1.14583 (accessed March 12, 2014).

9. Aristotle *On the Soul* 411a7–8, trans. Graham, *Texts of Early Greek Philosophy*, p. 35 (text 35).

10. Diogenes Laertius 1.24. See Graham, *Texts of Early Greek Philosophy*, p. 21 (text 1).

11. Plato *Theaetetus* 174a4–8, trans. Graham, *Texts of Early Greek Philosophy*, p. 25 (text 7).

12. Dante *Divine Comedy*, trans. BookCaps (BookCaps Study Guides, 2013), Kindle locations 10900–901.

13. Aristotle *Politics* 1259a5–21, trans. Graham, *Texts of Early Greek Philosophy*, p. 25 (text 8).

14. Diodorus of Sicily 1.39.1–3. See Graham, *Texts of Early Greek Philosophy*, p. 563 (text 84).

15. Aristotle *Meteorology* 342b25, trans. Demetris Nicolaides. See also Graham, *Texts of Early Greek Philosophy*, p. 303 (text 48).

CHAPTER 9. ANAXIMANDER AND THE INFINITE

1. Leon Lederman and Dick Teresi, *The God Particle: If the Universe Is the Answer, What Is the Question?* (Boston: Houghton Mifflin, 1993).

2. Werner Heisenberg, *Physics and Philosophy: The Revolution in Modern Science* (New York: Harper Torchbooks, 1962), p. 36.

3. Lederman and Teresi, *God Particle*, p. 56.

4. In the Kelvin scale the absolute zero is 0 degrees Kelvin, which is -273.15 degrees Celsius, which is -459.67 Fahrenheit.

5. Aristotle *On the Heavens* 295b10–16. See Daniel W. Graham, *The Texts of Early Greek Philosophy: The Complete Fragments and Selected Testimonies of the Major Presocratics* (Cambridge: Cambridge University Press, 2010), p. 59 (text 21).

6. Karl R. Popper, *Conjectures and Refutations: The Growth of Scientific Knowledge* (London and New York: Routledge, 1989) p. 138.

7. John Burnet, *Early Greek Philosophy* (London: A & C Black, 1920), chap. 1.

8. Charles Sherrington, *Man on His Nature* (Cambridge: Cambridge University Press, 2009), p. 302.

9. Erwin Schrödinger, *Nature and the Greeks and Science and Humanism* (Cambridge: Cambridge University Press, 1996), p. 66.

10. Heisenberg, *Physics and Philosophy*, p. 128.

11. Richard P. Feynman, *Six Easy Pieces* (New York: Perseus Publishing, 1963), p. 22.

12. Aëtius 5.19.4. See Graham, *Texts of Early Greek Philosophy*, p. 63 (text 37); Censorinus 4.7. See Graham, *Texts of Early Greek Philosophy*, p. 63 (text 38); Hippolytus *Refutation* 1.6.6. See G. S. Kirk, J. E. Raven, and M. Schofield, *The Presocratic Philosophers* (Cambridge: Cambridge University Press, 1983), Kindle Location 3598; Plutarch *Symposium* 730e. See Graham, *Texts of Early Greek Philosophy*, p. 63 (text 39); Ps.- Plutarch *Strom*. 2. See Kirk, *Presocratic Philosophers*, Kindle Location 3590.

13. Censorinus 4.7; Hippolytus *Refutation* 1.6.6; Plutarch *Symposium* 730e; Ps.- Plutarch *Strom*. 2.

14. Aëtius 5.19.4; Hippolytus *Refutation* 1.6.6.

CHAPTER 10. ANAXIMENES AND DENSITY

1. Simplicius *Physics* 24.26–25.1, Theophrastus frag. 226A. See Daniel W. Graham, *The Texts of Early Greek Philosophy: The Complete Fragments and Selected Testimonies of the Major Presocratics* (Cambridge: Cambridge University Press, 2010), p. 75 (text 3).

2. Aëtius 1.3.4, trans. John Burnet, *Early Greek Philosophy* (London: A & C. Black, 1920), chap. 1.

3. Hippolytus *Refutation* 1.7, trans. Burnet, *Early Greek Philosophy*, chap. 1.

4. Aëtius 3.3.2, trans. Burnet, *Early Greek Philosophy*, chap. 1.

5. This view is also expressed in Erwin Schrödinger, *Nature and the Greeks and Science and Humanism* (Cambridge: Cambridge University Press, 1996); Werner Heisenberg, *Physics and Philosophy: The Revolution in Modern Science* (New York: Harper Torchbooks, 1962).

6. Sextus Empiricus *Against the Professors* 7.135, trans. Schrödinger, *Nature and the Greeks*, p. 89.

7. Aristotle *Metaphysics* 985b4–20, trans. Graham, *Texts of Early Greek Philosophy*, p. 525 (text 10).

8. Ibid.

9. Carl Sagan, *Cosmos* (New York: Random House, 1980), p. 181.

10. Schrödinger, *Nature and the Greeks*, pp. 62–65, 84–86, 157–62.

11. Ibid., p. 160.

CHAPTER 11. PYTHAGORAS AND NUMBERS

1. Aristotle *Metaphysics* 987b22. See Erwin Schrödinger, *Nature and the Greeks and Science and Humanism* (Cambridge: Cambridge University Press, 1996), p. 35.

2. Diogenes Laertius 8.46. See G. S. Kirk, J. E. Raven, and M. Schofield, *The Presocratic Philosophers* (Cambridge: Cambridge University Press, 1983), Kindle Locations 9294–95.

3. Aëtius 2.1.1, trans. Demetris Nicolaides. See Greek book Βας. Α. Κύρκος, *Οι Προσωκρατικοί: Οι Μαρτυρίες και τα Αποσπάσματα τόμος Α* (Αθήνα: Εκδόσεις Δημ. Ν. Παπαδήμα, 2005), p. 247.

4. Aristotle *On the Heavens* 290b12, trans. Demetris Nicolaides. See also Kirk, Raven, and Schofield, *Presocratic Philosophers*, Kindle Locations 9131–33.

5. Johannes Kepler, *The Harmonies of the World*, quoted in George N. Gibson and Ian D. Johnston, "New Themes and Audiences for the Physics of Music," *Physics Today* 55, no. 1 (January 2002): 44.

6. Arnold Sommerfeld quoted in Gibson and Johnston, "New Themes and Audiences for the Physics of Music," *Physics Today* 55, no. 1 (January 2002): 43.

7. *The Elegant Universe: Part 1*, PBS, October 28, 2003.

8. Schrödinger, *Nature and the Greeks*, p. 122; Werner Heisenberg, *Physics and Philosophy: The Revolution in Modern Science* (New York: Harper Torchbooks, 1962), p. 46.

9. Schrödinger, *Nature and the Greeks*, p. 45.

10. Ibid.

11. Bertrand Russell, *The History of Western Philosophy* (New York: Simon & Schuster, 1945), p. 214.

12. Ibid., p. 217.

13. Ibid., p. 540.

CHAPTER 12. HERACLITUS AND CHANGE

1. Diogenes Laertius 2.22, trans. Daniel W. Graham, *The Texts of Early Greek Philosophy: The Complete Fragments and Selected Testimonies of the Major Presocratics* (Cambridge: Cambridge University Press, 2010), p. 181 (text 163).

2. Origen *Against Celsus* 6.42, trans. Graham, *Texts of Early Greek Philosophy*, p. 157 (text 58).

3. Aristotle *Eudemian Ethics* 1235a25–29, trans. Graham, *Texts of Early Greek Philosophy*, p. 157 (text 60).

4. Hippolytus *Refutation* 9.9.2. See Bertrand Russell, *The History of Western Philosophy* (New York: Simon & Schuster, 1945), p. 43.

5. Themestius *Orations* 5.69b, trans. Demetris Nicolaides. See Graham, *Texts of Early Greek Philosophy*, p. 161 (text 75).

6. Erwin Schrödinger, *Nature and the Greeks and Science and Humanism* (Cambridge: Cambridge University Press, 1996), p. 157.

7. Leon Lederman and Dick Teresi, *The God Particle: If the Universe Is the Answer, What Is the Question?* (Boston: Houghton Mifflin, 1993).

8. Planck's constant is a very small number equal to, 6.63×10^{-34} joules $\times$ seconds.

9. Plato *Cratylus* 402a8–10, trans. Graham, *Texts of Early Greek Philosophy*, p. 159 (text 63).

10. Clement *Miscellanies* 5.103.6, trans. Demetris Nicolaides. See Graham, *Texts of Early Greek Philosophy*, p. 155 (text 47).

11. Ibid., 5.104.3–5, trans. John Burnet, *Early Greek Philosophy* (London: A & C Black, 1920), chap. 3.

12. Plutarch *On the E at Delphi* 338d–e, trans. Graham, *Texts of Early Greek Philosophy*, p. 157 (text 55).

13. Clement *Miscellanies* 5.103.6. See Graham, *Texts of Early Greek Philosophy*, p. 155 (text 47).

14. Karl R. Popper, *Conjectures and Refutations: The Growth of Scientific Knowledge* (London and New York: Routledge, 1989) p. 147.

15. Werner Heisenberg, *Physics and Philosophy: The Revolution in Modern Science* (New York: Harper Torchbooks, 1962), p. 37.

16. Ibid., p. 45.

17. Heraclitus *Homeric Questions* 24, trans. Demetris Nicolaides. See Graham, *Texts of Early Greek Philosophy*, p. 159 (text 65).

18. Schrödinger, *Nature and the Greeks*, pp. 123–25.

CHAPTER 13. PARMENIDES AND ONENESS

1. Clement *Miscellanies* 6.23, trans. Erwin Schrödinger, *Nature and the Greeks and Science and Humanism* (Cambridge: Cambridge University Press, 1996), p. 27.

2. Aristotle *Metaphysics* 985b4–20, trans. Daniel W. Graham, *The Texts of Early Greek Philosophy: The Complete Fragments and Selected Testimonies of the Major Presocratics* (Cambridge: Cambridge University Press, 2010), p. 525 (text 10).

3. Ibid.

4. Einstein quoted in Joanne Baker, *50 Physics Ideas You Really Need to Know* (UK: Quercus, 2007), p. 165.

5. Werner Heisenberg, *Physics and Philosophy: The Revolution in Modern Science* (New York: Harper Torchbooks, 1962).

6. J. J. Sakurai, *Modern Quantum Mechanics* (CA: The Benjamin/Cummings Publishing Company, Inc., 1985), pp. 226–29.

7. Graham, *Texts of Early Greek Philosophy*, pp. 211–19.

8. Ibid., pp. 219–33.

CHAPTER 14. ZENO AND MOTION

1. Aristotle *Physics* 239b9–14, trans. Daniel Kolac and Garrett Thomson, *The Longman Standard History of Philosophy* (New York: Pearson, 2005), p. 33.

2. Elias *Commentary on Aristotle's Categories* 109.20–22. See Richard D. McKirahan, *Philosophy before Socrates* (Indianapolis/Cambridge: Hackett Publishing, 2010), p. 182 (Kindle ed.).

3. Aristotle *Physics* 239b14–20, trans. Demetris Nicolaides. See also Daniel W. Graham, *The Texts of Early Greek Philosophy: The Complete Fragments and Selected Testimonies of the Major Presocratics* (Cambridge: Cambridge University Press, 2010), p. 261 (text 18).

4. Ibid., 239b30–33. See Graham, *Texts of Early Greek Philosophy*, p. 261 (text 19); ibid., 239b5–9. See Graham, *Texts of Early Greek Philosophy*, p. 261 (text 20); Diogenes Laertius 9.72. See Graham, *Texts of Early Greek Philosophy*, p. 261 (text 21).

5. Robert J. Oppenheimer, *Science and the Common Understanding* (New York: Simon & Schuster, 1954), p. 40.

6. Aristotle *Physics* 209a23–25, trans. Demetris Nicolaides. See also Graham, *Texts of Early Greek Philosophy*, p. 263 (text 24).

7. Simplicius *Physics* 140.34–141.8. See Graham, *Texts of Early Greek Philosophy*, p. 255 (text 7); ibid., 140.27–34. See Graham, *Texts of Early Greek Philosophy*, p. 259 (text 13).

CHAPTER 15. EMPEDOCLES AND ELEMENTS

1. Leon Lederman and Dick Teresi, *The God Particle: If the Universe Is the Answer, What Is the Question?* (Boston: Houghton Mifflin, 1993), p. 340.

2. Ibid., p. 173.

3. Simplicius *Physics* 158.1–159.4. See also Daniel W. Graham, *The Texts of Early Greek Philosophy: The Complete Fragments and Selected Testimonies of the Major Presocratics* (Cambridge: Cambridge University Press, 2010), p. 251 (text 41).

4. Jeffrey Bennett, Megan Donahue, Nicholas Schneider, and Mark Void, *The Essential Cosmic Perspective 7th Edition* (Boston: Pearson, 2013), p. 450.

5. Aristotle *On Youth, Old Age, Life, Death, and Respiration* 473b9–474a6. See Graham, *Texts of Early Greek Philosophy*, p. 387 (text 127).

6. Simplicius *On the Heavens* 529.1–17, trans. Graham, *Texts of Early Greek Philosophy*, p. 361 (text 51).

7. Ibid.

8. Ibid., trans. Bertrand Russell, *The History of Western Philosophy* (New York: Simon & Schuster, 1945), p. 54.

9. Ibid., 586.12, 587.1–2, trans. John Burnet, *Early Greek Philosophy* (London: A & C Black, 1920), chap. 7 (frag. 57).

CHAPTER 16. ANAXAGORAS AND NOUS

1. Simplicius *Physics* 164.24–25, 156.13–157.4, 176.34–177.6. See also Daniel W. Graham, *The Texts of Early Greek Philosophy: The Complete Fragments and Selected Testimonies of the Major Presocratics* (Cambridge: Cambridge University Press, 2010), p. 291 (text 31).

2. Ibid., 164.23–24, trans. G. E. R. Lloyd, *Early Greek Science: Thales to Aristotle* (New York: W. W. Norton & Company, 1970), p. 44.

3. Ibid., 27.2–23. See Gregory Vlastos, *Studies in Greek Philosophy. Vol. 1: The Presocratics* (Princeton: Princeton University Press , 1993), p. 319.

4. John Burnet, *Early Greek Philosophy* (London: A & C Black, 1920), chap. 6; Vlastos, *Studies in Greek Philosophy*, p. 319.

5. Simplicius *Physics* 176.29, 175.12–14. See Richard D. McKirahan, *Philosophy before Socrates* (Indianapolis / Cambridge: Hackett Publishing, 2010), p. 194 (Kindle ed.).

6. Sextus Empiricus *Against the Professors* 7.140, trans. Demetris Nicolaides. See also Graham, *Texts of Early Greek Philosophy*, p. 309 (text 63).

7. Diogenes Laertius 2.6–15, trans. Demetris Nicolaides. See also Graham, *Texts of Early Greek Philosophy*, p. 275 (text 1).

8. Ibid.

9. Ibid.

CHAPTER 17. DEMOCRITUS AND ATOMS

1. Sextus Empiricus *Against the Professors* 7.135, trans. Erwin Schrödinger, *Nature and the Greeks and Science and Humanism* (Cambridge: Cambridge University Press, 1996), p. 89.

2. Plutarch *Against Colotes* 1110f–1111a, trans. Daniel W. Graham, *The Texts of Early Greek Philosophy: The Complete Fragments and Selected Testimonies of the Major Presocratics* (Cambridge: Cambridge University Press, 2010), p. 537 (text 28).

3. Aristotle *On Generation and Corruption* 315b6–15. See Graham, *Texts of Early Greek Philosophy*, p. 541 (text 41).

4. Ibid., 324b35–325a6, a23–b5, trans. Graham, *Texts of Early Greek Philosophy*, p. 529 (text 14).

5. Sextus Empiricus *Outlines of Pyrrhonism* 2.63, trans. Demetris Nicolaides. See Greek book Βας. A. Κύρκος, *Οι Προσωκρατικοί: Οι Μαρτυρίες και τα Αποσπάσματα τόμος Β* (Αθήνα: Εκδόσεις Δημ. Ν. Παπαδήμα, 2007), p. 255.

6. Graham, *Texts of Early Greek Philosophy*, pp. 579–95.

7. Sextus Empiricus *Against the Professors* 7.138–139, trans. Graham, *Texts of Early Greek Philosophy*, p. 597 (text 140).

8. Richard P. Feynman, *The Feynman Lectures on Physics* (Boston: Addison-Wesley Publishing Co., Inc., 1963), p. 1-2.

9. Leon Lederman and Dick Teresi, *The God Particle: If the Universe Is the Answer, What Is the Question?* (Boston: Houghton Mifflin, 1993), p. 340.

10. Aristotle *Metaphysics* 985b4–20. See Graham, *Texts of Early Greek Philosophy*, p. 525 (text 10).

11. Simplicius *On the Heavens* 242.15–26, trans. Graham, *Texts of Early Greek Philosophy*, p. 533 (text 23).

12. Bertrand Russell, *The History of Western Philosophy* (New York: Simon & Schuster, 1945), p. 71.

13. Plutarch *Against Colotes* 1108f–1109a, p. 527 (text 13).

14. Isaac Asimov, *Understanding Physics* (US: Dorset Press, 1988), p. 7.

15. Lederman and Teresi, *The God Particle*, p. 44.

16. Banesh Hoffman, *The Strange Story of the Quantum* (New York: Dover Publications Inc., 1959), p. 68.

17. Brian Greene, *The Elegant Universe: Superstrings, Hidden Dimensions, and the Quest for the Ultimate Theory* (New York: W. W. Norton & Company, 1999), p. 114.

18. Jeffrey Bennett, Megan Donahue, Nicholas Schneider, and Mark Void, *The Essential Cosmic Perspective 7th Edition* (Boston: Pearson, 2013), p. 479.

19. Werner Heisenberg, *Physics and Philosophy: The Revolution in Modern Science* (New York: Harper Torchbooks, 1962), p. 40.

20. Aëtius 1.3.18, S 1.14.1f. See Graham, *Texts of Early Greek Philosophy*, p. 537 (texts 31, 32); Cicero *On Fate* 20.46. See Graham, *Texts of Early Greek Philosophy*, p. 537 (text 33).

21. Aëtius 1.4.1–4. See Graham, *Texts of Early Greek Philosophy*, pp. 541–45.

22. Sextus Empiricus *Against the Professors* 7.135, p. 89.

23. Galen *On Medical Experience* 15.7, trans. Demetris Nicolaides. See also Graham, *Texts of Early Greek Philosophy*, p. 597 (text 139).

24. Sextus Empiricus *Against the Professors* 7.138–139, p. 597 (text 140).

25. Ibid., 7.140, trans. Demetris Nicolaides. See also Graham, *Texts of Early Greek Philosophy*, p. 309 (text 63).

26. "Astronomers Conclude Habitable Planets Are Common," Institute for Astronomy University of Hawaii, http://www.ifa.hawaii.edu/info/press-releases/HabitablePlanets Common/ (accessed March 23, 2014).

27. Bennett, *Essential Cosmic Perspective*, p. 511; Carl Sagan, *Cosmos* (New York: Random House, 1980), p. 300.

BIBLIOGRAPHY

Asimov, Isaac. *Asimov's Chronology of Science and Discovery*. New York: HarperCollins, 1989.

———. *Asimov's Chronology of the World*. New York: HarperCollins, 1991.

———. *The Greeks; A Great Adventure*. Boston: Houghton Mifflin, 1965.

———. *Understanding Physics*. US: Dorset Press, 1988.

Baker, Joanne. *50 Physics Ideas You Really Need to Know*. UK: Quercus, 2007.

Bertman, Stephen. *The Eight Pillars of Greek Wisdom*. New York: Barnes & Noble, 2007.

———. *The Genesis of Science: The Story of Greek Imagination*. Kindle ed. Amherst, NY: Prometheus Books, 2010.

Boardman, John, Jasper Griffin, and Oswyn Murray, eds. *The Oxford Illustrated History of Greece and the Hellenistic World*. Oxford: Oxford University Press, 1986.

Brunschwig, Jacques, and Geoffrey E. R. Lloyd. *A Guide to Greek Thought: Major Figures and Trends*. Cambridge, MA: Belknap Press of Harvard University Press, 2003.

———. *Greek Thought: A Guide to Classical Knowledge*. Cambridge, MA: Belknap Press of Harvard University Press, 2000.

Burckhardt, Jacob. *The Greeks and Greek Civilization*. Edited by Oswyn Murray. Translated by Sheila Stern. New York: St. Martin's Griffin, 1999.

Burkert, Walter. *Greek Religion: Archaic and Classical*. Translated by John Raffan. MA: Blackwell Publishing Ltd. and Harvard University Press, 1985.

Burnet, John. *Early Greek Philosophy*. London: A & C Black, 1920.

Dalling, Robert. *The Story of Us Humans, From Atoms to Today's Civilization*. New York, Lincoln Shanghai: iUniverse, 2006.

Davies, P. C. W., and Julian Brown. *Superstrings: A Theory of Everything?* Cambridge: Cambridge University Press, 1992.

Diamond, Jared. *The Third Chimpanzee*. New York: HarperCollins, 1992.

Economou, Eleftherios N. *A Short Journey from Quarks to the Universe*. Berlin and Heidelberg: Springer, 2011.

Einstein, Albert. *Relativity: The Special and the General Theory*. Kindle ed. Amazon Kindle Direct Publishing, 2011.

Feynman, Richard P. *The Meaning of It All*. New York: Basic Books, 1998.

———. *Six Easy Pieces*. New York: Perseus Publishing, 1963.

———. *Six Not So Easy Pieces*. New York: Perseus Publishing, 1963.

Freeman, Charles. *The Greek Achievement: The Foundation of the Western World*. New York: Penguin Books, 2000.

Freeman, Kathleen. *Ancilla to the Pre-Socratic Philosophers*. Cambridge, MA: Harvard University Press, 1996.

Graham, W. Daniel. *Explaining the Cosmos: The Ionian Tradition of Scientific Philosophy*. NJ: Princeton University Press, 2006.

Graham, W. Daniel, ed. *The Texts of Early Greek Philosophy: The Complete Fragments and Selected Testimonies of the Major Presocratics*. Translated by W. Daniel Graham. Cambridge: Cambridge University Press, 2010.

Graves, Robert. *The Greek Myths*. Canada: Penguin Group, 1955.

Greene, Brian. *The Elegant Universe: Superstrings, Hidden Dimensions, and the Quest for the Ultimate Theory*. New York: W. W. Norton & Company, 1999.

————. *The Fabric of the Cosmos: Space, Time, and the Texture of Reality*. New York: Vintage, 2005.

————. *The Hidden Reality: Parallel Universes and the Deep Laws of the Cosmos*. New York: Vintage, 2011.

Gregory, Andrew. *Eureka! The Birth of Science*. Cambridge: Icon Books, 2001.

Hamilton, Edith. *The Greek Way*. New York and London: W. W. Norton & Company, 1930.

Hardy, Alister. *The Biology of God*. New York: Taplinger Publishing, 1976.

Hawking, Stephen. *A Brief History of Time: From the Big Bang to Black Holes*. New York: Bantam Books, 1988.

Heisenberg, Werner. *Physics and Philosophy: The Revolution in Modern Science*. New York: Harper Torchbooks, 1962.

James, Renée C. *Seven Wonders of the Universe: That You Probably Took for Granted*. Baltimore: Johns Hopkins University Press, 2011.

Kirk, G. S., J. E. Raven, and M. Schofield. *The Presocratic Philosophers*. Cambridge: Cambridge University Press, 1983.

Kolac, Daniel, and Garrett Thomson. *The Longman Standard History of Philosophy*. New York: Pearson, 2005.

Lederman, Leon, and Dick Teresi. *The God Particle: If the Universe Is the Answer, What Is the Question?* Boston: Houghton Mifflin, 1993.

Lightman, Alan. *Great Ideas in Physics*. 3rd. ed. New York: McGraw-Hill, 2000.

Lindberg, David C. *The Beginnings of Western Science: The European Scientific Tradition in Philosophical, Religious, and Institutional Context, Prehistory to A.D. 1450*. 2nd. ed. Chicago: University of Chicago Press, 2008.

Lloyd, G. E. R. *Early Greek Science: Thales to Aristotle*. New York: W. W. Norton & Company, 1970.

————. *Greek Science after Aristotle*. New York: W. W. Norton & Company, 1973.

McKirahan, Richard D. *Philosophy before Socrates*. Indianapolis/Cambridge: Hackett Publishing, 2010.

Menzies, Allan. *History of Religion: A Sketch of Primitive Religious Beliefs and Practices, and of the Origin and Character of the Great Systems*. Memphis, TN: General Books, 2010.

Mourelatos, Alexander P. D., ed. *The Pre-Socratics: A Collection of Critical Essays*. Garden City, NY: Doubleday and Company, 1974.

Oppenheimer, Robert J. *Science and the Common Understanding*. New York: Simon & Schuster, 1954.

Pomeroy, Sarah B., Stanley M. Burstein, Walter Donlan, and Jennifer Tolbert Roberts. *A Brief History of Ancient Greece: Politics, Society, and Culture*. 2nd. ed. Oxford: Oxford University Press, 2008.

Popper, Karl R. *Conjectures and Refutations: The Growth of Scientific Knowledge*. London and New York: Routledge, 1989.

Randall, Lisa. *Knocking on Heaven's Door: How Physics and Scientific Thinking Illuminate the Universe and the Modern World*. New York: Harper Perennial, 2011.

————. *Warped Passages: Unraveling the Mysteries of the Universe's Hidden Dimensions*. New York: Ecco, 2005.

Ridley, B. K. *Time, Space and Things*. 2nd. ed. Cambridge: Cambridge University Press, 1984.

Rosenblum, Bruce, and Fred Kuttner. *Quantum Enigma: Physics Encounters Consciousness*. 2nd. ed. Oxford: Oxford University Press, 2011.

Russell, Bertrand. *The History of Western Philosophy*. New York: Simon & Schuster, 1945.

————. *Religion and Science*. New York and Oxford: Oxford University Press, 1997.

————. *The Scientific Outlook*. London and New York: Routledge, 2009.

Sagan, Carl. *Cosmos*. New York: Random House, 1980.

Schrödinger, Erwin. *Nature and the Greeks and Science and Humanism*. Cambridge: Cambridge University Press, 1996.

————. *What Is Life? & Mind and Matter*. Cambridge: Cambridge University Press, 1967.

Sean, Carroll. *The Particle at the End of the Universe*. New York: Dutton, 2012.

Sherrington, Charles. *Man on His Nature*. Reissue. Cambridge: Cambridge University Press, 2009.

Stark, Rodney. *Discovering God: The Origins of the Great Religions and the Evolution of Belief*. Kindle ed. HarperCollins, 2009.

Tattersall, Ian. *The World from Beginnings to 4000 BCE*. Oxford: Oxford University Press, 2008.

Thornton, Bruce. *Greek Ways: How the Greeks Created Western Civilization*. San Francisco: Encounter Books, 2002.

Vlastos, Gregory. *Studies in Greek Philosophy*. Vol. 1: The Presocratics. Princeton: Princeton University Press , 1993.

Waterfield, Robin. *The First Philosophers: The Presocratics and Sophists*. Oxford and New York: Oxford University Press, 2000.

Wrangham, Richard. *Catching Fire: How Cooking Made Us Human*. New York: Basic Books, 2009.

INDEX

'Fowler shocks and frightens, while making us laugh out loud . . . original, erudite and exciting'
Good Book Guide

'Fowler's fresh and unusual characters breathe new life into an established genre in which it's getting harder and harder to find anything genuinely fresh'
Booklist

'Madcap mystery . . . crazy and great fun for it'
Los Angeles Times

'This most unusual and impressive detecting duo . . . Fowler's wit and visual acuity combine for entertaining and thrilling results'
Chicago Tribune

'Christopher Fowler has offered his readership so much beyond a well-crafted British crime story . . . [he] will stretch your mind and leave you with a feeling of accomplishment after the final page is turned'
The Mystery Reader

'Places Fowler in the first rank of contemporary mystery writers'
Publishers Weekly

'Wartime London is conjured up with unique skill . . . Fowler's powers of description are enviable'
Independent on Sunday

Also by Christopher Fowler

Novels
ROOFWORLD
RUNE
RED BRIDE
DARKEST DAY
SPANKY
PSYCHOVILLE
DISTURBIA
SOHO BLACK
CALABASH
BREATHE
FULL DARK HOUSE
THE WATER ROOM
SEVENTY-SEVEN CLOCKS
TEN-SECOND STAIRCASE
WHITE CORRIDOR
THE VICTORIA VANISHES
BRYANT & MAY ON THE LOOSE
BRYANT & MAY OFF THE RAILS
BRYANT & MAY AND THE MEMORY OF BLOOD

Graphic Novel
MENZ NSANA

Short Stories
CITY JITTERS
CITY JITTERS TWO
THE BUREAU OF LOST SOULS
SHARPER KNIVES
FLESH WOUNDS
PERSONAL DEMONS
UNCUT
THE DEVIL IN ME
DEMONIZED

BRYANT & MAY AND THE INVISIBLE CODE

CHRISTOPHER FOWLER

BANTAM BOOKS

LONDON · TORONTO · SYDNEY · AUCKLAND · JOHANNESBURG

TRANSWORLD PUBLISHERS
61–63 Uxbridge Road, London W5 5SA
A Random House Group Company
www.transworldbooks.co.uk

BRYANT & MAY AND THE INVISIBLE CODE
A BANTAM BOOK: 9780857500953

First published in Great Britain
in 2012 by Doubleday
an imprint of Transworld Publishers
Bantam edition published 2013

Addresses for Random House Group Ltd companies outside the UK
can be found at: www.randomhouse.co.uk
The Random House Group Reg. No. 954009

The Random House Group Limited supports the Forest Stewardship
Council® (FSC®), the leading international forest-certification organisation.
Our books carrying the FSC label are printed on FSC®-certified paper.
FSC is the only forest-certification scheme supported by the leading
environmental organisations, including Greenpeace. Our paper procurement
policy can be found at www.randomhouse.co.uk/environment

Typeset in 11/13pt Sabon by
Kestrel Data, Exeter, Devon.
Printed and bound by
CPI Group (UK) Ltd, Croydon, CR0 4YY.

2 4 6 8 10 9 7 5 3 1

For Peter Chapman

'Money can't buy friends,
but it can get you a better class of enemy.'

Spike Milligan

'It started with me. It ends with me.'

**Unnamed teenager,
when asked about the history of London**

ACKNOWLEDGEMENTS

'Make your leading characters younger and put in more sex and violence if you want them to be a success,' a critic warned me as I embarked on the first Bryant & May mystery. Blithely ignoring his advice I ploughed on, determined to create a pair of intelligent Golden Age detectives who are forced to deal with the modern world. I knew I'd have fun just watching Arthur Bryant trying to use a smartphone.

Luckily, there were others who always agreed with me. Simon Taylor, my editor at Transworld, is so wonderfully enthusiastic that I sometimes doubt his sanity but never his *savoir faire*. Thanks too, to Lynsey Dalladay, who has restored my faith in publishing PR. Both she and Mandy Little, my charming agent, prove it's not all standing around drinking champagne and that we can also have fun going to secluded libraries on wet winter Wednesdays.

I really hope there are further Bryant & May adventures to come, as each book is more pleasurable to write than the last. Remember, the strangest parts of these tales are true. You can uncover lots more information at www.christopherfowler.co.uk

Peculiar Crimes Unit
The Old Warehouse
231 Caledonian Road
London N1 9RB

STAFF ROSTER FOR MONDAY 18 JUNE

Raymond Land, Acting Unit Chief
Arthur Bryant, Senior Detective
John May, Senior Detective
Janice Longbright, Detective Sergeant
Dan Banbury, Crime Scene Manager/InfoTech
Giles Kershaw, Forensic Pathologist (St Pancras Mortuary)
Jack Renfield, Sergeant
Meera Mangeshkar, Detective Constable
Colin Bimsley, Detective Constable
Crippen, staff cat

BULLETIN BOARD

Housekeeping notes from Raymond Land to all staff:

As you know, we now have a fully activated secure swipe-card entry system on the front door. It worked perfectly for two whole days, until Arthur Bryant accidentally inserted an old Senior Service 'Battle of Britain' cigarette card into the slot instead of his electronic keycard and somehow jammed it. The engineers hope to have the system working again by Thursday.

The new common room is to be used as a neutral zone for calm reflection and the sharing of information. It is not an after-hours bar, a videogame parlour or a place where you can stage chemical experiments, impromptu film shows or arm-wrestling matches for beers.

When the fire inspector came to test the smoke detector in the first-floor corridor last week, he found a box of Bryant & May matches wedged in place of the alarm battery. Obviously only a disturbed, selfish and immature individual would risk burning his colleagues alive in order to smoke a pipe indoors. I'm not mentioning any names.

I want to put the rumours to rest about our new building once and for all. While it appears to be true that a Mr Aleister Crowley once held meetings here (and decorated the wall of my office with inappropriate images of young ladies and aroused livestock), the building is most emphatically not 'haunted'. It's an old property with a colourful history, and has Victorian pipes and floorboards. The noises these make at night are quite normal and certainly don't sound like the 'death-rattles of trapped souls', as I overheard Meera telling someone on the phone. May I remind you that you are British officers of the law, and are not required to have any imagination.

There's a funny smell in the kitchen. It might be a gas leak. Our builders, the two Daves, are coming back to rip everything out. If I find one of you dropped a kebab behind the units, you'll be on unpaid overtime for a month.

Finally, I was under the impression that Crippen, our staff cat, was a neutered tom, but this appears not to be the case as she is clearly pregnant. Can someone please take care of this? I DO NOT want anyone unexpectedly giving birth in this unit.

PART ONE

—

The Case

I

CLOSE TO GOD

There was a witch around here somewhere.

The Fleet Street office workers who sat in the cool shadow of the church on their lunch breaks had no idea that she was hiding among them. They squatted in the little garden squares while they ate their sandwiches, queued at coffee shops and paced the pavements staring at the screens of their smartphones, not realizing that she was preparing to call down lightning and spit brimstone.

On the surface the witch was one of them, but that was just a disguise. She had the power to change her outward appearance, to look like anyone she was standing near.

Lucy said, 'She won't be somebody posh. Witches are always poor.'

Tom said, 'I can't tell who's posh. Everyone looks the same.'

He was right; to a child they did. Grey suits, black suits, white shirts, grey skirts, blue ties, print blouses, black shoes. London's workforce on the move.

Lucy pulled at her favourite yellow T-shirt and felt her tummy rumble. 'She'll have to appear soon. They often travel in threes. When a witch starts to get hungry, she loses concentration and lets go of her disguise. The spell will weaken and she'll turn back into her real self.'

She was crouching in the bushes and wanted to stand up because it was making her legs hurt, but knew she might get caught if she did so. The flowerbeds bristled with tropical plants that had spiny razor-sharp leaves and looked as if they should be somewhere tropical. A private security guard patrolled the square, shifting the people who looked as if they belonged somewhere else too.

'What does she really look like?' asked Tom. 'I mean, when she drops her disguise?'

Lucy answered without hesitation. 'She has a green face and a hooked nose covered in hairy warts, and long brown teeth and yellow eyes. And her breath smells of rotting sardines.' She thought for a moment. 'And toilets.'

Tom snorted in disgust as he looked around the courtyard for likely suspects. Nearby, an overweight woman in her mid-thirties was standing in a doorway eating a Pret A Manger crayfish and rocket sandwich. She seemed a likely candidate. The first of the summer's wasps was hovering around, scenting the remains of office lunches. The woman anxiously batted one away as she ate.

'It can't be her,' said Lucy.

'Why not?' asked Tom.

'Witches don't feel pain, so she wouldn't be scared of a stupid wasp.'

'Can a witch be a man?'

'No, that would be a warlock. It has to be a woman.'

Tom was getting tired of the game. Lucy seemed to be making up extra rules as she went along. The June sun shone through a gap in the buildings and burned the back of his neck. The sky above the courtyard was as blue as the sea looked in old films.

He was starting to think that this was a stupid way to spend a Saturday morning when he could have been at football. He had been looking forward to seeing the

Dr Who exhibition as well, but right at the last minute his dad had to work instead, and said, 'You can come with me to the office,' as if it was a reasonable substitute. There was nothing to do in the office. You weren't allowed to touch the computers or open any of the drawers. His dad seemed to like being there. He always cheered up when he had to go into the office on a Saturday.

The only other father who had brought his child in that morning was Lucy's, so he was stuck playing with a girl until both of their fathers had finished their work. At least Lucy knew about the game, which was unusual because most girls didn't play games like that. She explained that she had two older brothers and always ended up joining in with them. She didn't tell him they had outgrown the game now and spent their days wired into hip-hop and dodgy downloads.

'How about that one?' said Lucy, taking the initiative. Her brothers could never make up their minds about anything, and always ended up arguing, so she was used to making all the decisions.

'Nah, she's too pretty,' said Tom, watching a slender girl in a very short grey skirt stride past to the building at the end of the courtyard.

'That's the point. The prettier they look on the outside, the uglier they are inside. Too late, she's gone.'

'I'm bored now.'

'Five more minutes. She's here somewhere.' There were only a few workers left in the square, plus a motor-cycle courier who must have been stifling in his helmet and leathers.

'It's this one. I have a feeling. I bet she belongs to a coven; that's a club for witches. Remember, we have to get them before they get us. Let's check her out. Come on.'

Lucy led the way past a sad-looking young woman who had just seated herself on the bench nearest the church. She had opened a paperback and was reading it intently.

Lucy turned to Tom with an air of theatrical nonchalance and pointed behind the flat of her palm.

'That's definitely her.'

'How can we tell if she's a witch?' Tom whispered.

'Look for signs. Try to see what she's reading.'

'I can't walk past her again, she'll see. Wait, I've got an idea.' Tom had stolen a yellow tennis ball from his father's office. Now he produced it from his pocket: 'Catch, then throw it back to me in her direction. I'll miss and I'll have to go and get it.'

Lucy was a terrible actress. If the sad-faced young woman had looked up, she would have stopped and stared at the little girl gurning and grimacing before her.

'I'm throwing now,' Lucy said loudly, hurling the ball ten feet wide of the boy. Tom scrambled in slow motion around the bench, and the young woman briefly raised her eyes.

Tom ran back to Lucy's side. 'She's reading a book about babies.'

'What was it called?'

'*Rosemary's Baby.* By a woman called Ira something.'

'Then she's definitely a witch.'

'How do you know?'

Lucy blew a raspberry of impatience. 'Don't you know anything? Witches eat babies! Everyone knows that.'

'So she really is one,' Tom marvelled. 'She looks so normal.'

'Yeah, clever isn't it?' Lucy agreed. 'So, how are we going to kill her?'

2

DEATH IN THE WEDDING CAKE

Even though the presses of the Fourth Estate had been shifted to London's hinterlands by Rupert Murdoch, St Bride's Church was still known to many as the Printers' Cathedral. Tucked behind Fleet Street, it stood on a pagan site dedicated to Brigit, the Celtic goddess of healing, fire and childbirth. For two thousand years the spot had been a place of worship, and for the past five hundred it had been the spiritual home of journalists. Samuel Pepys, no mean reporter himself, had been born in Salisbury Court, right next to the church, and had later bribed the gravedigger of St Bride's to shift up the corpses so that his brother John could be buried in the churchyard.

St Bride's' medieval lectern had survived the Great Fire and the Luftwaffe's bombs. It still stood bathed in the lunchtime sunlight, barely registered by the tourists who stopped by to take photographs of just another London church. The building had been badly damaged in the firestorm of 29 December 1940, but had now been restored according to Wren's original drawings.

With the paperback in her hand, the sad young woman walked into the church and looked about. Amy O'Connor had been here many times before, but her visits had never brought her the satisfaction she'd hoped for. She knew little about the church except the one thing everyone knew: that the shape of a wedding cake came from its tiered spire. It was usually empty inside, a place where she could sit still and calm herself. Her encounter with the children in the courtyard had disturbed her. It was as if they had been slyly studying her.

Before her the great canopied oak reredos dedicated to the Pilgrim Fathers stood in front of what appeared to be a half-domed apse, but it was actually a magnificent *trompe l'oeil*. A striking oval stained-glass panel, like an upright eye holding the image of Christ, shone light down on to the polished marble floor, which was laid with black Belgian and white Italian tiles.

Amy looked around the empty pews with their homely little lampshades. If there had been any lunchtime worshippers here, they had all gone back to work now. The churchwarden was still on his break and had probably headed up the road for a pie and a pint in the Cheshire Cheese. Someone had taken over for him, and was manning the little shop selling books and postcards near the entrance.

Seating herself in one of the oak chairs arranged near the pulpit, she closed her eyes and let the light of God shine through the dazzling reds, blues and yellows of the stained glass on to her bare freckled arms and upturned face. It was like being inside a gently shifting kaleidoscope. The light divided her into primary colours. She swayed back and forth, feeling the changing patterns on her eyelids. She thought of lost love, wasted time and missed opportunities.

She was still furious with herself for losing the only man she had ever loved. She had been angry for more than two years now, and only coming to St Bride's could dull the ache of loss. If she had taken him more seriously and tried harder to help, she was sure he would still be with her.

His death had hastened the end of her trust in God, but here in the church he must have loved she felt a connection between the present and the past, the living and the dead. She could believe that angels were watching and guiding her thoughts.

But when she opened her eyes, she found that pair of children still peering through the door at her. Where were their parents, and why were they staring?

They looked as if they were waiting for something to happen.

The church's thick walls kept it cool even in the heat of summer. The chill radiated from the stones. But now, after just a few minutes, the interior started to seem hot and airless. The light from the windows hurt her eyes. She could feel her face burning.

Suddenly aware that she was perspiring, she wiped her forehead with the paper tissue she kept tucked in her sleeve, and looked up at the drifting motes of dust caught in the sunlight coming through the plain glass on either side of the nave. Perhaps it was her imagination, but today she really did feel closer to some kind of spiritual presence in here.

The sensation was growing, starting to envelop her. Perhaps God had finally decided to make himself known, and would apologize for screwing up her life. The colours in the oval window above the altar grew more vivid by the second. Even the oak pews that faced each other across the church seemed to give off waves of warmth.

It wasn't her imagination. The church was definitely getting hotter. The light streaming through the glass was tinged crimson. The floor was rippling in the heat. It was as if the entire building had divorced itself from its moorings and was sinking down to hell.

Suddenly she felt very close to a watchful being, but it wasn't God – it was the Devil.

She twisted her head to see the children leaning in from outside the church door, still staring at her intently. And someone or something no more than a stretched silhouette was behind them, dark and faceless, willing them on to evil deeds.

I am going to suffer, she thought. *This is all wrong. I can't die before knowing the truth.*

As the church tipped and she fell slowly from her chair, all she felt was frustration with the incompleteness of life.

3

HEALTH CHECK

'You need to start acting your age,' said Dr Gillespie.

'If I did that, I'd be dead.' Arthur Bryant coughed loudly, causing the doctor to tear off his stethoscope.

'Would you kindly refrain from doing that when I'm listening to your heart?' he complained. 'You nearly deafened me.'

'What?' asked Bryant, who had been thinking about something else.

'Deaf,' said Dr Gillespie. 'You nearly deafened me.'

'Yes, I'm quite deaf, but don't worry, it's not catching. You're a doctor, you should know that. I've got a hearing aid but it keeps picking up old radio programmes. I put it on yesterday morning and listened to an episode of *Two-Way Family Favourites* from 1963.' He coughed again.

Dr Gillespie coughed too. 'That's not possible. How long have you been coming here?' he asked, thumping his chest.

'Forty-two years,' said Bryant. 'You ought to cut down on the oily rags.'

'The what?'

'The fags. The snouts. Gaspers. Coffin nails. Lung darts.'

'All right, I get the picture.'

'The doctor I had before you is dead now. He was a smoker, too.'

Dr Gillespie coughed harder. 'He was run over by a bus.'

'Yes, but he was on his way to the tobacconist.'

'You smoke a pipe.'

'My tobacco has medicinal properties. Is there anything else wrong with me?'

'Well, quite a lot, but nothing's actually dropping off. It's mostly to do with your age. How old are you, exactly?'

'My date of birth is right there in your file.' Bryant reached forward and slapped an immense sheaf of yellowed paperwork.

Dr Gillespie donned his glasses and searched for it. 'Good Lord,' he said. 'Well, I suppose, all things considered, you're doing all right. Mental health OK?'

'What are you implying?'

'I have to ask these things. No lapses of memory?'

'Well of course there are, all the time. But I know if I'm at the park or the pictures, if that's what you mean. It proves quite convenient sometimes. Birthdays, anniversaries and so on.'

'Jolly good. Well, you should make sure you get adequate rest, take a snooze in the afternoons.'

Bryant was apoplectic. 'I can't suddenly go for forty winks in the middle of a case.'

'Yes, but a man of your age . . .'

'Do you mind? I am certainly not a man of my age! I'm running national murder investigations, not working for the council,' Bryant bellowed.

'Well, there's nothing wrong with your voice.' Dr Gillespie made a tick on his list. 'You could always take up a hobby.'

'What, run the local newsletter or work in a community puppet theatre? Have you met the kind of busybodies who do that sort of thing? I'm not interested.'

'That's not what I heard.' Dr Gillespie coughed again and blew his nose. 'I think I'm coming down with something. What was this about you thinking someone had been murdered by a Mr Punch puppet recently?'

'Where did you hear about that?'

'Your partner Mr May is one of my patients too. He's in very good nick, you know. Takes care of himself. He's got the body of a much younger man.'

'Well, he should give it back.'

'He's wearing much better than you.'

'Thank you very much. I'm so pleased to hear that. We solved the Mr Punch case, by the way. Beat people a quarter of our age.'

'Well done. Good appetite? Bowels?'

'I'm sorry?'

'Are they open?'

'Not right at this minute, no, but they will be if you keep me here much longer.'

'I'm almost through. How's your eyesight?'

'It's like I'm living in a thick fog.'

'You should try cleaning your glasses occasionally.' Dr Gillespie's cough turned into a minute-long hack. 'God, I'm dying for a cigarette.'

'If you need one that badly, I'll wait.'

'Can't,' Dr Gillespie wheezed, 'no balcony.'

Bryant absently patted him on the back, waiting for him to catch his breath. 'You don't sound too good. Ciggies just bung up your lungs. I bet your chest feels sore right now.'

'You're right, it does.' The doctor hacked again.

'Like a steel strap slowly tightening around your ribs. Hands and feet tingling as well, no doubt. You're probably heading for a stroke.'

'I've tried to give up.'

'Lack of willpower, I expect.'

'I know, it drives me mad.'

'Perhaps you should think about retiring.'

The doctor bristled. 'Don't be ridiculous, I'm perfectly capable of doing my job.'

'There, now you know how I feel.' Bryant was triumphant. 'Let's call it quits.'

'Fair enough. Put your – whatever that is – back on.'

'It's my under-vest. Then I have my vest, my shirt and my jumper.'

'Aren't you hot in that lot? It's summer.'

'Ah, I thought the rain was getting warmer. I need these layers. They keep my blood moving around.'

'I saw a case that was right up your street the other day,' said Dr Gillespie as Bryant dressed. 'Young woman, Amy O'Connor, twenty-eight, pretty little thing, dropped dead in a church on Saturday.'

'Where was this?'

'St Bride's, just off Fleet Street. It was in the *Evening Standard*.'

'Why do you think that's a case for us, then?'

'You run the Peculiar Crimes Unit, don't you?' said Dr Gillespie. 'Well, her death was bloody peculiar.'

After the doctor had outlined what he knew about the case, Arthur Bryant left the GP's scruffy third-floor office situated behind the Coca-Cola sign in Piccadilly Circus and set off towards the Peculiar Crimes Unit in King's Cross, to check out the case of a lonely death in a City of London church.

Bryant ambled. In Paris he would have been a *boulevardier*, a *flâneur*, but in London, a city that no longer had time for anything but making money, he was just slow and in the way. Accountants, bankers, market analysts and PR girls hustled around him, cemented to their phones. The engineers and artists, bootmakers, signwriters and watch-menders had long fled the centre. Who worked with their hands in the City any more? The ability to make something from nothing had once been regarded with the greatest respect, but now the Square Mile dealt in units, its captains of industry preferring to place their trust in flickering strings of electronic figures.

Bryant would not be hurried though. He was as much a part of London as a hobbled Tower raven, a Piccadilly barber, a gunman in the Blind Beggar, and he would not be moved from his determined path. He was, everyone agreed, an annoying, impossible and indispensible fellow who had long ago decided that it was better to be disliked than forgotten.

And over the coming week, he would find himself annoying some very dangerous people.

4

STRING

'Why did I have to hear about this from my doctor, of all people?' asked Bryant petulantly.

'It's not our jurisdiction,' replied John May, unfolding his long legs beneath the desk where he sat opposite his partner. 'The case went straight to the City of London Police. They're a law unto themselves. You can't just cherry-pick cases that take your fancy, they'll come around here with cricket bats.'

Bryant was aware that the City of London's impact extended far beyond its Square Mile inhabitants. Marked out by black bollards bearing the City's emblem and elegant silver dragons that guarded the major entrances, it contained within its boundaries more than 450 international banks, their glass towers wedged into Palladian alleyways and crookbacked Tudor passages. As the global axis of countless multi-national corporations, it demanded a bespoke police force equipped to protect this unique environment with special policies and separate uniforms.

'If there's a reason why we should take over the investigation we can put in a formal request,' he suggested.

'True, but I can't think of one.'

'How did you know about it?'

'I picked up the details as they came in,' said May. 'It was kept away from us because Faraday wanted it to be handled by the City of London.'

Leslie Faraday, the Home Office liaison officer charged with keeping the Peculiar Crimes Unit in line, was under instruction from his boss to reduce the unit's visibility, and therefore decrease their likelihood of embarrassing the government. His latest tactic was to starve them of new cases.

'But you made some notes, I see.'

'Yes, I did, just out of interest.'

'Well?' asked Bryant, peering over a stack of old *Punch* annuals at May's papers like an ancient goblin eyeing a stack of gold coins.

'Well what?' May looked innocently back across the desk, knowing exactly what Bryant was after.

'The details. What are the details of the case?' He waved his ballpoint pen about. 'There, man, what have you got?'

'Look at you, you're virtually salivating.'

'I have nothing else to concern myself with this morning, unless you happen to know where my copy of *The Thirteen Signs of Satanism* has got to.'

'All right.' May pulled up a page and held it at a distance. Vanity prevented him from wearing his newly prescribed glasses. 'It says here that at approximately two twenty p.m. on Saturday, a twenty-eight-year-old woman identified as Amy O'Connor was found dead in St Bride's Church, just off Fleet Street. Cause of death unknown, but at the moment it's being treated as suspicious. No marks on the body other than a contusion on the front of the skull, assumed by the EMT to have been incurred when she slipped off her chair and brained herself on the marble floor.'

'So what did she die of?'

'It looks like her heart simply stopped. There was a lad running the church shop, but he left his post to go for a cigarette a couple of times and didn't even notice her sitting there. She was found by one of the wardens returning from lunch, who called a local med unit. The only note I have on the initial examination is an abnormally high body temperature. The building has CCTV, which the City of London team requisitioned and examined. They know she entered the building alone, and during the time that she was in there nobody else came in. That's about all they have.'

'Where was she before she entered St Bride's?'

'She was seen sitting on a bench in the courtyard outside the church. A lot of the area's local workers go there at lunchtime. Quite a few work on Saturdays. O'Connor was alone and minding her own business, quietly reading a book.'

'Was she working in the area?'

'No. She had a part-time job as a bar manager at the Electricity Showroom in Hoxton.'

'Why would an electricity showroom have a bar?'

'They kept the name from the building's old usage. It's a popular local hostelry. There aren't any electricity showrooms as such any more, Arthur, even you must have noticed that.'

'What about her movements earlier in the morning?'

'Nobody's too sure about those. She was renting a flat in Spitalfields, had been there a couple of years. Her parents live on the south-west coast. She'd never been married, had no current partner, no close friends. There, now you know as much as anyone else.'

'Where was her body taken?'

'Over to the Robin Brook Centre at St Bart's, I imagine. They handle all the cases from the Square Mile. But you can't go near the place.'

'Why not? I know the coroner there. We used to break into empty buildings together before my knees packed up.'

'Why did you do that?'

'Oh, just to have a look around. I think I'll pop over.'

'No, Arthur. I absolutely forbid it. You can't just walk into someone else's case and stir things up.'

'I'm not going to, old sport. I'll be visiting an old friend. There's a big bowling tournament coming up. He's a keen player. I think I should let him know about it.' Bryant rose and jammed a mouldy-looking olive-green fedora so hard on his head that it squashed his ears. 'Want me to bring you anything back?'

The hospital and the meat market occupied the same small corner of central London, the saviours and purveyors of flesh. In Queen Square, the doctors lurked like white-coated gang members, grabbing a quick cigarette before returning to their wards to administer health advice. Not far from them, in Smithfield, the last of London's traditional butchers did the same thing. Both areas were at their most interesting before 7.00 a.m., when the doctors were intense and garrulous, the butchers noisome and amiably foul-mouthed.

Dr Benjamin Fenchurch's parents had been among the first Caribbean passengers to dock in Britain from the SS *Empire Windrush* in 1948. He had spent his entire working life in the St Bartholomew's Hospital Coroner's Office. Over the decades, he had become so institutionalized that he hardly ever left the hospital grounds. He owned a small flat in an apartment building that was so close to his office he could see into it from his kitchen window. He ate in the St Bart's canteen and always volunteered for the shifts that no one else wanted. Perfectly happy to cover every Christmas, Easter, Diwali and Yom Kippur, he actively avoided the living, who were loud and messy and unreliable, and always let you down. Bodies yielded their secrets with far more grace.

It seemed to Arthur Bryant that this was not a healthy way to live, and yet in many ways he was just as bad, preferring the company of his staff to the world beyond the unit. Working for public-service institutions had a way of making conscientious people feel as if they were always running late. They spent their lives trying to catch up with themselves, and Fenchurch was no exception.

Threading his way through a maze of overlit basement corridors, Bryant reached the immense mortuary that served both the two nearby hospitals and the City of London Police. In the office at the farthest end, Fenchurch was at his lab desk, hunched over his notes, lost in a world of his own.

Bryant cleared his throat.

'I know you're there, Arthur. You don't have to make that absurd noise. I know the sound of your shoes.' Even after all these years, Fenchurch had retained his powerful Jamaican accent. He removed his glasses and raised a huge head of grizzled grey hair.

Bryant was surprised. 'Really? My Oxford toecaps?'

'Nobody else I know still wears Blakey's.' He was referring to the crescents of steel affixed to Bryant's toes and heels that saved leather and ruined parquet floors. 'I haven't seen you since that disgusting business with the Limehouse Ratboy.'

'Yes, that was rather nasty, wasn't it?' Bryant looked around. 'All by yourself today?'

'Do you see anyone else? My assistant's off having a baby. I mean it's his wife who's having the baby. Why he has to be there as well is a mystery to me. It's a simple enough procedure. So, what have I done to deserve a visit?'

'Amy O'Connor.'

'Oh yes. Thought you might be sniffing that one out. Very interesting.'

'That's just what I thought.'

'Pity it's not your jurisdiction.'

'It should have been. She died in a church. Part of our remit is to ensure that members of the general public aren't placed in positions of danger. If people can't trust the sanctuary of a church, what can they trust? But I'm not here in an official capacity. I thought you might like some company. Here, I brought you some sherbet lemons.'

Bryant rustled the corner of a paper bag. Fenchurch sniffed. 'Not much of a bribe, is it?' He fished inside and took one anyway.

'We're playing the Dagenham Stranglers at the Hollywood Lanes Saturday week. I'll put you on our team.' For some peculiar reason, bookish Bloomsbury was the home of two decent central London bowling alleys.

'I thought you'd been banned after that incident with the nutcases.'

'New ownership. Don't think you should call them nutcases.' Bryant sucked ruminatively on a sherbet lemon, clattering it loudly against his false teeth. Last year he had fielded a team of anger-management outpatients to play in a bowling tournament against a group of Metropolitan Police psychotherapists. The outpatients had proven to be sore losers. One of them had tried to make a psychotherapist eat his shoes before knocking him unconscious with a bowling pin. 'Have you carried out a post-mortem yet?'

'Last night. I'm afraid it's going to be an open verdict.'

'Why so?'

'You know I'm not allowed to tell you.'

'Oh come on, Ben, who am I going to tell? I'm old. Most of my friends are either dead, mad or on the way out.'

'How's John?'

'Well, he's fine, obviously. And he's not a friend; he's the other half of my brain. I'd discuss it with him, I admit, but it would go no further.'

'Promise?'

'Cross my hardened heart.'

'To be honest it's a bit of a puzzler, and I could do with some feedback. She had a slight contusion to the orbital frontal region, but was otherwise clean of any marks.'

'You mean falling from the chair and banging her head wasn't enough to kill her?'

'Our bodies are a little tougher than that, Mr Bryant. Otherwise we'd be smashing ourselves to bits like bone-china teacups.'

'Then what else could it have been?'

'With the heightened body temperature it *feels* like toxicosis – systemic poisoning of some kind – but there's no agent present that I could trace. No oesophageal trauma, so she hadn't ingested anything severe. Stomach's fine. That's the thing with City of London workers – you always find the same gut contents, courtesy of our friends at Pret A Manger. The City workers tend to favour the crayfish and rocket sandwiches.'

'She wasn't a City worker. She had a job in a bar in Hoxton. No other marks on the body at all?'

'None that I could see. The admitting officer says they checked the CCTV and she'd been alone outside the church and inside it. There was a boy working in the shop, but he was in and out – a smoker – and went nowhere near her. There's a witness report from him that's the blankest document I've ever seen. She was completely alone except for a couple of kids.'

Bryant's ears pricked up. 'What kids?'

'The officer said it looked like she had an argument with two small children a few minutes before going into the church. She was trying to read. They were playing ball near her, annoying her apparently. Not hoodies – well dressed.'

'Has anyone tried to track the children down?'

'I wouldn't have thought so. What would be the point? What could a small child do?'

'You never know these days. Nothing else unusual at all? Clothes, personal belongings, mobile, handbag?'

'You'd have to ask someone else about that. I'm only dealing with the physical remains. Wait a minute – there was one thing.' He rose and went over to a stack of steel drawers labelled alphabetically. 'Hang on, it's gone.'

'Not "C",' said Bryant. 'Try "O" for "O'Connor".'

'Don't know my own filing system.' Opening the lower drawer, Fenchurch pulled out a clear plastic bag and held it up. 'I hung on to this because I had to cut it off her body. She had a piece of red string knotted around her left wrist.' He threw it over to Bryant. 'My first thought was Kabbalah.'

'No,' said Bryant. 'A Kabbalah string is usually a single strand of red woollen thread, and it's associated with Judaism. It's called a *roite bindele* in Yiddish. With a name like O'Connor she certainly wasn't Jewish, and it wasn't her married name because she'd never had a husband. St Bride's is the church of St Bridget of Ireland, so I daresay it attracts Irish worshippers.'

'Then maybe it was just decoration.'

Bryant turned it in the light, thinking. 'St Bride's. An interesting place to die. It's one of the oldest churches in London, at least the seventh to have stood on that site. Wedding cakes.'

'I'm sorry?'

'A baker called William Rich saw the spire from his window and had the idea for the shape of his daughter's wedding cake. Oh, and journalists always used it. A few old ones still do. Suggestive, don't you think?'

'No, I don't, Arthur. My mind doesn't store things up for later use like yours does. I prefer to have a brain, not a shed.'

'A couple of things,' said Bryant. 'You had to cut the string off, yes?'

'Yes, the knot—'

'Precisely. Not the sort of knot you could do up by yourself. So someone else tied it on for her.'

'What's the other thing?'

Bryant had second thoughts. 'Well, the colour is indicative – but I'll have to do some research on it.'

'Except that it's not your case.'

'I know, everybody keeps saying that.' Bryant jammed his hat back on and walked to the door. 'Ben, will you do me one favour for old times' sake? Don't file your conclusions for a couple of days. Say the printer ran out of paper or something. I want to try and get the investigation transferred to the unit.'

'All right,' said Fenchurch. 'There's no one pressuring me, and we're short-staffed. I suppose I can sit on it for forty-eight hours without too much trouble.'

'You're a pal. Saturday night, weekend after next, bowling, you're playing for us. Eight p.m. sharp for warm-up drinks at the Nun and Broken Compass.'

'I won't do it,' said Raymond Land, shaking his head angrily.

'I don't see why not,' said John May. 'The City of London's on a high alert because of the banking protests, their resources are overstretched and I'm sure they'd appreciate the offer of help.'

'You just don't get it, do you?' Land hissed. 'They hate us. All of them, from the Commissioner downwards. Not just us. They especially hate Arthur. He makes them look bad. He swans in and nicks all the high-profile work, solves the cases and gets the column inches, and accidentally forces up their targets. Why should they give him a case that's already been assigned? He's been in to see me about it and I said no. Absolutely not. We have to keep our noses clean for a while.'

'Fair enough,' said May, raising his hands. 'The others wanted me to ask.'

'Wait, what others?'

'Everyone. Janice, Jack, Meera, Colin, all of them.'

'Are you telling me you've been going around canvassing support behind my back?'

'Of course not. But you know when Arthur gets a hunch it usually turns out to be right.'

Land caught sight of himself in the mirror and saw the usual mix of puzzlement, frustration and anger stirred together like a pudding in a bowl. The little hair he had left was turning grey. He wanted to show authority, but how could he when his detectives defied him at every turn? 'Look, it's bad enough having to fight everyone else in the police service without internal divisions as well. Bryant is a detective, not a mystic. He chases these cases because he fancies having a crack at them, not because he has some strange psychic ability to know exactly when—'

Land's office door opened and Bryant shambled in, his hands thrust deep in the pockets of a shapeless, patched corduroy jacket, his unlit pipe jutting from the side of his mouth. 'Wind's changed direction. It's in the east,' he said meaningfully. 'Looks like there's a storm coming.'

'Where have you been?' asked Land, annoyed.

'Ah. I was on my veranda having a quiet smoke and a think.'

'You haven't got a veranda. This is King's Cross, not New Orleans. It's a rickety old loading platform and it's unsafe. Please don't stand on it.'

Bryant gave a derisive snort. 'It doesn't matter at my age. These days I'm amazed if I just wake up in the morning. Senior citizens should take more chances, not less. Teenagers sleep all the time and us oldies manage four hours a night. Life is upside down. I have a hypothesis about how Amy O'Connor died.'

'You can't possibly know anything about her,' Land protested as a faint but ominous roll of thunder rattled the windows. He glanced out at the seething grey skies above the station, unnerved.

'The old insurance office,' said Bryant, removing his pipe. 'They were tearing down a Victorian building in Salisbury Court, right behind the bench where O'Connor was sitting, but work stopped while they excavated a Roman floor in the basement. Some very nice mosaics. I've just been over there. I looked down into the ruined brickwork and saw something lying in the shadows. It might have been the reason for her death.' The raising of his eyebrow was a study in Stanislavskian method acting.

Land was dumbfounded. His attempts to show leadership were always undermined by his utter amazement at the abilities of others. As a student of human nature he would have made a fine pastry chef. 'Are you telling me that she was murdered?'

'I didn't say that. But I can see how she might have died. I need to find the children who were playing ball in the courtyard.'

'Well you can't, it's not your case.'

'No,' said Bryant, 'but it soon will be.'

'So you're some kind of clairvoyant now?' said Land, exasperated.

'Answer the phone,' said Bryant, pointing to the desk. 'It's your wife.'

The phone suddenly rang, making Land jump. He gingerly raised the receiver. 'Raymond Land. Oh, Leanne, it's you. Yes, I know. I won't be late. All right.' He put the phone down. 'How did you . . . ?'

'The same way I know that you've developed a fear of

rats, that you think you're undergoing a mid-life crisis and you've recently started to believe in the supernatural,' said Bryant.

'You can't possibly – who have you been speaking to?'

Bryant rolled his eyes knowingly and grinned, exposing an amount of white ceramic not seen since the reduction of the East Midlands Electrification Programme had resulted in a surfeit of semi-conductors on the London black market.

'I know everything about you, Raymondo, even things you don't know yourself.' Bryant gave a lewd wink as Land stared at him in ill-disguised horror.

Suddenly, the eerie sound of a theremin started up, the *oooo-weee-oooo* call sign of a hundred old monochrome science-fiction films. 'That's my mobile,' said Bryant, 'I must take this call. If anyone wants me, I shall be in my boudoir.'

'You haven't got a boudoir,' Land called after him helplessly, 'you've got an office!'

'All right, what's with the Sherlock Holmes stuff?' asked May, closing the door behind him. 'You're really getting up Raymond's nose.'

'Oh, it was a dreadfully cheesy trick, I know,' said Bryant airily, 'but I couldn't resist getting him back for refusing to let me try for the case. He's so adorable when his mouth is hanging open, like a spaniel trying to understand house-training instructions.'

'How did you know all that stuff about him? Or did you just make it up?'

'It's easy. His wife just called me by mistake and I rerouted it. He left a card from a rodent exterminator on his desk. We had rats at the old headquarters in Mornington Crescent and they never bothered him, but ever since Janice mentioned she's heard noises in the walls in this building late at night, he's been on edge.'

'The mid-life crisis?'

'He found out about his wife's affair, yes?'

'Only because you told him.'[1]

'Now she's talking about divorce and he's suddenly realized he'll be back on the dating scene, hence his recent purchase of several appallingly unsuitable shirts. Oh, and that horrible aftershave he's starting pouring over himself. You must have noticed that he's smelling like a perfumed drain. And before you ask, he's started to believe in the supernatural because I can see that he's borrowed some books from my top shelf, notably *Psychogeographical London*, *Great British Hauntings* and my 1923 copy of *Mortar and Mortality: Who Died in Your House?* He's been upset ever since he discovered that Aleister Crowley ran a spiritualism club in our attic. Nearly every London house has been lived in by somebody else, and Crowley was all over this town like a cheap suit. It's hardly anything to get upset about.'

'You could try being nice to him for a change,' said May. 'He's been very supportive lately. I feel sorry for him, stuck in a job he hates, having to look after us lot. He can't understand how you think.'

'I should hope not,' said Bryant indignantly. 'I would be most offended if he could. But perhaps you're right. I'll make it up to him.'

'No.' May hastily held up his hand. 'Don't do anything unusual. Just do what he says for a while.'

'You mean don't push for the Amy O'Connor case.'

'Exactly.'

'All right,' said Bryant, 'but don't say I didn't warn you.'

'What do you mean?'

'"By the pricking of my thumbs, something wicked this

---

[1] Bryant used his newly rediscovered ventriloquism skills to inform Land of Leanne's affair with her flamenco instructor. See *Bryant & May and the Memory of Blood*.

way comes."' He sauntered to the door. 'I'm going to the terrace for a pipe of St Barnabas Old Navy Rough Cut Shag. But I'm telling you, there's more to Amy O'Connor's death than meets the eye.'

'Because of what you saw in a Roman excavation?' asked May.

'That, and because of the string that was tied around her wrist.'

5

THE ENEMY

'You're not going to be happy about this,' warned John May. 'Home Office Security has backed up the City of London. They won't let you have the O'Connor case.'

'Why not? What's it to them?' Bryant asked, as he and May made their way across Bloomsbury's sunlit garden squares towards the Marchmont Street Bookshop.

'Your pal Fenchurch has already tipped someone off about his likely verdict, although he seems to be holding back the full official report. Once that's been filed, the case is technically closed unless you get Home Office dispensation, and they won't grant it.'

'That's odd. I was with him this morning and he said he'd delay the process by forty-eight hours. Why would he have told someone?'

'You weren't supposed to go there. Maybe he's being pressured.'

'That makes no sense unless someone at the Home Office thinks the case is more important than it looks. Amy O'Connor was a low-paid bar manager. Apparently she studied biology at Bristol University, but dropped out. She's not connected to anyone important. Unless there's something in her past. I could take a look at her employment records and see if—'

'Arthur, maybe she really did just black out and fall.'

'Without a cause of death? Next you're going to tell me she was struck down by the hand of God. Nobody dies without a reason, and no reason has been found. If I can just go back through her history . . .'

'But it's not your—'

'Don't say it again, all right? Here we are.' Bryant stopped in front of the bookshop and pointed proudly at the window. 'Sally's given me pride of place.' Bryant's wrinkled features peered up from the cover of a slim volume entitled *The Casebook of Bryant & May*, by Arthur Bryant, as told to Anna Marquand. Beside it, a joss stick protruded from the head of a green jade Buddha, as if in funereal remembrance.

'It's just the first volume, as you know, but it covers quite a few of our odder investigations, from the Leicester Square Vampire and the Belles of Westminster, to the Billingsgate Kipper Scandal and the hunt for the Odeon Strangler.'

'And you honestly think the public wants to read this stuff? People aren't interested in the past any more. The young want to get on and make something of their lives. They don't want to wallow about in ancient history.'

'I didn't write it for the ambitious young,' said Bryant primly. 'I wrote it for the mature and interested. And, if you don't mind, it isn't ancient history, it's my life. Yours, too.' Privately, though, Bryant had to admit that the events of his life were receding into history. Last Christmas the milkman had come in for a warm-up and had asked his landlady if she collected art deco. 'No,' Alma had replied, 'this happens to be Mr Bryant's furniture.' Yesterday's fashions were today's antiques.

The owner of the small bookshop greeted Arthur.

Now in her early fifties, Sally Talbot was an attractive blue-eyed blonde with the natural freshness of someone raised on a warm coastline. John May was a great appreciator of beautiful women, and his pride required him to smooth his hair and pull in his stomach.

'Nice to see me in the window,' Bryant commented. 'I'm not sure about the incense, though. It looks as if I've died.'

'Oh, we've got damp,' said Sally. 'It's better than the smell of mildew. Thank you for coming by to sign the stock. You only went on sale this morning but we've already sold a few copies.'

'One of them wasn't to a man who looks like a vampire bat, was it?' asked Bryant. Oskar Kasavian, the cadaverous Home Office Security Supervisor, had made it publicly known that he objected to Bryant writing his memoirs, and had been trying to get hold of the manuscript so that he could vet it for infringements. The Peculiar Crimes Unit was the flea in his ear, the pea under his mattress, the ground glass in his gin, but at least he had lately abandoned his attempts to have it closed down. So long as the unit's strike rate remained high, there was little he could do to end its tenure. He was not against the idea of the place so much as its method of operation, which defied all attempts at rational explanation, beyond a vague sense of *modus vivendi* among its staff.

'No, they went mostly to sweet little old ladies who love murder mysteries,' said Sally.

Bryant dug out his old Waterman's fountain pen, uncapped it and shook it, splodging ink about. 'How many do I have to sign?' he asked.

'Well, five if you don't mind.'

'Is that all you have left?' Bryant beamed at the bookseller. 'How many did you sell?'

'Three.'

'Oh. What's your bestselling biography?'

'*Topless* by Katia Shaw,' said Sally. 'She's a glamour model.'

Bryant turned to his partner in irritation. 'You see? This is what's wrong with the world. A young lady with bleached hair, an estuarine accent and unfeasible breasts can outsell a respected expert with decades of wisdom and experience.'

'She's human interest,' replied May. 'You're not. People reading her story will feel that if she can make it without talent, maybe they can.'

'Well, I find that phenomenally depressing.' Bryant's theremin call sign sounded once more. 'Well, speak of the Devil,' he said, checking the number, 'it's Mr Kasavian himself. I bet I know what this is about. I'd better take it outside.'

Ten minutes later, the detectives had hailed a taxi and were heading south towards Victoria. 'My guess is he wants an explanation about the memoir,' said Bryant.

'Then why would he ask to see me as well?'

'You're mentioned in the title of the book, John. You're as involved in this as I am. I think he might have found something unpalatable in one of the chapters and taken objection.'

'I wonder if it's the part where you refer to MI7 as a secure ward for the mentally disenfranchised, or the bit where you describe his department as a hotbed of paranoid conspiracy theorists with a looser grip on reality than a stroke victim's hold on a bedpan handle?'

'I'm impressed you remembered that,' said Bryant, pleased. 'There's nothing in the book that breaches the Official Secrets Act, and that's the only thing he can get me on. Anna triple-checked it.'

'Yes, but Anna Marquand is dead.' Bryant's biographer had supposedly died of septicaemia in the South London home she shared with her mother, but she had passed away shortly after being mugged by an unknown assailant. The case remained unsolved.

'You know my feelings about that,' said Bryant. 'I'm sure Kasavian's department is implicated somehow. He might not have been directly involved, but I bet he knows who was.'

'I'm not so convinced any more,' said May. 'You honestly think the Home Office found something in your memoirs that was so damaging they would commit murder to cover it up? They're part of the British government, not the Vatican.'

'I think they might have gone as far as condoning an unlawful killing, if it involved the Porton Down case.' Bryant sucked his boiled sweet ruefully.

Porton Down was a military science park in Wiltshire, the home of the Ministry of Defence's Science & Technology Laboratory, DSTL. The executive agency had been set up and financed by the MOD to house Britain's most secretive military research institute. Three years ago there had been a rash of suicides at a biochemical company outsourced by the DSTL. The project leader at the laboratory had turned whistle-blower, and had been found drowned. At the time, Oskar Kasavian had been employed as the head of security in the same company. It might have been coincidence – government defence officials moved within a series of tightly overlapping circles – but the absence of information made Bryant suspicious.

'Why do something so dramatic?' asked May. 'Why not simply slap an injunction on the book?'

'That would be the best way to draw attention to it, don't you think? Do you honestly imagine governments can't make people disappear when they want to? Looks like we're here.'

The taxi was pulling up in Marsham Street, the new Home Office headquarters. The building had won architectural awards, but to Bryant's mind its interior possessed the kind of anonymous corporate style favoured by corrupt dictators who enjoyed picture windows in the boardroom and soundproofed walls in the basement.

'A word of advice, Arthur,' May volunteered. 'The less you say, the better. Don't give him anything he can use as ammunition.'

'Oh, you know me, I'm the soul of discretion.'

May's firm hand on his shoulder held him back. 'I mean it. This could go very badly for us.'

'That's fine, John, so long as you remember that he is our enemy. Anna Marquand was more than just my biographer, she was fast on her way to becoming a good friend; someone I trusted with the secrets of my life. And she may have paid for it with her own.'

In the immense open atrium, the detectives appeared as diminished as the figures in a Lowry painting. A blank-faced receptionist asked for their signatures and handed them plastic swipe cards.

Three central Home Office buildings were connected from the first to the fourth floors by a single walkway. This formed part of a central corridor running the length of the site, commonly known as the Bridge. Kasavian's new third-floor office was in the only part of the building that had no direct access to sunlight. As the detectives entered his waiting room, they felt the temperature fall by several degrees.

Kasavian's assistant looked as if she hadn't slept for months. 'Perhaps he drains her blood,' Bryant whispered from the side of his mouth. She beckoned them into an even dimmer room. Kasavian was standing at the internal window with his back to them, his hands locked together, a tall black outline against a penumbra of dusty after-noon light. In this corner of the new century's high-tech building it was forever 1945.

May glanced across at his partner. Arthur Bryant had

no interest in what others thought of his appearance. His sartorial style could most easily be described as 'Post-war Care Home Jumble Sale'. It was usually possible to see what he had been eating just by glancing at his front. John May prided himself on a certain level of elegance, although his police salary did not run to handmade suits. When Kasavian turned, May instantly recognized the Savile Row cut of charcoal-grey cloth, the lustrous gleam of Church's shoes, the dark glitter of Cartier cufflinks, and felt a twinge of jealousy.

'Take a seat, both of you. I'm sorry about the light. Sometimes when I'm stressed my eyes become hyper-sensitive.'

Bryant shot his partner a meaningful look. *Something's wrong here.* Kasavian never revealed anything that could be interpreted as a human flaw; it wasn't in his DNA to do so.

Kasavian sighed and absently ran a palm against the side of his oiled black hair. As yet he had looked neither of them in the face. He stalked around his chair, picked up an onyx-handled letter opener and set it back down, then suddenly seemed at a loss. Searching about in vague confusion, he eventually planted himself on the edge of the desk and carefully studied each of them in turn.

'This isn't about your memoir,' he said finally. 'Our legals gave it a cursory glance when it was still proofing.'

'That's odd,' said Bryant. 'The galleys were locked away.'

Kasavian waved the implication aside. 'We let you off because it appears your early cases weren't covered by actionable security regulations, and the department has resolved not to take a stance on your more provocative jibes. We like to think we can take a joke, and besides, your personal opinions don't matter to us. This – well, it's about something else entirely. And it occurred to me that you might be able . . . that is, you might be the only ones . . . who could help me out.'

'What do you mean?' asked May.

'Perhaps I'm not making myself clear,' said Kasavian, rising and starting to pace about. 'I want to hire your services.'

6

PERSECUTED

Bryant was taken aback by the tone of Kasavian's voice. The civil servant he knew had a steely grimness that turned the lightest remark into the pronouncement of a death sentence, Judge Jeffreys with a gastric complaint. Now he sounded unsure of himself and almost human.

'I know there has been a certain level of . . . dissension between us in the past, but I want to put that behind us.'

'Fine,' said Bryant, 'but could you sit down? You're making me nervous.'

Kasavian went behind his desk and sat. Bryant was amazed. It was the first time the security chief had ever heeded one of his requests. Steepling his long, crab-leg fingers, Kasavian thought for a moment. 'I don't want this to go beyond my office, do you understand?'

'Of course,' May readily agreed, inching forward on his chair. Bryant shot him a jaundiced look.

'I have a problem. It has nothing to do with the antagonism between your unit and my department. This is a purely personal matter.'

Bryant was clearly fighting to suppress a grimace. He

liked his enemies cold and bitter, like his beer. Anything less weakened them in his eyes.

'I don't know if either of you has ever met my wife?'

'Certainly not,' said Bryant. 'I never pictured you being married to a—' He was thinking of *human being*, but hastily ended the sentence.

'She's – very beautiful. Very young. Perhaps too young.' He lifted a framed photograph from his desk and showed it to them. 'In this job one expects to be vetted for many different degrees of security clearance, but one thing they don't do is decide whom you fall in love with. Perhaps they should do. How can I describe Sabira? She's wilful and easily bored. Rather like a Christmas tree: beautifully adorned but likely to burn the house down if left unattended.'

For Bryant, this was intolerable. The last thing he wanted was to know about the private life of his archnemesis, but the framed photograph was extraordinary. It showed an extremely attractive woman with a heart-shaped face, an absurdly flat stomach and cantilevered breasts, lying in a cheesecake pose on a sun lounger in a candy-striped bikini. It looked less like she was absorbing the rays of the sun than radiating them. She did indeed appear to be very young. If Kasavian had to be married at all, surely his wife should have had a face that could send a dog under a table?

Meanwhile, May was taking another look at the security supervisor and trying to imagine how on earth women could find him attractive. There was, he supposed, power and gravity in his bearing, authority in his saturnine features. A wife would be able take shelter if not comfort, and some were more concerned with finding a safe harbour than igniting passion.

'She is the light of my life. Everything changed after I met her. But now there's something wrong between us. It's hard to explain. In the last six weeks her personality has undergone an extraordinary transformation. She is angry all the time – very angry. Not just with me but with everyone around her.'

'We're detectives, not marriage-guidance counsellors,' said Bryant. 'Have you tried taking her to the pictures occasionally?' May fired off a warning look.

'I'm not explaining myself very well. I'm not used to having this kind of conversation. Let me tell you more.'

Bryant's intestines cringed. He forced out a staggeringly insincere smile.

'We've been married for almost four years. Sabira is eighteen years younger than me.'

May gave a low whistle. Kasavian glared at him before continuing. 'Yes, I know it's a big gap. And on the surface, we have few common points of interest. She left school at fifteen to work in a biscuit factory; I went to Eton and King's College, Cambridge. Her parents used to manage an industrial aluminium smelting plant in Albania and still live on the contaminated site; mine are landowners in Herefordshire and breed horses. She was raised a Muslim, I was High Anglican.'

'But you love each other very much,' said May, leaning further forward in his seat. Disgusted, Bryant unwrapped a Hacks cough sweet and crunched it noisily.

'Our affection for each other is beyond question. I don't want you to think this is some kind of lovers' quarrel – there's been a tangible and dangerous psychological change in her.'

'Can you give us an example of her behaviour?'

'I knew Sabira was raised in a religious household, but by the time she left home she was no longer a practising Muslim. I discovered she was superstitious when she began covering all the mirrors in the house.'

'Why did she do that?'

'She said there was an evil presence nearby.' Kasavian shook the thought aside, impatient with its absurdity. 'She said she could feel something following her around, intending harm. Believe me, I am fully aware how ridiculous this sounds.'

'Not at all,' said May. 'Please go on.'

'At first it just seemed like another of her quirks. Albania is one of the most religious countries in Europe, and she was raised in a small village, the kind of place about which they have a saying: "In order to live peacefully here, you must first make war with your neighbours." Coming to London must have been a profound shock for her. Lately her belief in this so-called evil presence has escalated. She started ransacking the house, but wouldn't tell me what she was looking for. One evening I came home and found her burning books and letters in the garden. She talked about devils taking human form, about satanic conspiracies and witchcraft and a plot against her and God knows how many other crazy notions. It's not as if this happened slowly; the change occurred over two, perhaps three weeks.'

'How did you cope with this?'

'I was very busy here. I had just taken over the development of the new UK border-control directive, so I wasn't at home much. It's by far the largest project the department has ever undertaken, a pan-European initiative designed to curtail the movement of members of terrorist organizations within the EU. As the head of the UK delegation I'm representing the wishes and intentions of Her Majesty's Government.'

'What does that mean exactly?' asked Bryant.

Kasavian levelled ebony eyes at him. 'It means I don't get home in time for supper.'

'Does your wife have many friends here?'

'Hardly any. We met during her second London visit, and soon after she moved here to be with me.'

'You said the psychological change in her was dangerous – what do you mean by that?'

'Mr May, my job is to establish a rational explanation for why things go wrong and come up with practical solutions. But what do you do when your wife suddenly announces that she is being chased by demons? She swears there's someone in the grounds of the house at night, someone who watches her all the time and wishes her harm. She believes she's the victim of a witch-hunt. She says she only feels safe in a place of worship, so she spends more and more time in mosques and churches. One night she dragged me into the garden to look at a pattern carved on a tree and said it was a satanic sign, that she had been marked as a victim. I can't talk to any of my colleagues about this, and I certainly couldn't go to the police without any evidence. There's no proof, no consistency to these absurd stories. I thought if anyone could understand, it would be you. You seem to know a lot of abnormal people.'

'I'll take that as a compliment. Has she seen a doctor?'

'She absolutely refuses to do so, and I can't force her. She tells me there is a history of mental instability in her family, and fears becoming like her grandmother or her aunt, both of whom were sectioned after years of aberrant behaviour. She thinks a doctor will look at her past and make assumptions about her mental health.'

'But if she's delusional it seems she needs a therapist, not a detective.'

'You don't understand,' said Kasavian. 'I and my entire department operate within the confines of the Official Secrets Act, and although I've told her virtually nothing about my work over the time we've been together, she is my wife. Within a marriage there can be no absolute guarantee of privacy. And now she is running around talking to complete strangers, telling them people are casting spells on her. I have no idea what else she's saying to them. My position here is being compromised. It's as if she's two people, ecstatic one minute, suicidal the next.

If I thought she was going mad I would force her to seek psychiatric help, but I suspect there's something more to it.'

'Why do you think that?'

'Because this cult of Devil-worshippers she imagines lurking behind every car and tree – I have a feeling she thinks I'm their leader.'

'Well, you must admit you do look—' Bryant began, but once again thought the better of it. 'What do you imagine brought on this sudden change in her behaviour?'

'I can only think something happened around six weeks ago – perhaps she met someone unsavoury, or did something foolish. Got herself into some kind of trouble. She won't give me a straight answer.'

'Then what do you expect us to do?'

'I need you to find out if there's anything behind these fantasies of hers,' said Kasavian. 'Obviously I wouldn't be able to grant the case official status, but if you get to the root of the problem I think I can promise a very agreeable recompense.'

'What did you have in mind?' asked Bryant.

'A full exoneration for the unit, an amnesty on your memoirs and a permanent guarantee of official status within the City of London Police structure. You'd no longer face challenges from the Met or the Home Office.'

'We'd be reinstated and officially recognized?' asked Bryant, staggered.

'I just want my wife back,' said Kasavian, looking suddenly pitiful. 'Please, find a way to make her sane again.'

'I haven't felt this revolting since we wormed Crippen,' said Bryant as they headed towards Victoria Station. 'Dracula seeks our services and asks us to sort out his barking wife's persecution complex? The very unit he's spent the last few years trying to close down?'

'You heard him,' said May. 'He has no one else to turn to.'

'Of course not, everybody hates his guts. But we're not experts on mental health. Quite the reverse, if anything. Besides, if I know Kasavian he's less concerned about his wife's sanity than he is about making sure his department isn't brought into disrepute.'

'That's understandable. He's about to represent British interests in Europe. The last thing he needs right now is something that will break his concentration and damage his reputation. And we might be able to deal with the problem quickly. It sounds as if Sabira's parents live on a chemically contaminated site, and presumably she was raised there as well, which might explain her mental problems now. We have nothing to lose by taking on the case.'

'Oh no? What if we fail? He'll have the perfect ammunition against us.'

'You heard him say that our involvement would be kept strictly off the record. He won't be able to blame us if we fail. What have we got to lose?'

'Do I have to remind you?' asked Bryant. 'Anna Marquand may have been murdered because somebody wanted to destroy the notes she made from my interviews.'

'It would help if you could remember what you told her about your past cases.'

'The sessions took place over a two-year period. I have no idea what I might have said. You know what my memory's like. I can't even remember where I'm living.'

'Good Lord, I'd forgotten you were moving at the weekend. How is it?'

'I don't know, I haven't been there yet.'

'But you must have seen the place.'

'No, I left it all to Alma.' Alma Sorrowbridge had been

Bryant's long-suffering landlady for over thirty years, and had arranged to find a new flat for them after their old home had received a compulsory purchase order. 'That's not important right now. The important thing is . . . I've forgotten the important thing.'

'Anna Marquand.'

'Anna, yes. Her attacker had been to Oskar Kasavian's office. There was a Home Office slip in his pocket with Kasavian's department named on it.'

'But their new premises doesn't use entrance slips any more – it operates on a swipe-card system.'

'You're right. I hadn't thought of that,' said Bryant. 'A swipe card wouldn't have led us to Kasavian.'

'Exactly – so maybe somebody was trying to implicate him.'

Bryant groaned. 'This is dreadful. Not only do I lose my only suspect, I gain Dracula as a client.'

'I thought it would be right up your street – his wife thinking she's being hunted down by a satanic cult.'

'Well, it is actually. Except that Kasavian should turn out to be the culprit, not the client. He looks like the sort of man who'd try to drive his wife mad, doesn't he? He's a cold-blooded bureaucrat. He can make children cry just by staring at them.'

'But he'd gain nothing by coming to us. His career is about to come under the microscope, and the last thing he'd want is to draw attention to his marital problems. He's giving us the power to wreck his career, Arthur. That's not the action of a man who wants his own complicity uncovered. It's the act of someone who's desperate and has been forced to turn to his enemies for help.'

'Fair enough, but it's Occam's razor, if you ask me. When a man looks like Christopher Lee with irritable bowel syndrome, it's hard to suddenly imagine him buying flowers and patting puppies. Never mind, I shall set aside my personal antipathy while we figure out what's behind it all. Just don't ask me to be friends with him afterwards.'

The detectives descended into the muggy tunnels of Victoria tube station.

7

THE ENGLISH DISEASE

Like most of the venerable institutions in London, the Guildhall was more impressive than beautiful. It had been the corporate home of the City of London for eight hundred years, and tonight was illuminated to welcome six European heads of state, from Finland, the Czech Republic, Spain, the Netherlands, Poland and Italy.

The chevrons and monograms of previous reigning monarchs and Lord Mayors shone down on the Great Hall's assembled guests, who were seated beneath monuments to Nelson, Wellington, Chatham, Pitt and Churchill. It was in this room that Dick Whittington had entertained Henry V, paying the delicate compliment of burning His Majesty's bonds on a fire of sandalwood. Gog and Magog, the short-legged, flame-helmeted giants who founded London before the time of Christ, glowered over the oblivious diners, who were finishing their desserts and moving on to coffee.

Sabira Kasavian was growing more upset by the minute. Sandwiched between a moth-eaten City alderman and a twitchy, crow-faced woman named Emma Hereward, she tried to spot her husband. Oskar was seated on the top table between the Deputy Prime Minister and the Chief of the Metropolitan Police. The intense conversation did not allow him time to look up and offer her a complicit smile. She had been relegated to an unimportant table because regulations prevented her from being within earshot of private ministerial conversations.

After almost four years of marriage she should have become used to such snubs, but each one still came as a shock. Her security clearance for visiting her husband at his place of work was one of the department's lower grades, because other wives had longer-serving spouses. She was not permitted to call him between certain hours, nor ask Oskar any questions about his work. There were rooms in the house she was forbidden from entering because they contained sensitive documents; she was not allowed access to any of his electronic devices. In times of a security crisis her friends, family members and correspondents were vetted, and sometimes, during those periods when the city was on the highest level of security alert, a guard was posted outside their flat.

She had thought she was freeing herself from her lunatic ex-boyfriend and her own impossible family, from the endless financial worries and the painful peculiarities of Albanian life, but instead she had stepped into a secretive gilded prison.

So she drank. Downing the dregs of her red wine, she snatched the brandy bottle away from the alderman, filling her water glass from it. She looked around at the other guests: the florid businessmen and their badly dressed wives; the desiccated accountants and corporate lawyers; the frumpy horse set; the charmless couples who couldn't wait to get back to their dogs and their mock-Tudor Thames Valley houses; the so-called cream of the *nouveaux riches*. They all treated her as if she was a fool and a foreigner, as if those states were synonymous. None of them had bothered to get to know her. If they had, they would have found out that she was well read and intelligent, and spoke flawless, if accented, English. The other wives could speak only their own language plus a smattering of restaurant French.

She knew the real reasons for their enmity. She was young and attractive, and liked to dress glamorously. She was wearing a red dress edged with silver bugle-beads, a look none of the other wives would have dared to try and pull off, and she simply wasn't apologetic enough about not being English.

'Of course, we've largely stopped going to Capri because these days it's full of the most ghastly people.' Emma Hereward was talking across her. 'The budget airlines all fly into the region now.' As usual, Emma was seated with Anastasia Lang and Cathy Almon. The three of them were hardly ever apart and were as poisonous as scorpions. Cathy was the plainest and therefore the most picked on. All were married to men in Oskar's department.

'We go to a marvellous little island off Sicily—'

'Isn't that where Giorgio Armani has his villa?' asked Ana Lang, also talking across Sabira. 'The Greek islands are ruined, of course. Do you still have the place in Tuscany?'

'We gave it to the children. Better for the tax man.' Emma noticed Sabira listening. 'Does your husband have a bolt-hole?'

'I'm sorry?' She didn't understand the question. What was *bolt-hole*?

'A second home – you know . . .' She walked her fingers. 'Somewhere you can whizz off to in the school holidays.'

'He has a house in Provence,' Sabira replied, 'but I have never been there.'

'Why ever not? I mean, the French are frightful, obviously, but you must get so bored being stuck in London.'

'I go home to see my family in Albania, but Oskar is usually too busy to accompany me.'

'Of course it's different for you, not having any children,' said Ana. 'But what would Oskar do in Albania?'

'We go to the beach there.'

'You have beaches?' Ana exclaimed. 'How extraordinary.'

'Yes, we have very nice ones.'

'That's a surprise. I always assumed the country was mainly industrial. We have a Polish chap – you must have met him, Emma. He built our patio. A terrible one for the vodka, but then all Eastern Europeans drink like fish.'

Sabira dropped out of the conversation and refilled her glass.

The speeches dragged on. Someone from the Animal Procedures Committee was talking about a new initiative, but he had a habit of moving his face away from the microphone, and whole sentences dropped out of earshot. The Deputy Prime Minister, a fair, faded little man who might easily have been mistaken for the manager of a discount software firm, was whispering in her husband's ear. She was so far away from the speaker's table that she could barely see who was talking.

'These initiatives are a waste of time,' Cathy Almon was saying. 'Democratic governmental procedures are hopeless. People respond better to a benign dictatorship; it saves them having to take responsibility.'

'I don't agree,' said Sabira, jumping in. 'Surely the key to any democratic process is representation.'

Cathy stared at her as if she expected frogs to start falling from her mouth. 'I'm sorry,' she said. 'Remind me who you are again?'

'I'm Sabira Kasavian. We have met a dozen times.'

'Goodness, of course, you must forgive me. I have absolutely no memory for faces. You must be very proud of your father.'

'I am, but Oskar Kasavian is my husband.'

'Then you must be more mature than you look.' She meant it as a compliment, Sabira decided, a very English kind of compliment, the sort that offended as it flattered.

'No, I am not,' she said in a louder voice than she intended. 'He is forty-five and I am twenty-seven. There is an eighteen-year age difference between us.'

Ana Lang laid a beringed claw on her arm. 'There's no need to take offence, dear. You mustn't be so sensitive.'

'But I do take offence,' said Sabira hotly. 'You know where Giorgio Armani has his holiday villa but seem unaware that Albania has a coastline. That one, Mrs Almon, likes to pretend we've never met, and makes me introduce myself again. And you just accused my countrymen of being alcoholics. You've been patronizing and condescending to me ever since we sat down.'

'I think you're overreacting,' said Ana, who could only cope with indirect criticism. 'There's no need to get so overwrought. This is simply dinner conversation. How long have you been married to Oskar?'

'Nearly four years,' Sabira replied.

'Then I'm sure you must be familiar with at least some of our social customs by now, just as we are with yours. For example, your drinking habit could hardly go unnoticed, and while you might consider it part of a noble heritage there are others who could misconstrue it as intemperance.' Ana bared her teeth in a mirthless smile, daring her to answer back.

'Then you'll know that, according to my noble heritage, when someone is insulted custom requires them to take revenge,' said Sabira.

She felt her hand going towards her full water glass. She intended to take a sip of brandy to steady her frayed nerves.

'The girl has some spirit, Ana. I think Oskar's done rather well for himself.' Emma Hereward laughed.

'I think you should stop picking on her,' said Cathy

Almon, who knew what it was like to be constantly bullied.

'If you think I'm beneath him, you should say so to my face,' said Sabira. 'Hypocrisy is the English disease, isn't it?'

'I imagine dear Oskar probably woke up on an Albanian fact-finding mission and found you beneath him,' said Ana Lang, chuckling softly with the others.

Sabira's grip on the brandy-filled glass tightened.

8

SABIRA

'You've got to admit it's a great photo.' Detective Sergeant Janice Longbright threw the newspaper at Jack Renfield. 'Look at her, she's a real wildcat.'

'Blimey, that'll sell a few copies.' Sergeant Renfield grinned approvingly. 'I wonder why they stuck a blurry box over the top of her thighs.'

'According to the *Daily Mail* she didn't have any knickers on,' said Longbright. 'She said she took them off before the dinner began because it was too hot in the room. It looks like a very tight dress. She probably didn't want a VPL.'

'Typical of a woman to think it was about fashion. Perhaps she was just feeling horny.'

'She was attending a dinner to welcome heads of state at the Guildhall, not hitting on guys in a Nottingham nightclub, Jack. It says there that she threw a glass of brandy in some old bag's face.'

'That "old bag" is Lady Anastasia Lang,' said John May, snatching up the paper as he entered on Tuesday morning. '"Sabira Kasavian was arrested for being drunk and disorderly last night, and was taken to Wood Street

Police Station." The arresting officer told me that Oskar tried to get her off the hook, but they had no choice but to run her in. Ana Lang was ready to press charges.'

'Wait a minute,' said Longbright, 'it's "Oskar" now? Since when did you switch to first-name terms?'

'Since he hired us to investigate his wife,' said May. Everyone in the common room turned to look at him. 'What can I say? I know. He's always been the enemy, and now he's the client.'

'After all the terrible things he's done in the past, I'm amazed he would trust us with something so personal.'

'Kasavian didn't have anyone else he could turn to. He managed to get his wife released from Wood Street a short while later, but by that time the damage had been done.'

'I wonder what upset her so much?' Longbright asked. 'It says here the Finnish Minister for Finance was forced to stop his speech.'

'This is a big deal,' said May. 'Three months ago our Deputy PM made a speech in Finland that was halted by hecklers halfway through, so now everyone's saying this was payback. But Sabira Kasavian says it wasn't planned; she'd been insulted all evening and finally had enough of it.'

'OK, she was drunk, but she must have known it would reflect badly on her husband. Kasavian's in line for one of Europe's top security posts, isn't he?'

'He may not be after this. Check the rest of the online press; see if there are any more details. I bet they're having a field day. Then fix up an appointment with the wife this morning. If she refuses to meet with us, I'll get Oskar to call her.'

'He's on the line right now,' said their detective constable, Meera Mangeshkar, covering the phone. May took the call with a certain amount of trepidation.

'I suppose you've seen the news this morning,' said Kasavian.

'I could hardly have missed it.'

'My wife was carried from the Guildhall kicking and screaming last night. *The Guildhall*. She smashed a tray of glasses and threw a shoe at one of my colleagues' wives, then swore at the arresting officer and tried to run off down the street.' He sounded exhausted.

'But you got the charges dropped.'

'Yes, but I can't keep her locked up at home. I'm not putting her under house arrest: I'm her husband, not her jailer. I don't know what to do. In an ideal world I'd take her away for a holiday, but this border-control thing is taking up all my time. And I can't send her home to Albania. Imagine how that would look just ahead of the talks.'

'Then I suggest you concentrate on your work and allow us to take care of her,' said May. 'You know our methods are unorthodox, but you'll simply have to trust us. We're going to need a level of access that may cause problems for you.'

'I've a stack of reports on your past activities from Leslie Faraday. I'm fully aware of the lines you cross to get results, Mr May. But in this case, I need you to do whatever you can for my wife. Go and see her, and I'll get you any other access you need. I have half a dozen important social occasions this month, and Sabira is expected to accompany me to them. If she suddenly stops turning up, my opposite numbers will be quick to make capital of it. The trouble is, I no longer know what she's likely to do.'

'First we'll look at your calendar and take her out of the more sensitive events.'

May prided himself on his understanding of women, but he felt uncomfortable knowing that if he failed to get to the cause of Sabira Kasavian's problem, her husband would have good reason to come down hard on the unit. Her behaviour could derail his career and wreck a European-wide initiative. The Americans would be watching, and would step in fast.

He swung into the office he shared with Bryant. 'Get your hat and scarf on, Arthur,' May instructed. 'Kasavian's granted us clearance. Let's catch his wife by surprise and find out what she's up to.'

'They have a house in Henley, but his London apartment is in Smith Square,' said May, opening the badly rusted door of Victor, Bryant's leprous yellow Mini. 'I'll drive.'

'You'll need this,' said Bryant, handing him an apostle spoon.

'What am I supposed to do with it?'

'Stick it down the side of the gear stick. It seems to hold it in place.'

May gave up trying to move the seat back, and set off into the traffic, heading towards the river. There was something wrong with Victor's gears. 'I'm surprised this thing passed its MOT,' he said as the car leapfrogged across the Euston Road.

'It passed under certain conditions,' replied Bryant vaguely. 'I think one of them was that I must never drive it anywhere.'

'Then it's an illegal vehicle.'

'No – I'm not driving, am I?'

Smith Square, just south of the Palace of Westminster, was dominated by the immense white frontage of St John's, a baroque church now used as a concert hall. Surrounding it were the offices of the Department for Environment, Food and Rural Affairs, the Local Government Association and the headquarters of the European Parliament. Sandwiched between these grandly appointed workspaces were a number of elegant flats.

'I wouldn't want to live here,' sniffed Bryant, pulling his scarf tighter as he gazed up at the grand buildings.

'Why not?' asked May.

'The noise.'

'There isn't any.'

'Not now, but whenever there's a government crisis the BBC sends its outside-broadcast vans over here, and they're so full of electronic equipment that the technicians have to leave their air-conditioning units running all night, and they keep everyone awake.'

'You're a mine of useless information, do you know that? Come on.' May trotted up the stairs and rang the doorbell.

A porter admitted them into a hallway chequered with black and white diamond tiles. 'Janice texted me to say that she'd cleared the way, but leave the talking to me for once, OK?' May instructed.

The second-floor front door opened to reveal a slender, delicate-boned young woman with large, expressive eyes, her blonde hair knotted in a graceful chignon. She was wearing a black and silver T-shirt that read 'Wild Girl', very tight jeans and high heels. For a moment, Bryant assumed it was the maid. Then he remembered the photograph.

She studied the detectives in puzzlement. 'I'm sorry, I was expecting you to be more, well, Scotland Yard, you know? At home we used to have an old English television programme with a detective, always doing crossword puzzles and breaking secret codes.'

'Ah, you were expecting someone in a gabardine mackintosh with a pencil moustache and a pipe,' said Bryant. 'Possibly wearing a bowler hat. Actually, I'm very good at breaking codes and I do have a pipe.'

'No,' said Sabira, 'I just meant he was younger.' Her blue eyes widened and her hand rose to her mouth. To their surprise, she started giggling. 'Oh God, I've done it again,' she said, horrified and amused in equal measure. 'Lately I seem to have offended every English person I've spoken to.' She ushered them into a narrow painting-filled hall that led to the drawing room.

'It's quite all right,' said Bryant, revealing a crescent of bleached false teeth. 'I don't suppose there's any chance of a cup of tea?'

'Of course – the tea, always the tea!' Settling the detectives, she ran to the kitchen, her heels clicking on the oak floor, and called back over her shoulder: 'Do either of you know a cure for a hangover? I feel terrible this morning.' She sounded unrepentant.

Bryant had already found an armchair. He pointed back at the wall mirror behind him; a bolt of black velvet material had been thrown over the glass. 'She's behaving very oddly,' he whispered. 'You'd better go and see to her.'

May raised his hand. 'Leave this to me.'

He joined Sabira in the kitchen. 'Mrs Kasavian—' he began.

'Oh, call me Sabira, I can't bear being so formal.'

'I have something that might sort out your head. But I'll need—'

'Just dig around in the cupboards for anything you want. I have a tiny man with a road-drill behind my eyes.' She was racing around in the tiny space, boiling water, spilling milk, rattling cups, nearly dropping them.

May found what he needed. Filling a tumbler with milk, he added a dash of Worcester sauce, chilli sauce and black pepper, and cracked an egg into the mixture. 'You must drink it straight down without breaking the yolk,' he explained. 'The egg contains cysteine, which helps fight the free radicals in your liver.'

Sabira gave the tumbler a mischievous sidelong glance, and then grabbed it and downed it in one, slamming the empty glass into the sink. '*Gëzuar!*' she shouted. She wiped her lips with the back of her hand, flicked a loose blond curl away from her eyelashes and grinned. 'That was truly – disgusting.' She laughed again.

Bryant had settled so deeply into the armchair that he looked as if he came with the room. 'You were a long time,' he complained, helping himself to biscuits.

'We were getting rid of an annoying little man,' said Sabira, dropping on to the sofa opposite. 'I suppose you're here to tell me off.'

'I think it goes beyond that,' said May. 'I'm sure I don't need to remind you that you're married to a very high-powered official, and he has many enemies. They watch and wait for incidents like last night's, and use them against your husband's department.'

'Oh, the woman was rude, the speeches were long and boring, and I got drunk. In my country such a thing is not important. We laugh because there is so much pain in our lives. Sometimes there is nothing else to do but laugh – you understand this?'

'But your situation is very different now. Your husband is a very important man.'

'I know! Everyone keeps telling me about the important man! Don't you think I know that I have shamed him? Of course I know! But there are things you don't know.' She stabbed a painted nail at both of them in turn.

'Perhaps you'd like to tell us,' said Bryant.

'How do I know I can trust you?'

'I'm too old for games,' said Bryant. 'If you can't trust a man of my advanced years, who can you trust?'

'That is a fair point,' Sabira conceded. 'I'll try to answer your questions.'

'Was this the first time such an incident has occurred?'

'In public, yes. I've been upset for a while now.'

'What about? Why are you so upset?'

Sabira leaned forward with her head in her hands, trying to compose her thoughts. 'This is the fine English society I heard so much about. When I married Oskar, I knew things would not go easily for me, but I did not think I would be shut out so completely. Right from the start, I would walk into a room and feel it go cold. The women are the worst. At least the men fancy me. The women look at my clothes, my face and go – *poof*!' She flicked up her nose, imitating their disdain. 'They ask who are my people, where do they live, what do they own and I tell them with complete honesty. I say I was born Sabira Borkowski, and I grew up with the smell of a smelting plant in my nostrils. Oskar always said I would have to be less honest, but it's not in my nature. About a year ago I came to the . . . understanding? Is that the word? . . . that this was how it would always be from now on. I would be a social outcast.' She looked from one of her guests to the other, anxious to make them understand. 'I thought my marriage would open the doors, not slam them in my face. They think I'm stupid, common, a gold-digger, a whore. I was largely self-educated, but I am a clever woman. Since coming here I have studied English literature and art history as well as the history of London. I am better than these dried-up snobs, but perhaps not as confident.'

'Why not?' asked May, 'You seem to know your own mind.'

'I'll never be accepted by the people closest to my husband, and I really don't care. The old ones with their inherited furniture, their horses and boat races and seasons at Glyndebourne – boring, boring. Who cares? They talk about breeding, they trace their history back through the centuries but it's really just about who owns the most. Many of these people are Oskar's colleagues. It's as if they all belong to some big private club that no one else is allowed to enter. I smile and keep my mouth shut. I dress nicely and behave well and I outsmile every last one of them.'

'You didn't last night.'

'No, a demon came out of the brandy bottle.' She laughed again.

'But something else happened, didn't it? Tell me what occurred six weeks ago.'

'Did Oskar tell you that?' For the first time the pair saw a flicker of fear in her eyes.

'He says he saw a change in your behaviour from that time.'

'I don't want to talk about it. Some things are personal.'

'Then we can't help you,' said Bryant, putting down his cup. 'Ta for the biscuits.' He made to get up, but couldn't get out of the armchair.

'No, wait, please. Ask me something else.'

'All right. How do you get on with your husband's colleagues?'

'Which ones? The people in his department?'

'Yes, Edgar Lang, Stuart Almon and Charles Hereward,' said May. He saw Bryant mouthing '*Who?*' at him. 'They're all in Mr Kasavian's division, and they're also his business partners. That's correct, Sabira, isn't it?'

'Yes, they own a company together. Oskar is very careful about declaring his interests. He places great value on honesty.'

'Pegasus Holdings provides intelligence to the scientific community,' May told his partner. 'They check for security leaks and make sure data doesn't get passed to the wrong parties. It's part-funded by British and American homeland security interests.'

'There's no conflict with the ministry?' Bryant asked.

'It's the kind of public–private initiative this government loves. There are guidelines governing the running of such companies. The Home Office isn't allowed to outsource to Pegasus without holding an open tender.'

'Stuart Almon fell out with Oskar and is now just doing the books,' said Sabira. 'They are colleagues but not friends.'

'You didn't answer my question,' said Bryant. 'You must meet these people socially. It was Edgar Lang's wife you threw the drink over, wasn't it?'

'I don't care for Edgar or his wife. I find them insufferable. Charlie seems less pompous. I don't think he went to Eton. Stuart is simply invisible. I've met him dozens of times but can't even remember what he looks like.'

'And their wives?'

'They don't like me, of course. They spent their lives being groomed to marry powerful men, and along I come and steal their husbands' boss. I hate them all. But it doesn't matter what I think. I suppose they are all very clever men, and their wives – well, they do what such wives are trained to do.'

'Why did you cover up the mirror?' asked Bryant.

'There are bad things here.'

'What kind of things?'

'Devils. In my country we call them devils. I don't know what you call them.'

'You mean spirits?'

'They can be in many forms. They can be the ghosts of the dead, or people who are not what they say they are.'

'Who are these people?'

'They take different shapes,' Sabira warned. 'Some of them are Oskar's friends. Some of them can walk through the walls.'

'Walk through walls?'

'Yes, in places where they should not be.'

May felt a growing sense of frustration. Each time he thought they were getting somewhere, Sabira's answers became abstract.

'Let's see if we can cut through some of the mystery,' said Bryant impatiently. 'You covered the mirrors because you didn't want to see these spirits? Or you didn't want them to see you?'

'That spirit waits for me in the dark. He glares over my shoulder. He will kill me if he can.'

Bryant clambered to his feet and walked over to the

mirror. With a flourish, he whipped away the bolt of black cloth. Sabira gasped and turned her face aside. To May, it seemed like a piece of terrible overacting.

Bryant stepped back and examined the mirror's surface. 'See? There's nothing.'

'He's not there now.'

'Your husband says you talk to strangers in churches.'

'Certainly. Why not? I feel safe there. When I was a little girl, if I ever felt sad or frightened I would go to the mosque and the feeling would go away.'

'So why do you go to churches?'

'What, you think because I was once a practising Muslim I cannot enter a church? It is a sanctuary to me, nothing more. A mosque is where my thoughts can be heard, but a church will do almost as well.' She laughed. 'I'm glad my parents can't hear me say that.'

'But your husband also says you believe there is some kind of . . . satanic club—'

'You have met Oskar's colleagues. They all belong to clubs, Boodle's, the Devonshire, White's, but sometimes there are clubs inside of clubs and this – this' – she stamped her palms together – 'is where they plan their evil.'

'But you don't honestly mean they're *satanic*?'

'Well – perhaps this is the wrong word.'

'Do you have many friends of your own age?' asked May, changing tack. 'Anyone in whom you can confide, have a good honest conversation?'

'Only in Albania. No English. My husband does not approve of my Albanian friends because they are low class.'

'You're not wearing any jewellery,' said Bryant, cutting in. 'Do you normally?'

The question took Sabira by surprise. 'Sometimes, for formal occasions only. But not like the other women. You hang baubles from a straggly tree to distract from the meanness of its branches.'

Bryant laughed but May could see they were not going to get any further. 'I think that's all we have to ask you today,' he said, rising. 'I hope we'll meet again.'

'I hope so too,' said Sabira, smiling warmly. 'My head is feeling much better now.'

'Well, I thought she was delightful,' said May as they headed back across the square. The sky had clouded over and a strand of grey shadow was massing above the church. 'But highly strung. All the paranoid stuff, it's just in her mind. She feels cut off from her friends, she hates the circles she's forced to mix in, and when she picks a fight I imagine her husband refuses to take her side.'

'I think it's something more than that,' murmured Bryant. 'Come over here. Children don't use this square. Hardly anyone cuts across it because the back gate is kept locked, and they certainly don't deviate from the path if they do. Take a good look at the grass.'

He wandered over to a patch of green within the boundary of the church and poked at it with his walking stick. Then he looked back at the Kasavians' second-floor apartment.

'This is the area of the street she sees reflected in the mirror. That's why she keeps it covered. Look.' He directed his stick at a lamp-post on the path. 'She sees a man standing under the lamplight at night, watching her.'

He bent and examined the flattened area. 'It rained on Saturday night. Someone stood here on the wet grass.' There were several cigarette butts tightly grouped in among the crushed blades. 'I think he stood here and watched her. And she's terrified of him.'

9

PERMISSIBLE MATERIAL

Alma Sorrowbridge dragged the last of the cardboard cartons inside the front door and kicked it shut with her slippered foot. When the removal men refused to pack up Bryant's chemistry experiments and transport them, citing health-and-safety regulations, her church group had kindly undertaken the task.

Now everything from his reeking Petri dishes to his mummified squirrels and the stuffed bear inside which Kensington Police had once discovered the body of a gassed dwarf had been shifted into the new flat's spare room, in an almost perfect replica of Bryant's old study.

Alma picked up a book and checked its spine: *Intestinal Funguses Volume 3*. None of Mr Bryant's books seemed to have been arranged alphabetically, but were grouped by themes and the vagaries of his mind. She set the tome between *A User's Guide to Norwegian Sewing Machines* and *The Complete Compendium of Lice* and hoped it would eventually find its place. After setting his green leather armchair behind his stained old desk and arranging what he referred to as his 'consulting chair' before it, she satisfied herself that everything was in its rightful place, gave the shelves a final flick of her duster and sat down to await her lodger's arrival. Bryant had been sleeping in his office, and had yet to see his new home.

The move to number 17, Albion House, Harrison Street, Bloomsbury, had been delayed because the council painters had decorated the wrong flat, but as she checked each of the rooms she saw much that was to her liking. The windows were large and let in plenty of light. The oven had already been put to good use and the kitchen was filled with the smell of freshly baked bread. Best of all, her bedroom was at the far end of the corridor away from Mr Bryant, so she wouldn't be disturbed by his appalling snoring.

The impatient knock at the door suggested that he had already mislaid his keys. 'You never told me we were on the third floor,' he complained before she had even managed to open the door wide.

'There's a lift. Why didn't you take it?'

'It smells of wee.'

'Don't be ridiculous, I just bleached it.'

'Aha, then it *did* smell of wee.'

'Of course not, I just knew you would make a fuss.'

'I really didn't realize we'd be all the way up here.' Bryant sniffed and peered about himself in vague disapproval. 'There are lots of bicycles chained to the railings downstairs, and there's an Indian man in a string vest watering some kind of vegetable patch. He offered me a turnip.' He unwound his moulting green scarf and took a tentative step inside. 'Hm. Nice paintwork. Did you do that?'

'No, the council sent someone round.'

'What, they paid for it?'

'Yes, they pay for maintenance and upkeep.'

'That's a good wheeze. I don't know why we didn't think of this years ago. And you say the rent's very low?'

'You're classed as an essential worker, Mr Bryant, although I can't imagine why.'

'Where's my study?'

Alma pushed open the study door with a little pride, although pride was technically regarded as a sin by her church. 'Here we are,' she said, stepping out of his way.

Bryant walked around his desk, shifting books and ornaments by an inch here, an inch there. 'Where's my Tibetan skull?'

'Exactly where it always is,' said Alma. 'In your office at work.'

'And my Mexican Day of the Dead puppets?'

'You gave them to Mr May's sister's children the last time you went down to see them. She confiscated them from her boys after one of them cut himself on a crucifix and came up in boils.'

'Just testing. My books are out of order.'

'Well, that will give you something to do when you're home, won't it?'

'And where's my marijuana plant?'

'This is a council block. You can't keep it here any more, the police have dogs.'

'I am the police, you silly woman.'

'I sent it to your office. Honestly, I thought you'd be pleased. It took half a dozen of us to move all your stuff in and lay it out correctly. There's a nice southerly light.'

Bryant sniffed. 'I suppose it'll have to do.'

'It'll have to do,' Alma repeated. She was a large, cheerful woman predisposed to a kind smile, but right now the smile was fading to a scowl. '*It'll have to do?* You ungrateful, miserable old man! You didn't help me in any way. I had to attend the court hearings and deal with the compulsory purchase order of our old place, then search for accommodation and apply for the flat and deal with the council, a job I wouldn't wish on a dog, then move everything by myself and reinstall it here without a single thing broken, missing or out of place, and all you had to do was walk out of your old home and into this one with nothing more than the clothes on your back. I still have relatives in Antigua; I could have left you and gone home to live somewhere happy and sunny, but I stayed here. If I wasn't a good Christian I'd smack you around the head until your ears rang.'

'All right, you've made your point,' Bryant mumbled. 'It's very nice. What's for tea?'

'There's ginger cake and banana bread laid out in the kitchen, and a spiced chicken salad later.' She stood with her hands on her hips and resisted the temptation to give him a whack on the ear as he passed.

Longbright was staying late at the unit, transferring John May's interview notes. Downloading all the images she could find of Sabira Kasavian, including those in her social-networking profiles, she reassembled them by date and location. *She has a hell of a clothing allowance,* thought Longbright. *Skinny women can wear anything.* There were hardly two photographs where she was in the same outfit.

Longbright dreamed of a clothing allowance, although that would have been a slippery slope. She would have soon lavished it on impractical corsetry and 1950s sheath gowns.

The next thing she noticed was how closely Sabira stayed by her husband's side. In the few photographs that showed her seated with other people at government dinners, she appeared to be mutely listening. Her attitude was demure, as if she had been advised not to speak by her husband.

Around the end of the last week of May there was a noticeable change in the pictures. Sabira was rarely photographed without a drink in her hand, and appeared introspective, sullen, even startled. In the few Facebook shots she had put up from public events she looked flushed and nervous. *Perhaps her drinking just got out of control*, thought Longbright. *It happens.*

She spotted something else: a uniformity of style in the official press pictures. Checking the provenance of the images, she found that the same news agency had taken them, which probably meant that Sabira had been targeted by one specific photographer. She called the agency but it was shut for the night, so she checked PhotoNet's list of clients and found *Hard News* at the top of the supply list. She rang the editor, Janet Ramsey, on her mobile.

'Janice, you'd better have a damned good reason for calling me on my private number,' Janet warned. The unscrupulous editor was well known to the staff of the PCU.

'Do you have someone at PhotoNet permanently assigned to cover Sabira Kasavian?'

Janet sounded as if she was in a crowded cocktail bar. 'I wouldn't say he's permanently assigned. We have a special-interest list of public figures and their partners, as I'm sure you're well aware.'

'I guess you've been waiting for her to screw up.'

'Of course. We all want a good story, darling. And she has, hasn't she? I assume that's why you're calling me.'

'Who took the shot of her being carried out of the Guildhall last night?'

'You'd have to take that up with PhotoNet.'

'Obstruction, Janet. You know how that goes.'

'OK, it's no secret. His name is Jeff Waters. He likes taking shots of her. She's a very photogenic girl. She brightens a page on a slow news day.'

'Do you or Mr Waters get any instruction on the taking of photographs? I mean, from the Home Office. Her husband's—'

'—in the security department, I know. There are guidelines. The smudgers aren't usually allowed inside the ministerial venues, and if they are, they have to stay within specifically defined spaces. It's implicit that we don't take shots if they're tipsy, but negotiable. There are other protocols which you'd have to speak to the HO about.'

'How did you get away with last night's shot?'

'You're slipping, Janice. Take a careful look at the sequence. She's off the front step of the building. That's pavement under her shoes. Technically public space. She'd been cautioned by the police, which made her fair game. Permissible material.'

Longbright rang off. Sabira Kasavian must have noticed that she was being targeted by the same photographer every time she appeared in public. Perhaps that was part of the reason why she thought there was a conspiracy against her.

The detective sergeant pushed back from her desk and rubbed her eyes. It didn't seem like much of a case. But there was something else in Sabira's photographs – a certain look in her eyes, a certain angle of the head. Unable to put her finger on it, she closed down her screen for the night and decided to head home.

Jack Renfield stuck his head around the door. Although the room was cold, he was sweating. 'I've been whacking the punchbag upstairs. Didn't realize you were still here. I wouldn't get too close to me if I were you. D'you fancy a quick beer?'

'Not tonight, Jack,' said Longbright.

'You don't have to worry, I'm not going to jump on you or anything. At least, not without a shower. Joke.'

'No, I'm really not up for it.'

'A drink is just a drink, y'know. You look like you're going out later in that clobber anyway.'

Longbright toned down her look for work, but there was still a touch of the nightclub hostess about her. She had long ago decided that she would die in high heels. 'I'm just in a bit of a weird mood tonight.'

'Gonna start thinking you're avoiding me soon.'

'I'll take you up on the offer, I promise.' She knew how much Jack liked her, and was slowly getting used to his rough-and-ready manner, but while she thought of him as a tree or a fence-post, something strong you could lean on or shelter under, he seemed a bit too rooted to the soil. She still dreamed of achieving something beyond her work on murder investigations. 'What do you think about Kasavian's wife?' she asked.

Renfield shrugged. 'She's rich and bored and feeling neglected. Give it a few years, she'll start studying horoscopes and supporting cat charities; it's what they all do.'

'Thank you for your rich insight into the female psyche.'

'I mean it; she's going against the grain to try and assert some power, to tell him she's still got free will and that he can't take her for granted.'

'John and Arthur say she thinks she's being sort of – hunted.'

'Maybe someone has decided that she's a security risk and is keeping an eye on her movements.'

'Perhaps you're right.'

'See you in the morning, beautiful. Did I tell you that you look beautiful?'

'Jack, bugger off.'

As he headed out, Longbright turned back to her computer and reopened the photo file, to try to understand what she might have missed.

10

THE INVISIBLE CODE

Sabira Kasavian hated Fortnum & Mason.

The department store's pastel shades, its veiled windows, the airless, hushed old-world atmosphere that was meant to be charming felt merely repressive to her. She imagined wandering around its food hall with her mother, who would marvel at the jars of apricots in brandy, the tins of caviar and miniature hat-boxes of champagne truffles that cost as much as her family's weekly food bill. It seemed to be a store for people who disliked the simple pleasures of eating.

The wives of the Home Office officials met in the Fountain Restaurant every third Wednesday to boast about their holidays and luncheons and to complain about their husbands. Lately attendance seemed to have become compulsory, if only to ensure that no one was talking about you behind your back. They would have preferred to meet in the more expensive fourth-floor St James's Restaurant, but its bloated sofas and armchairs squatted around the tables like sumo wrestlers, and prevented the ladies from being seated closely together.

There were eight wives today, presided over by the

impermeable Anastasia Lang. Everyone had turned up to see if further sparks would fly. Sabira had decided to attend because she needed to show that she was unrepentant about her behaviour on Monday night. The first few minutes were made all the more awkward by the wives' determination to act as if everything was normal between them all.

There was no air-kissing and no cocktails; they were not footballers' wives. Instead they went straight to the table and ordered the lightest and most complex starters, less salads than fragile ecosystems of exotic greenery. The wine list was considered with the gravity a juror might reserve for convicting a rapist.

'That is, if you're drinking wine?' Ana asked her foe pointedly as the other wives fell silent.

'I'll have a glass of anything red and full-blooded, a large one,' said Sabira, fixing the group with an open I-dare-you-to-argue smile.

Lunch progressed through the usual roster of subjects, dinners and weekend trips, charity work and the difficulties of finding reliable nannies and gardeners, but inevitably it arrived, as it always did, at husbands.

With little to contribute, Sabira listened and drank. She heard the chatter of birds on a fence: creatures noted more for their colourful plumage than their songs. If Sabira had known that she would be required to give up her voice, she might not have been persuaded to move to England and get married.

'I don't mind that he comes to bed with his laptop,' one of the women said, 'but he puts it between us. He might as well stick an electrified fence down the middle of the duvet. I get the message; he's working. He doesn't need to point it up.'

'Perhaps you should consider having an affair, darling,' said Cathy Almon, already bored by the conversation.

'That's easy for you to say. Your husband is an accountant.

He's not required to think about anything but totting up numbers.'

'Actually he heads up the Home Office's Workforce Management Data System, and has a number of very important functions within the ministry,' Mrs Almon recited.

'Aren't there websites that arrange affairs for you?' asked the woman opposite her.

'God, you couldn't leave a data trail. It would have to be the butcher's boy or someone like that.'

'Someone uncouth and inarticulate.' They all laughed.

Ana Lang turned to Sabira. 'But, of course, we should ask you, shouldn't we? I hear you already have someone uncouth waiting in the wings, don't you?' Her smile was as venomous as ever.

'I don't know what you mean,' Sabira replied, sipping her wine.

'Oh come on, darling, your little secret is safe with us,' said Emma Hereward, looking around the table. 'Spill the beans.'

'There is only my husband.'

'It's perfectly understandable. You're *au mieux de votre forme*. You must have at least fift— ten years on the rest of us. You're among friends here. You can tell all.'

'There's nothing to tell. I would never dream of being unfaithful. I was raised a Muslim.'

'We know all about the Muslims, sweetie, they're the worst. Half of them operate double standards. I don't know why you insist on pretending that you're more virginal than anyone else. We all heard you weren't wearing any underwear at the Guildhall dinner.'

'We know who he is,' said Ana. 'We've all seen him.'

'Have I?' asked Cathy, confused.

'Yes, darling, outside the Spanish Embassy – dark-haired, rather dishy.'

'You mean . . . He's just a photographer,' said Sabira, colouring.

'Yes, but we've all seen the way he looks at you.'

'He means nothing to me.'

'So you are seeing him!' Ana was triumphant.

'No, of course not!'

'It's funny how you always seem to be wearing more make-up when he's around.'

'If I wore as much make-up as you I'd suffocate,' Sabira replied, draining her wine. She had yet to master the subtle art of English sarcasm. The joke fell flat.

'You're supposed to sip that,' said Ana, pointing to the glass. 'It's a Montrachet, not potato vodka.'

Sabira meant to slap her face, but in the process Ana's earring came loose somehow and gashed her left cheek. Ana yelped as she saw the smear of blood on her fingers. As the others came to her aid, Sabira realized that none of them would take her side. Dipping her napkin in her water glass, she tried to help but Ana slapped her hand aside. 'Get away from me,' she hissed. 'Go back to the pig farm where you were raised.'

'If I'd been raised like a pig I'd still have better manners than any of you,' she said, rising sharply. Her chair went back and the shoulder bag she had draped over it fell to the floor, spilling papers.

Ana looked down, clutching at her face. 'What the hell are you doing with my husband's private correspondence?' she said.

And that was when all hell broke loose.

'She's done it again,' said Meera Mangeshkar, running into the office. 'And this time it's on Sky News.' She reached over and opened a fresh screen on Longbright's computer. 'She attacked the same woman over lunch. Two security guards just took her away.'

'That looks like Fortnum and Mason,' said Longbright. 'I made a shoplifting arrest there once.' As she watched, a red banner rolled across the footage: 'Home Office official's wife in restaurant brawl'. 'They'll have taken her to West End Central in Savile Row.'

'No,' said Mangeshkar. 'I called them. She's been driven to the Home Office. This is serious.'

'One of us needs to be there as an independent observer,' said Longbright. 'You'd better tell John.'

May arrived in Victoria forty minutes later and found that Sabira Kasavian had been taken to a private room on the ground floor of the building opposite her husband's department. The trapezoid of grey concrete in which he found himself was far less welcoming than the airy glass atrium it faced. The staff security passes had jumped a couple of grades.

'You can't see her at the moment,' warned the scrubbed young man who came out to find him. 'This is out of your jurisdiction.'

'You're Andy Shire, aren't you?' said May, squinting at the laminated ID pinned on the security official's breast pocket. 'We met with the Police Commissioner a few months back.' May was owed a favour after he had helped Shire locate a suspected arsonist.

'I remember,' said Shire, 'but I still can't give you access.'

'I can obtain written permission from Sabira's husband if need be. Why wasn't she taken to West End Central?'

'I think you know the answer to that one, John. In matters of national security we override the police.'

'The PCU isn't part of the Met, it's a Home Office unit, so we're working on your side. I'm just trying to understand why she was brought here. I don't want to have to ask Oskar. I know he's got a lot on his plate right now.'

Shire knew that his boss would complain if his staff failed to shield him from unnecessary interruptions. 'All right,' he said. 'It appears a number of sensitive papers were found on her when she was removed from the restaurant. We don't yet know if they were taken from this building.'

'You mean she's going to be held on a spying charge?'

'We're trying to ascertain the importance of the documents at the moment. If she took them without authorization, it looks as if this is going to fall under the Terrorism Investigation Act. You interviewed her, so I assume you know she's still in contact with her Russian ex-boyfriend.'

'No, I didn't,' he admitted. 'How is she?'

'She came in here kicking and screaming, having a real panic attack. It looked like a full-blown psychotic episode to me. Her doctor gave her a sedative and she's feeling a little better now, but she keeps calling for her husband. He's still in a meeting.'

May knew that Kasavian was preparing to present the UK's case for the European border-security initiative in just over a week's time. As it was likely that his future career relied upon driving the deal through, he wouldn't take kindly to being disturbed, even to help his wife. May dreaded to think what his reaction would be when he heard the details of Sabira's latest outburst.

'Andy, I'm working with her husband to try and sort this out,' he explained. 'When do you think I can get to see her?'

'Obviously we don't want to hold her any longer than is necessary, but we need an explanation as to why she had a classified file in her handbag. This isn't half an hour in the cop shop and a slapped wrist. We're following protocol now, and you have to realize it could lead to a prosecution.'

'It could be a set-up, Andy. Maybe Kasavian has an enemy who's using his wife to get at him.'

Shire gave a cold laugh. 'Are you kidding? Oskar has

nothing but enemies. They're required to sustain the department. Just don't ask us to narrow down the suspects.'

The South Bank Centre and the Royal Festival Hall were bedecked with fluttering blue and red flags, turning the promenade into an urban seaside town. Artificial sandbanks had been placed against the embankment walls and topped with beach huts as part of an arts festival.

'What a mess,' said May, leaning on the cool stone balustrade of Waterloo Bridge.

'I don't know, I quite like it,' said Bryant, screwing the pieces of his pipe together and digging around for his tobacco pouch. 'The juxtaposition of sand and grubby old buildings. It's a bit like Margate.'

'I mean the investigation. We didn't ask any of the right questions. It's my fault. I let her charm me. It was completely inappropriate behaviour.'

'I'm glad you pointed that out. You've always been a sucker for a pretty face. She played you like a Stradivarius, matey. We should have run a thorough background check on her first. I didn't know about the ex-boyfriend or her past mental-health issues. I'm losing my touch.'

'I thought it was strange to find her in such an upbeat mood the morning after she'd made a spectacle of herself. I asked Janice to find out if she's on prescription medication, but she says apparently not. She's being assessed by her doctor later this afternoon.'

'Have you ever read Henry Mayhew?' asked Bryant, looking out over the olivine water as he drew flame into his pipe bowl. '*London Labour and the London Poor*, 1851. Fascinating stuff. Conversations with ordinary working-class Londoners. Of course we still have a tremendous class divide, but the disenfranchised weren't always outsiders. If anything, they knew more about what was really going on. Mayhew met rat-catchers and fire-jugglers, pickpockets and sewer-hunters, and the thing you notice most when you read their accounts is this: a poor man will tell you everything, and someone in society will tell you nothing.'

'That's true enough,' May agreed. 'It seems the further up you go, the less you find out.'

'There are only a handful of major landowners in London, but we never fully discover the truth about them. It's all part of the invisible code of English conduct that baffles foreigners.'

'So you think we're wasting our time even talking to Sabira?'

'No, I believe she feels she's been genuinely ill treated.'

'Isn't that just naivety on her part?' asked May.

'Maybe. When you reach a certain level it's not about how much money you have, but your background. Knightsbridge and Notting Hill may be home to wealthy New York bankers and Dubai businessmen, but even they have trouble reaching the inner circles of power. It's certainly not something that's automatically conferred upon you by marriage. The Foreign Office and the Home Office have always been run by men like Kasavian. I'll bet you his entire family moves in government circles. He may well worship the ground she walks on, but he'll never let her in.' Experience had encouraged Bryant to hold bleak views about the British class system.

'So it's his fault she's behaving like this? You don't think she just has anger-management issues?'

'Don't say "anger-management issues". What's wrong with the word "temper"? I'm not saying it's a conscious act, John. A job at the most senior level of government is a Mephisophelean deal. In return for power you surrender your peace of mind.'

'Are you sure you're not just siding with her because of your own working-class background?'

'Look, I remember standing at the end of Petticoat Lane with my father one freezing Sunday morning. All around us were men selling chickens and canaries and skinned rabbits, and my old man – who was sober for once – lifted me up so I could look into the window of Arditti's restaurant, and he said, "There'll always be a pane of glass between you and these fine gentlemen. Even when you think it's gone, it will still be there."' As if to illustrate the point, a cyclist passing behind them waved two fingers at the diplomatic vehicle that had just cut him up on the bridge.

'Oh, you've always had a chip on your shoulder about your background. You enjoy being an outsider.'

'I made the best of it because I never had a choice,' Bryant pointed out. 'Your father used to take you to the Wigmore Hall for classical concerts when you were a little boy.'

'Because he was a musician and wanted me to appreciate the finer things in life. But he never had any money. Our family was always hard up.'

'But your parents gave you ambition. Mine just wanted me to have a job. The fear of poverty is never far away from the working-class mind, and all the plasma TVs, PlayStations and iPhones are just talismans warding off that darkness. I think Sabira Borkowski married to free herself from the fear of poverty, and now she's paying the price.'

'Fine, but in her husband's eyes the situation is worsening and we're not helping. If you take her side, you'll be setting us against the government.'

'You know diplomacy has never been my strong point. I think everyone already assumes we're against the government.'

'Then what do you suggest we do next?'

'We go to this address.' Bryant held up the crumpled piece of paper he had used for wrapping up his sherbet lemons.

'I can't read your writing.'

'Neither can I, but don't worry, Janice put it in my phone. The Home Office is looking into the possibility of the spying charge. Let's talk to a more rational enemy – the woman Sabira Kasavian assaulted in Fortnum and Mason.'

11

THE GLASS

Edgar Lang and his wife lived in a redbrick Edwardian house with a wrought-iron veranda overlooking a wide, calm section of the Thames at Barnes, just past Hammersmith Bridge. Anastasia Lang was not at all pleased to find a pair of detectives standing in her porch, and reluctantly invited them in. She had covered the cut on her cheek with taped cotton wool, but her left eye was now turning a lurid shade of mauve.

'I'm fully prepared to press charges; I won't let anybody talk me out of that,' she said with ice in her voice. 'Not for the physical attack, but for stealing private documents.' She waved them into an immense glass-roofed kitchen that had been added to the rear of the already substantial house. 'I can't offer you anything, I've sent the maid home.'

'Nice gaff,' said Bryant, walking to the wall-sized window with his hands in his pockets. 'A very popular look, this. Classic at the front, modern at the back. Architectural hypocrisy. Should I call you Lady Anastasia?'

'Mrs Lang will do.'

'We're not here to ask you about the argument. What

do you think Mrs Kasavian was doing with your husband's property in her handbag?'

'She must have taken it from the Pegasus offices in Great Portland Street. Edgar never keeps documents anywhere else.'

'You have children?'

'No, we have a dog.'

'Do any of the directors have kids?'

'Yes, Cathy and Emma do. I can't see what that has to do with—'

'Do you think Sabira was being paid to spy?' It was Bryant's technique to keep his witness wrong-footed.

'She's hardly short of money, the number of new outfits she wears. No, I don't think she was being paid to spy. I think she did it because she's jealous. She wishes she had my husband. Do you know how they met? My husband and Oskar were in a wine bar in the city. She started talking to Edgar first and, being a married man, he turned her down, so she went after Oskar. She was on a mission to find a successful man, operating with a fairly limited arsenal and a tight time limit on her sex appeal. Eastern European girls blow up like zeppelins when they hit thirty. Edgar said no, so she switched her attention to Oskar, who at that point had been divorced for over three years and was vulnerable to a pretty face.'

May could not imagine Kasavian ever being vulnerable. 'You're saying Mr Lang turned her down over four years ago, so she stole papers from his office? Doesn't that seem a little pointless to you?'

'I don't understand the point of anything she does,' said Ana Lang, touching her face lightly as if checking that nothing had shifted.

'How well do you and Mrs Kasavian know each other?'

'I meet her at social events, but we have nothing to say to one another. Her every utterance is a mystery to me. All I know is that she drinks too much and has an uncouth personality. You'll take her side, of course. You're a policeman. But what you must understand is that women of our social standing remain by our men, no matter how wrong we think they might be. It's part of the deal, it's what we signed up for.'

A tinkle of metal made them both turn around. Bryant had pulled the head off a small but rather valuable sculpture. Mrs Lang ploughed on with determination. 'When I want to know what a British politician really thinks, I ignore what he's saying and talk to his wife, Mr May. Neil Hamilton and Jeffrey Archer both had strong women at their sides to support and further their careers.'

'Archer went to jail, Mrs Lang, and Hamilton got into a fair bit of trouble himself.'

'I was merely illustrating the point. Sabira could have had it all, and now she'll have nothing. No one will have anything to do with her after this. I suppose you know she was having an affair?'

'I understand you accused her of having one, yes,' said May. 'We called your fellow lunch guests.'

'I did not "accuse" her, I merely stated what everyone already knew. It's the Australian photographer who always takes her pictures.'

'We know about him.'

Ana Lang was surprised. 'How?'

'He has an exclusive deal with a magazine called *Hard News*, Mrs Lang. He's assigned to follow Mrs Kasavian to social events. Do you have evidence that they're having an affair?'

'You only have to look at the way he photographs her.'

'So no actual proof.'

May heard the front door open and shut. A broad-bodied man in his late forties came in and set down his briefcase. With his slicked grey hair and pinstriped blue suit, he had the appearance of a stockbroker or an auction-house expert. 'Who's this?' he asked. 'What on earth's that on your face?'

'They're detectives,' Edgar Lang's wife explained. 'I was attacked today, but of course I couldn't get hold of you.'

'You should have pulled me out of my meeting. Why are they here?'

Bryant resented being discussed as if he was invisible. 'We needed to ask your wife a few questions,' he said.

'Not without a lawyer present,' warned Lang. 'I think you'd better leave now. I'm a very good friend of the Commissioner, and he'll have something to say about this.'

'No, it was better to make you leave right then,' said May as they walked along the footpath that ran beside the river. Ahead of them, a pair of swans swooped down to the water and folded their wings, looking like funfair love boats. 'I could see you were about to open your mouth. It seems Mr Lang's first concern was the impropriety of our presence and not his wife's health. You realize this is impossible, don't you? They've built a wall around themselves. How are we supposed to find out anything? Anyway, how can you help a woman who behaves so irrationally? It's as if Sabira deliberately set out to wreck her life. Why would she risk throwing her marriage away by stealing classified documents?'

'Oh, they weren't classified,' said Bryant cheerily. 'There's nothing in the paperwork of any value whatsoever.'

'What are you talking about? Why is the Home Office holding her if she's not suspected of spying?'

'Well, they don't yet know that there's nothing of value in the papers.'

'But you do.'

'Yes.'

'Would you care to explain how you know?'

'I made a few inquiries.'

'Then why didn't you tell me?'

'I wanted Mrs Lang to think they were important for now. If we admitted they weren't, she wouldn't have talked to us at all. We need to find out what happened to Sabira six weeks ago.' He patted his partner's broad back. 'Don't worry: I'll fill you in as we go along. But we have to act fast. I think something very bad is about to happen.'

'If you're trying to convince me that you're clairvoyant, it won't work,' said May. 'I've shared an office with you for most of my adult life. I know how you think.'

'Well, I wish you'd tell me,' said Bryant. 'I have absolutely no idea how my brain operates.'

'I think it's a sort of intelligent threshing machine. It chews up bits of information and spits them back out in a different order. They should pickle it when you die.'

'I'd quite like to end up in a glass jar in the Wellcome Institute.'

'Yes, I thought you would.' With a despairing sigh, May led his partner back to Barnes Bridge Station.

At the unit, Longbright had summoned all members of staff to the common room. Raymond Land sulked in his office for a few minutes, upset that he hadn't been in charge of calling the meeting, but then, worried that nobody would miss him, he reluctantly attended.

'I just had a call from the Home Office,' said Longbright. 'There was nothing sensitive in the paperwork. It was just a folder of Edgar Lang's taxi receipts and dinner expenses. Sabira says she doesn't know how the file got into her bag. She's agreed to be placed in a private clinic. Based on her past history her doctor feels she's at risk, and is admitting her tonight.'

'Can we get our mitts on her medical records?' asked Renfield.

'No, they're off limits.'

'I wonder if this is for her health or because she's become a major embarrassment,' said May. 'Which clinic?'

Longbright checked her notes. 'Somewhere in Hampstead. It's called the Cedar Tree Centre, just off Fitzjohn's Avenue. She's not allowed any visitors tonight.'

'She's at risk, but not from herself,' said Bryant. 'Bring her smudger in. I want to meet him.'

Jeff Waters arrived in the doorway of Bryant and May's office a little over an hour later. The handsome Australian was in his late thirties, unshaven and long-haired, still slung with cameras. He plucked at his lapel and grinned. 'I don't need the photographers' jacket now that we're fully digital, but I can't bring myself to give it up. I'm on my way to work.'

'I suppose you keep late hours,' said Bryant.

'It's mostly night assignments, and when I've not got a schedule I make sure I'm outside the Ivy by eleven p.m. Then I do the rounds of the clubs to see if anything's going on.'

'How do you know who's going to be there?'

'There's a network of tip-offs. We bung some of the maître d's.'

'You've got some misdemeanours on your record, I see.'

'Small stuff. In this job it happens.'

'Grab a seat, Jeff,' said May. 'We need to know just how well you know Sabira Kasavian.'

'Janet Ramsey appointed me to tag her. I cover about fifteen women for PhotoNet. Sabira photographs like a dream. You get to know your clients pretty quickly.'

'Do they want to know you?'

'Most of them act like they don't care about having their photos taken, but they love it. I can always tell the ones who want to get their faces in the press. They find excuses to slow down when they walk past us, stop and talk to their partners, turn and laugh about nothing. If a woman adjusts her dress as she passes you, you know she wants her shots done. But they never want to speak to you. I'm careful, I only have one chance to get the right shot, so with some of them I stick to "Over here, love, turn to your left" – that sort of thing. Sabira Kasavian isn't like that. She's always happy to be photographed. She loves the camera; the camera loves her.'

Bryant watched the photographer's hands. He was glib, fast, hard as nails, but there was something else. He was smoothly moving the conversation on, trying to control it.

'So the two of you never get to talk?'

'No, not at all, you can't when you've been railed into a ten-by-eight with a dozen other paps, security all around, and you've got maybe ten seconds for each target. Over-step your mark and you risk being blacklisted.'

'Have you ever spoken to Mrs Kasavian privately?'

'No, not so much as a single word. I'm sure she'd be fine with it if I did, though. She seems honest and friendly, a bit more fun, not like the others.'

'But you do form some kind of relationship with your subject?'

'What do you mean?'

'You fancy her – she appeals to you.'

'No, nothing like that.' Waters laughed. 'We're not in the same class, are we? Christ, I used to push a vegetable barrow in Melbourne. I mean, I know her background but even so . . . there might as well be bullet-proof glass between us. It's all over this city, the glass.'

'Is she always with her husband?'

'No. One evening outside a conference centre in Canary Wharf he had to take the driver and it took fifteen minutes to find her another car. She was standing there with a friend – they were speaking to each other in Albanian. Not many people speak it, so I guess she's glad when she finds someone who does.'

'Do you know any Albanian?'

'I know a little of every language. In this job you have to. I got talking to the friend while they waited. Sabira was standing off to one side, a bit aloof. Then I realized she was shy. It was raining hard, so I lent the pair of them my umbrella. The friend told me how much Sabira hated going to the embassy dinners. She said she'd rather go and eat pizza in a café.' He smiled, but it faded with the memory. 'I thought Sabira was very – nice.' Both of the detectives could see that it wasn't what he had been about to say.

'Do you remember the name of the friend?' Bryant asked.

'I would have written it down. You always do; it's a habit.' He pulled out a BlackBerry and thumbed through his notes. 'Edona. I guessed the spelling. Didn't get the last name – probably too many consonants for me to handle. I took a picture of her for fun. I wasn't going to use the shot.'

'What did you talk about?'

'Nothing much, we were just filling in the time.' He suddenly rose to his feet. 'Is that it? Can I go now?'

'One last thing,' said May. 'When was this?'

Waters checked his BlackBerry again. 'I made the note in early June. So, nearly a month ago.'

'Mr Waters,' said Bryant sharply, 'did you really not talk to Sabira Kasavian? The woman your camera loved so much?'

'I told you, no.' He did not catch Bryant's eye. As he left, he passed Detective Constable Fraternity DuCaine in the passageway. DuCaine looked back as he entered.

'Who was that?'

'A photographer who knows Sabira,' said May. 'Why?'

'So that's Waters. I've seen him around. He's always outside West End clubs, chatting up women. A real eye for the ladies. Did you find a connection?'

May went to the window and watched Waters crossing

the street. 'No,' he said, puzzled. 'According to him, he stood next to her for fifteen minutes and they never exchanged a single word. He's stretching the truth, but I have no idea why.'

12

THE ENGLISH HEART

Hampstead had always prided itself on being a cut above other London areas. The homes of Byron, Dickens, Keats and Florence Nightingale had now been usurped by financiers who had turned the village into one of the most expensive places in the world. Its street names were printed in elegant reverse text, white lettering out of black tiles, its avenues were sumptuously leafy, its houses gabled and slightly suburban, set back from the sight of vulgar vehicles. It had lakes and the largest open heathland in London, and looked down on everyone else from a windswept peak where the city temperatures cooled, and on a summer day like this you could almost believe you were deep in countryside until you saw the high-street prices.

May wondered aloud who lived there, and Bryant was delighted to enlighten him; in 1951, he explained, the Church Commissioners, who owned most of Hampstead, were advised by their estate agent to sell everything off as prices were about to plunge. They sold, prices soared, and Hampstead Man, the pipe-smoking chap who wrote books that didn't sell and supported nuclear disarmament, moved down the hill to shabbier postcodes, leaving Hampstead to rapacious property developers. Even the politicians moved to cheaper areas.

The Cedar Tree Clinic was founded in a house formerly owned by an English composer who had chosen the spot for its tranquillity. In 1937 it was bought by a wealthy American benefactress who came to paint and stayed to heal. The clinic's gardens sloped to manicured woodlands and had provided a sheltered spot for officers recuperating from a devastating war. Now the main house was used by burned-out musicians and detoxing media executives, but the east wing was for more troubled souls, those with recurring addictions and nervous disorders.

'Put that out,' John May instructed. 'The last thing they want is the smell of tobacco drifting over their lawns.'

'Lightweights.' Bryant knocked out his pipe, unscrewed the stem and without checking for embers dropped the bowl into his jacket pocket. He stamped his feet on the porch steps. 'Bloody English summer, my feet are frozen.'

'You should try wearing thicker socks.'

'I shouldn't have to. I've already got the linings from my carpet slippers tucked inside my boots. It's not like this on the continent. Everyone's in flip-flops and Bermuda shorts. They smile at each other and eat vegetables that still have earth on them. They're happy. If I lived over there I'd have retired by now. I'd be living a life of luxury and deceit. Instead I'm stuck here with a pension that wouldn't buy a beach hut. They have sunlight. What do we have? Sublight. D'you know, I accidentally caught sight of myself when I shaved this morning. I looked like a very old apple. Slightly green with a wrinkled skin, probably full of worms. I need a suntan.'

'If you don't stop complaining, I'll leave you here.'

'You couldn't afford to.'

'If you dressed smarter, you'd feel better about yourself.'

'You can't pour a pint of bitter into a cocktail glass.'

May sighed. 'I don't know where you get these sayings.'

The door was opened by one of the senior nurses, who introduced herself as Amelia Medway. She led them through to the wing where Sabira Kasavian had been settled. 'The main thing is to ensure that you don't upset her,' she said as they passed the empty dining room. 'We try to avoid medication wherever possible, preferring to encourage our guests to participate in holistic programmes and natural therapies.'

'Is she allowed out?' asked Bryant.

'Sabira isn't a prisoner, Mr Bryant. She is free to come and go as she pleases, although she is required to keep us informed of her whereabouts at all times.'

'What's to stop her from doing a runner?'

'Her access to money is limited. Her credit cards have been put away, and she is provided with a daily cash allowance. There's a wardrobe restriction, and for bigger trips she's required to wear a bracelet that allows us to track her movements, although we find that's rarely necessary. Most of our guests only go as far as the heath or the high street. They like to return in time for meals and special events; there's always something going on. Many of our guests soon find they enjoy having a structure imposed on them, and try their best to maintain the regime, but it can prove challenging for some.'

'You mean they either become institutionalized or they get stroppy with the staff. I suppose getting them into a load of smells-and-bells therapy is better than doping them up.'

May could see the nurse was trying not to be defensive. Considering his partner believed in all manner of bizarre alternative practices, it surprised May that he was so sceptical about private clinics, but consistency had never been Bryant's strong point.

Inside, the atmosphere was calm, relaxed and low-key.

There were no locks or bars, just carers and guests and a full schedule of daily activities with which to occupy fretful minds. Banning definitions like 'nurses' and 'patients' was meant to speed the journey to recovered health.

Sabira Kasavian was sitting in the morning room wearing a thick white towelling robe. Without make-up she looked diminished and ghostlike. 'They're bringing you some herbal tea whether you want it or not,' she whispered. 'I'm going to be a model patient.' She raised her hand to her mouth. 'Sorry, I said the forbidden word – I keep forgetting we are all *guests* here, except guests don't usually have to pay their hosts five hundred pounds a day.'

'How are you feeling?' asked May.

'Fine for a woman who has just been locked away in a madhouse by her husband.'

'It's hardly a madhouse. It was your doctor who recommended your admission, and you agreed that it was a good idea.'

'I met some of the inmates this morning and they are pretty crazy. There's a TV talent-show winner who has to be kept out of the kitchens because he knows how to make drugs from household items, and some PR woman who had a screaming fit this morning because they took her mobile away from her. But it's fine, I am safe here.'

'Safe from what?'

She looked blankly at them. It was as if she disconnected from their questions when faced with anything uncomfortable. 'My mind. Being poisoned. I think they are putting spells on me. It's some form of witchcraft, but I don't know how it works. Do you think there are books on witchcraft in the local library? Why did you come to see me?'

'Why do you think Mrs Lang accused you of having an affair? Is that why you attacked her?'

'How is she – have you seen her? Does she have a scar? I hope so. I think they're deliberately trying to manufacture a scandal to humiliate me and drive me back home.'

'Who are *they*?'

'The establishment. The members of the club.'

'Including your husband?'

'Of course. He has no choice but to go along with the others. If he refuses, his initiative will fail and he won't get his promotion. Without a vote of confidence, his career will be over.'

May leaned forward. 'Mrs Kasavian . . . Sabira, this makes no sense. I understand the anger you feel at being excluded, but why would anyone go to so much trouble?'

'Because I know what's going on.'

'And what is going on?'

'I can't tell you that. At least, not yet, until I have more proof.'

May tried another tack. 'Why did you steal papers from Edgar Lang's office?'

'I didn't steal them. They were important documents.'

'You didn't steal them? Then how do you know they were important?'

A cloud of doubt crossed Sabira's features. 'I'm – I get confused. They were taxi receipts.'

'I don't think you're telling us the whole truth,' said May.

'It is true. I saw them for myself. I am the victim. You cannot make this sound as if it is my fault!'

'What do you mean, your fault?'

'The press says I am trying to damage my husband's reputation, but they don't know what he's like.'

'Has he ever hurt you, threatened you, subjected you to mental cruelty?' May asked.

'No, of course not. He has always been wonderful to me, but don't you see, that is all part of the trick.'

Bryant could tell that his partner was losing patience, and made a rare attempt to be the voice of reason. 'Sabira, we know there's no truth in the rumour that you were having an affair with your photographer. But I think you spoke to him. He knows a little of your language. I'd like to know what you talked about.'

She looked away, watching a crow hopping about on the lawn. 'I asked him if he had ever had to photograph a dead body. I asked him to do something for me. Something it was impossible for me to do.'

'And what was that?'

'I asked him to explain what was in the English heart. He didn't know the answer, and that is why I must die.'

'If it wasn't for the fact that the man in charge of our future personally requested the investigation, I'd have walked away from this by now,' said May when they reached Hampstead Tube station. 'I told you what I thought of her at the outset. She's one of those women.'

'One of what women?'

'You know, the ones who always need an audience, an attention-seeker.'

'She's overwrought and imaginative, I agree. But what if she really is being victimized?'

'Over *what*, Arthur? We haven't had one word from her that's made a lick of sense.'

'That's because she's frightened. I think she honestly believes someone is going to try and harm her. And we know someone's been watching her. Renfield talked to the neighbours in Smith Square. A couple of dog-walkers remember seeing someone standing on the grass at night, but they didn't get a good look at him. She's safer where she is, but I don't want her leaving the place and wandering about. I'll have a word with the nurse.'

'If she's that scared, why won't she tell us anything that will help catch him?'

'Because she doesn't trust us. She thinks we're part of the problem, and for all I know she may be right.'

'What do you mean?'

'If Sabira's life is being threatened, it can only be by someone she knows through her husband. She has no friends of her own here except this Albanian woman Waters saw her with. We should try and talk to her.'

'If she can't tell us what's wrong I don't see how anyone else will be able to help.'

'That's the trouble. We answer to the same people Sabira hates. It's a closed circle. I think Jeff Waters was given information that holds the key to her behaviour. If she won't tell us, we have to get the truth from him.'

'What makes you think he has any inkling of what she's talking about?'

'He lied to us, didn't he?' said Bryant.

13

IN CORAM'S FIELDS

The motorcycle rider had been following the photographer all day. Now he watched as Waters slipped between the tables outside Carluccio's, and wondered what the hell he was doing. He seemed to be searching for someone, but it was hard to tell from here. The enclosed concrete rectangle of Brunswick Square was crowded with shoppers and diners enjoying the sunny afternoon.

He saw Waters clearly for a moment, and spotted the object of his search: the small girl in a yellow T-shirt and jeans. He only glanced away briefly, but when he looked back the pair had disappeared.

Pushing his way through the pedestrians, he headed into the walkway beside the Renoir Cinema. It was the only exit on the east side. Waters was a hundred yards ahead, hurrying down the street with the girl in his arms, and looking as guilty as hell. The rider followed at a safe distance, trying to see where they were going, then realized they were heading for Coram's Fields. Waters would be able to enter, but he would be turned away.

Smart idea, he thought, *but it's not going to keep you alive.*

'Christ, I'm going to be arrested for abducting a bloody minor,' Jeff Waters muttered as he looked down at the little girl. He already had quite a few arrests on his police file. Adding *Suspected Paedophile* to the list wasn't going to do him any favours when it came to getting visas for overseas assignments.

'It was just a game,' said Lucy. 'We didn't mean to hurt anyone.'

'I know you didn't, darling, but something went wrong and we have to see if we can put it right. This is the best place to talk, trust me.'

Setting her down, he led her across the busy road to the entrance of the playground. Coram's Fields was a unique seven-acre park for children in the centre of the city, constructed on the site of the old Foundling Hospital. At the gates a sign read: 'NO ADULTS UNLESS ACCOMPANIED BY A CHILD'. Right now, it was the safest place for Waters to be.

He had followed the child all morning, starting at her school in Belsize Park, then to her father's office off Fleet Street, heading across to Hamley's toy store in Regent Street, where Lucy was bought a talking pink poodle ('Comes with Built-in Wi-Fi!' said the box) and finally to the crowded farmers' market that filled the central courtyard of the Brunswick Centre in Bloomsbury.

He had used his Nikon zoom to track the girl in the yellow T-shirt and jeans as she trotted behind her father with the toy poodle under her arm. His name was Mansfield; Waters called the office just after the pair had left and asked a few questions; he was good at teasing out answers from suspicious receptionists. Apparently Mansfield was taking care of his daughter today because she had an appointment at Moorfields Eye Hospital to have her new glasses fitted, which meant that either the wife worked or he was divorced.

Waters had kept the Nikon trained on the yellow shirt. It darted behind the Portuguese food stall, reappeared briefly by a woman selling iced cupcakes, and then slipped between a fence of dark blurs that proved to be a line of Chinese tourists taking photographs of London litter bins.

The crowd was denser on this side of the square because it was directly in the sun, and people were sitting on the edge of the fountains that never seemed to be working, eating sausage baguettes, waiting for friends, talking on their phones.

Lucy's father had released her hand and was walking over to a bookstall, where he turned his back for a moment to examine a hefty volume of New York City photographs.

It was long enough. Waters had lifted the girl off her feet and made a run for it before she could cry out, slipping away down the steps and beneath the raised concrete platform of the precinct. To his surprise Lucy didn't cry out. 'You again,' was all she said. Waters's great advantage was his face, handsome, wide and friendly. Girls turned to him and smiled even before they realized that he was a photographer.

He had been following Lucy for days, and they had reached the point of smiling and tentatively waving, but Waters was under no illusions – Lucy was likely to turn and shriek in the way that only little girls could if he put a foot wrong now.

He figured he had less than ten minutes before her father thought of searching the park and all hell broke loose.

'Why are we here?' Lucy asked, clutching her pink poodle. 'You said you knew about the witch.'

'I do, Lucy, I just wanted to ask you something very quickly before we go back to your daddy.' He crouched beside her, reducing his height to something more manageable and safe. 'About the lady who went into the church. The one on Saturday.'

'It wasn't my fault,' Lucy warned him. 'Tom agreed with me. His father works with my father. I'm nine months and seven days older than him, and I know all the rules of the game because my brothers used to play it, but they got bored with it and gave it to me.'

'What game? Is this the game you were playing on Saturday morning?' Waters checked over his shoulder, watching the plaza steps, expecting to see Mansfield appear on them at any minute.

'Yes,' said Lucy loudly and clearly in her best explaining voice, which you had to do because adults were slow. 'It's called *Witch Hunter* and you have to find the witches and kill them. And me and Tom looked for a witch and found the lady who was one, and we put a curse on her to make her die.'

'How did you put a curse on her, Lucy?' Waters's sightline remained fixed on the steps, watching for a distraught father.

'You have to make her pass a test,' said Lucy. 'The man showed me how to do it.'

'What man?'

'He works with my dad but I don't know his name. He brings the food.'

'What do you mean, he brings food?'

'You know, pizzas. He has a big bike.'

'What exactly happened?' asked Waters. 'I mean from when you saw the lady?'

'She was sitting eating a sandwich and she was reading a book about how to eat babies.'

There was a sudden movement across the road. Mansfield's Ray-Bans flared in the sunlight. He was running down the steps, taking them in pairs, watching the traffic, seeing when he could cross the road to the park.

'Shit,' said Waters under his breath, rising.

'You mustn't say that,' said Lucy.

'I have to go. You mustn't mention this to anyone, do you understand? It has to be our secret. Like your game.'

Lucy remembered the rules of *Witch Hunter* and smiled. 'All right.'

He turned and checked the park for cover. It was a bright, clear afternoon, but there was deep green shade beneath the immense plane trees and oaks that lined the path to the petting zoo.

'That's him,' said Lucy softly, 'he's here.' But Waters didn't hear her.

A young man in a black motorcycle jacket and black jeans was shifting out of the shadows, moving swiftly towards Waters. Judging by the bulk of his chest he'd either been in jail or spent his life on a bench press.

Waters was still checking Mansfield's progress across the road. He stepped back from the little girl and waved her away. 'Lucy, I can see your daddy, he's coming to get you right now, and it's very important you don't say anything about us being—'

He didn't finish the sentence.

The motorcycle rider was up close and turning Waters to him. A slender blade found an entry point between the photographer's ribs, slicing directly into the chambers of his heart. Waters tried to finish his warning but hot coppery blood filled his throat and he was frightened of spitting it on to her, so he dropped as quietly as he could to his knees, trying not to fall on his injured side. It was important to him that the girl didn't see there was something wrong. She was safe; she had her back to him now.

The knife was smoothly extracted and reinserted. The searing heat appeared further up, then again to the right, and all he could think was *She didn't see, she got away*, because he could hear the girl running back to her father, back into the sunlight where she belonged.

14

CONNECTIONS

The mock-Gothic windows of the St Pancras Mortuary and Coroner's Office peered out on to a Victorian graveyard gilded with scrolled gates. The Regent's Canal wound around it, sparkling in the milky evening sunlight. Beyond was an Edwardian crescent of terraced houses, a third-century church, giant blue cement tanks preparing to create a new town square and several six-floor blocks of council flats, crammed into a messy collage so typical of the capital city that Londoners never noticed its strangeness.

Inside the coroner's office, Giles Kershaw was thinking about knife wounds. 'There's a mandatory minimum four-month prison sentence for sixteen- and seventeen-year-olds found guilty of aggravated knife offences now,' he said, checking over his new arrival. 'Every Tory government returns to the old "lock 'em up" policy eventually, just as every Labour one tries to introduce a more liberal penal attitude to stabilize the prison population.'

'Will either of them stop kids from tooling up?' asked John May.

'Unlikely. The anti-knife campaigns are endless and well meaning but they don't make it any easier for a kid to walk down a street at night, staying out because his mum's got a new boyfriend.'

On the steel tray before Kershaw was the stripped body of Jeffrey Martin Waters, a grey plastic mesh sheet arranged above his hips. He was lying face down, so his wounds were not visible from this side. It looked as if he was waiting to have a massage.

'We were about to bring Waters back in,' said May. 'We interviewed him yesterday but didn't get very far. He knew more than he was willing to tell us.'

'So he knew his attacker?'

'It looks that way.'

'Before we get into the question of how you managed to pre-empt a murder victim, John, let me quickly outline what happened,' said Kershaw. 'I can turn him over – do you want to see?'

'Not unless the killer signed his work,' said May.

'Good. He's a big lad and I put my back out last week playing squash. There are five narrow but very deep puncture wounds over and around the heart, no serrations on the blade. At first I thought the weapon had penetrated so deeply because it had been incredibly well sharpened, but then I found traces of oil inside the wound. The blade had been sprayed with WD40 and all the cuts were pointing towards the heart itself. Waters was wearing a baggy T-shirt and a jacket with lots of pockets, so stabbing should have been a hit-and-miss affair. This was someone attacking with a decent knowledge of anatomy and an intent to kill, not wound. That's pretty rare. Knives are kept to be brandished, to ward off, to mark territory. This one was . . . well, you remember that business with Mr Fox and his sharpened skewer? I don't suppose he's out on the streets again.'

'He's safely behind bars,' said May.

'OK, but it's someone like that. I'd say he set out to remove your witness and did a very neat job. How did Waters get into Coram's Fields? You're not allowed inside the perimeter without a child.'

'He had a little girl with him,' said May. 'Coram's Fields has several CCTVs around the outer railing. Unfortunately, Mr Waters was standing behind a very large plane tree when he was stabbed.'

'Then how did his attacker get in?'

'He vaulted the fence covered by some bushes – the council had been due to cut them back – and made his way straight towards Waters. He knew his target.'

'Waters was with – who, his daughter?' Kershaw's interest always extended beyond the bodies on his table.

'We don't know. We've got a muddy shot of a girl running away, maybe nine or ten years old, that's all. We've only just started piecing together the witness reports. She ran off moments before he was attacked and carried on until she reached the far side of the park railing. It looks like Waters warned her away. We've got a brief shot of his arms outstretched, then we lose him.'

'Did you get a description of the killer?'

'It's useless,' May said. 'Black motorcycle helmet, black leathers, boots, broad build, young and obviously fit. Thanks to the helmet we don't even know if he was Caucasian. No branding on the jacket, which is unusual. Probably removed it to avoid identification, which also suggests intent.'

'Well, I think our crime scene manager is probably going to disappoint you on particle evidence, assuming we can afford any proper tests. There isn't much to go on. I get the feeling our man stuck his right arm out, gripped with the left, hauling Waters into close contact by keeping him off balance. You can't see anything on the CCTV?'

'Not a sausage. My guess is he knew where the cameras

were positioned, and avoided walking on the grass, so there are no prints to speak of.'

'About the intent to kill. I'd say Waters was targeted and dropped as neatly as a bull at a corrida. There's a fresh abrasion on the left knee.' Kershaw picked up his telescopic indicator – a bequest from his predecessor – and flicked it at the corpse's leg. The Victorian device served no real purpose but was a trade accessory, like a journalist's pencil. 'It's a textbook army hit.'

'So you think it was a professional job?'

'It seems the likeliest scenario. Do you want to tell me what this is all about? You interviewed a – what, suspect, witness? – who was then murdered. And you don't have the case, because it came to me direct. So what on earth's going on?'

'You remember Oskar Kasavian?'

'Of course. Is he still trying to close the unit down?'

'He commissioned us to investigate his wife. She's been suffering from behavioural problems and has turned into a security risk. She was being shadowed by Waters here, who was commissioned to take photos of her, but we think she told him something that got him killed. Oh, and his killer matches the description of the man who mugged Arthur's biographer, Anna Marquand.'

'Well, that's as clear as creosote,' said Giles, covering the body. 'It's a bit of a tenuous link, isn't it?'

'Not at all. There were two murders in London today, in a city of eight million people. One was the victim of a gang stabbing on a Tower Hamlets estate and the other was Waters, who appeared in my office just a day before he was killed. I'd say we have a link, wouldn't you?'

'Then where does the kid fit in?'

'No idea yet. There's no reason to assume there's a connection between her and Mrs Kasavian, although I'm sure Arthur is looking hard. I've sent Dan Banbury over to Waters's apartment to retrace his final day on earth. He enjoys jobs like that. I have to say, it feels like we're pulling on threads that may unravel something big.'

'Like what?' asked Giles.

'I don't know,' May admitted gloomily. 'Something that'll probably come down and crush us all.'

Dan Banbury was the only member of the unit who still knocked on the door of Bryant and May's shared office, or at least he knocked on the lintel, as the door had been removed by the decorators because it was sticking and had yet to be put back because they had lost the screws between the floorboards. 'I've got Waters's movements for the full day,' he said. 'Do you want to come and see?'

'Why can't you just print them out?' asked Bryant, looking over the top of his spectacles. He was completely surrounded by loose pages with scrawled-in margins, Prospero marooned on his island of books.

'Because nobody uses paper any more.'

'Well, I do.'

'You mean you want me to create a document and print it so that you can read it, screw it into a ball and then throw it away? That's very old-fashioned and wasteful.'

'So am I. Just do it.'

Banbury sighed and returned a minute later, setting the sheet on Bryant's desk. 'Waters wasn't driving to assignments, because they're usually all in the centre of town and he hates paying the congestion charge. He took the Tube, and touched in and out with his Oyster card. So we get eight fifty a.m. out of Belsize Park, then just after one p.m. back in at Belsize Park, touching out at Blackfriars. The cameras picked him up in Fleet Street, Salisbury Court, then St Bride's Church—'

'He went into St Bride's?'

'He was inside for about five minutes. Cameras show

him waiting around but you can't see much in the courtyard because of the trees. It's a problem at this time of the
year, most of them should have been trimmed back but
the weather—'

'Get on with it.'

'OK, the next one is Piccadilly Circus, then back in at
Oxford Circus, suggesting he walked up Regent Street,
but I can't get access to those cameras at the moment
because the Met's using the footage to find a gang of
Ukrainian shoplifters, then out at Russell Square at four
fifteen p.m., which points him in the direction of Coram's
Fields. I think he spent the day looking for the kid or
waiting to get her alone, and it might not have been the
first time he did that. His Tube card has similar times and
destinations on other days.'

'Hm. I imagine he was so busy watching out for the girl
that he didn't notice someone was following him. Amy
O'Connor died in St Bride's Church. This has to be connected.'

'You don't know that.'

'O'Connor spoke to two small children before she
entered the church. What if the little girl Waters met
was one of the kids who were there that day? I want the
O'Connor case.' He picked up the phone and called Oskar
Kasavian.

'Are you absolutely sure this is relevant to my wife's
situation?' Kasavian asked after Bryant had put forward
his argument.

'I think we're going to find there's a clear chain of events
linking her to the earlier death,' Bryant replied.

'You mean you don't have evidence yet.'

'Not quite. Does your wife like children?'

'Not especially. Why?'

'I can't see how she would get to know a little girl. One
doesn't tend to come across them in central London.'

'I'm sorry, you're losing me. What little girl?'

'Amy O'Connor spoke to two children shortly before she died.'

'What has that got to do with her death? Were they related to her? Did she tell them she wasn't feeling well or something?'

'No, I don't think she knew them. But nobody else came near her, and something made her die. Healthy young women don't just drop dead. That's why I need the case. I want to look into O'Connor's background, her medical records, her employment history, and I need Home Office sanction to do it.'

'All right,' said Kasavian finally. 'I'll do what I can. The City of London Police are bound to kick up a fuss, but I'll see if we can get things moving from this end. If you honestly reckon it will help Sabira I'll do whatever it takes, but you may have to leave it with me for a few days. Things don't move as quickly here as they do in your world.'

Bryant ended the call. 'It looks like he's going to get us O'Connor,' he said. 'Dan, how soon can you tackle Waters's apartment?'

'I'll go right after this.'

'Good. I need to find out what else connects Waters to Sabira Kasavian. Check his computer and his cameras, look for the pictures he was taking on the night they met. I'd like to know if she sent him to St Bride's Church.'

'Wouldn't it be easier just to ask her?' said Dan.

'I'd get clearer answers from the cat.' Bryant tapped his false teeth with a chewed biro, thinking. 'Speaking of which, ask Meera to see if Crippen has given birth yet. I don't want to be treading on kittens. You know, it would be better if we can find a link, because it doesn't sound as if Kasavian's department is going to come up with anything overnight. The more we're delayed, the more we risk.'

'You say there's risk,' said Dan. 'I don't understand.'

'I think Sabira Kasavian is right to believe that her life is in danger,' said Bryant. 'If we lose her, we'll never get to the truth. And we have to act before she finds out that the only person she trusted is dead.'

15

GHOST IMPRINT

'John wanted me to see how you work,' Jack Renfield explained as he watched Dan Banbury attempting to open the front door. 'He thought it would help me understand your thinking. For God's sake give me the bloody key.'

Jeff Waters lived in one of the bland new high-rises that surrounded Swiss Cottage. Clearly his photographic business paid well. The top floors had glass walls that faced south, overlooking the city.

'We're in the wrong bloody jobs,' Renfield grunted. 'Look at this place, a million plus, easy.'

'I'm going to tell you this once,' said Banbury. Now that they had entered the flat, they were on his turf. 'I have my own way of working and I need you to follow my instructions. You remain behind me, don't deviate to the left or right unless I clear an area first. I work to a grid, but I'll create two cleared access paths through the site. After that—'

'This is boring,' said Renfield. 'Just go in and stop pissing about.'

'I need to explain this because Mr Bryant fails to

understand the concept of site contamination. He's been known to leave sweet wrappers by a body. He can't resist *touching* things.' Banbury grimaced. 'And bits seem to fall off him. He sheds foreign material like a dog. I once picked up trace liquids at a murder site and followed them through three rooms before I realized that he'd made himself a cup of cocoa and dripped it through the flat.'

'What are you expecting to find here?'

'Waters said he never spoke to Sabira, never saw her anywhere except from behind the paparazzi barrier, but Mr Bryant thinks otherwise.'

'Why would Waters have lied?'

'Presumably because she confided something of importance to him, and instructed him not to mention their conversation to anyone.'

'Pillow talk.'

Banbury held up his tweezers. 'The woman Sabira had the fight with in Fortnum's accused her of having an affair with Waters. A long blond hair would be a good start. The last thing her husband will want is to be confronted with proof of her infidelity. But Mr Bryant reckons it might shock her into giving some honest answers.'

'Has it occurred to anyone that she might just be having a nervous breakdown?' Renfield asked. 'Birds do, you know. It's not easy living in the public eye, as my sister can tell you after she got done for shoplifting at Ikea.'

'What did she take?'

'She put an occasional table up her kaftan. Now that she's gained weight she could probably get away with a lawn chair.'

'I hardly think Sabira's change of lifestyle can be compared to your sister's light-fingered habits.' Banbury eased himself down on to his knees, opened his forensics box and began taping the floor. 'Besides, her husband is clear about the date of her personality dysfunction. He says it started six weeks ago. If I can find something that approximates that date, we'll be able to give him a reason for her behavioural change.'

'It may not be something he wants to hear.'

'I'm going to do the bathroom first.' Banbury cleared a path to a bare, white-tiled corner shower room and began checking the toiletries cabinet. 'There's no woman residing here,' he said. 'Not recently, anyway.'

'How do you know that?' asked Renfield.

'Single men hardly ever remember to clean the insides of their bathroom cabinets. It's special territory, like your shed. How long has he been living here?'

Renfield checked his notes. 'Three years.'

'Cleaning lady?'

'Janice spoke to the neighbour. She says no.'

'Overnight guests?'

'I can nip next door and ask her.'

'Don't bother, I'll soon tell you. I'm only going to grid the seating area in the lounge. I can see where he's been. Singles form more regimented patterns than couples. Let's do the bedroom.'

The apartment had been recently painted in soothing shades of grey, offset with lime-washed light oak floors, thick cream rugs, white walls and hidden downlighters. The bedroom was elegant and minimalist.

Renfield noticed that the forensic pathologist had a habit of peering about himself like a cat venturing into a stranger's flat. 'No clutter to deal with, no knick-knacks, all very masculine.' Banbury opened another of his cases and set aside a packet of brown paper bags.

'What are they for?' Renfield asked.

'Best way to avoid evidence contamination. Should get some nice Cinderellas off those rugs. I can tell if there are only his shoes in the wardrobe from checking the angle of the footfall, rubbed spots, weight distribution, stuff like that. I can do that without going to Forensics.'

'And by looking at the sizes,' said Renfield sarcastically. 'Unless she had massive plates of meat.'

'What if there are two males living here with the same shoe size? He works strange hours, could be subletting without the neighbour even noticing. Hang on.' Banbury lowered himself beneath the bed and emerged with a tiny fragment of broken glass in his tweezers.

'Blimey, how did you spot that?'

'Practice. Normally I'd send this off to the GRIM room at Lambeth FSS.'

'Grim room? What's that?'

'A Glass Refractive Index Measurement room. If we'd found a glass fragment from Waters's clothing, Lambeth would stick the fragments in a special oil, heat it, then cool it down until the point when the bits refract light at the same point as the oil. So the glass vanishes in the oil, giving its refractive index. If this bit and the recovered sample refract light at the same point then they're probably from the same source, and you could say he was killed here and dumped there. But in this case we know he was killed where he was found because of the CCTVs and witness reports. Anyway, I wouldn't be able to use the Forensic Science Service now. The government's closing it down.'

'Why would they do that? I thought it was supposed to be the best in the world.'

'It's the best, but it's also losing two million a month, so they're going to outsource to private firms. A total disaster, in my opinion. The FSS built its rep on shared information, the very thing private companies don't do.' He rose and stretched his back. 'No one else has been here. We're lucky Waters had a hairy chest.'

'Why?'

'Hairy blokes can't help shedding as they move about. You wouldn't believe the amount of stuff that comes off the human body. There's not been anyone else in his bed.

Look at that.' Banbury had folded back the cotton covers and was pointing to tiny curls of hair on the bottom sheet. 'Heavy sleeper.'

'How do you know that?'

'He sleeps on the left but deliberately keeps the alarm clock on the right-hand table. It's so he has to cross the bed to turn it off in the morning. If it was on the left he knows he would just hit snooze and go back to sleep. So there was no one to the right of him.'

'I can see why you like this job now,' said Renfield. 'It gives you a chance to have a right old nose around.'

'I like it', said Banbury, waggling a dusting brush between his fingers, 'because it allows me to build a picture of someone without me ever having met them. A ghost imprint, if you will.'

'All right, then,' said Renfield, folding his arms. 'Tell me what you know about Waters that you didn't when you came in.'

'He's tall, around six three. The apartment's bespoke, and he's had the cupboards, sink and counters raised above normal height. That fits with the size twelve trainers, comfortable shoes for standing around. He's a night person; he drinks brandy alone, which no one does early unless he's French, in my experience, and his TV viewing history will back that up. Never eats at home; likes to think he's fit: the cupboards have protein shakes and there's a note in the kitchen reminding him to renew his gym membership. He probably put his back out two years ago – there are old packets of diazepam and tramadol in the bathroom, strong doses.'

'That's easy stuff. Tell me something I don't know.'

'All right. He likes his women young. He's got serious commitment issues because of his brain-damaged four-year-old daughter.'

'You're reading that from his apartment?'

Banbury shrugged. 'He decorated it alone. This isn't

the kind of flat women would hang around in without altering something. There's no bath, for a start. Have you ever met a woman who could live without a tub? Plus there are some internet sites on his laptop that tread a bit close to the legal age limit.'

'You got the computer?'

'In his bag.'

'You're thinking about him talking to the little girl? The porn doesn't make him a pervert.'

'I agree, I'm just pointing it out. His hours are unsociable and he likes to sleep in late, so he never brings anyone back because he'd have to talk to them in the morning. He hardly ever sees his daughter. From the number of Plaxo reminders about doctors' appointments and a few of his emailed replies my guess is that she suffered some kind of brain-trauma, maybe meningitis, and the stress wrecked his relationship with his partner. There are a few pictures of her up to the age of three, but they're in a drawer. She was healthy then, everything was fine. Then he put his past away, a coping mechanism. I've got an address for the partner; you can check it out. He's obviously still involved and concerned because there are over a dozen books in the lounge on the subject of coping with serious child illnesses, so I'd say he was still handling the fallout. He's got no enemies because he has no friends. All he ever does is work.'

'But he knew something that could get him killed.'

'If the information's not in his laptop it has to have been in his head.'

'And he never has sex?'

'I didn't say that. He brought a woman back here eight days ago.'

'How do you know that?'

Banbury had lifted something from the top of a pillow. 'Here's your long blond hair. I've got another one.' He held up a clear plastic pouch. 'The laundry hamper holds a pile of identical T-shirts, his working uniform. This was between the bottom shirt and the one immediately above it. To my eye it matches the one from the bed. We need to get them under a microscope, and get a sample from Sabira Kasavian. And find out her whereabouts the Monday before last.' Banbury grinned. 'I worked back through the number of shirts. You see? The old-fashioned methods are still the best.'

'You don't think he followed the girl because she reminded him of his daughter, do you?'

'Interesting idea. People sometimes do things they don't understand. Ask someone to explain their actions and they usually find a justification, but often I think they make it up to cover for the fact that they don't know themselves. If Waters was just randomly following girls around it would throw everything out of order, and I wouldn't like that. I'm a very orderly person.'

'I bet you keep a special stick for stirring paint with,' said Renfield. 'The unit must drive you nuts.'

'Mr Bryant lives in a whirlwind of filth and chaos,' muttered Banbury. 'If anyone murdered him in his office we'd never find out who did it, that's for sure.'

16

WATCHING

Meanwhile, John May was briefing the remaining assembled staff in the common room. 'We need to find the girl,' he said. 'We have to build this case solidly, step by step. Otherwise we can't connect Waters's death to Sabira Kasavian. We have CCTV shots of the children in Salisbury Court. We can match the footage against the entrance camera in Coram's Fields. That gives us two co-ordinates. Waters spent half of his last day in Belsize Park, so there's a chance the girl lives in that area.'

Bryant dug through a stack of ragged notes. 'Janice, did you talk to the chap on the gate at Coram's Fields?'

'He saw Waters and the girl enter and assumed they were father and daughter. He thinks Waters was holding her hand but admits he wasn't paying attention. A big crowd of Chinese children was coming in, and he was checking to make sure they were supervised. Kasavian called back. We've definitely got the O'Connor investigation. They're sending the files over right now.'

Bryant rubbed his hands together. 'Good, I thought he'd drag his feet. Where's Raymond?'

'He had to go home,' said Longbright. 'Bad news, apparently. He wouldn't tell me what was wrong.'

'Right, you two.' Bryant turned to his PCs, Colin Bimsley and Meera Mangeshkar.

'No bins,' said Meera. 'I'm not doing rubbish duty.'

'No, you'll like this one: warm weather, a leafy square. I want you and Colin to stake out the courtyard and surrounding alleyways tomorrow morning, starting before the offices start filling up, so you'd better be there by seven. You're on the lookout for someone who knows this little girl.' He handed them an enlarged still taken from the courtyard camera. 'There are no schools in the area. You hardly ever see kids around that neck of the woods, so she and the boy must have been with someone, probably waiting for a parent.'

'It shouldn't take long to check all the offices in the square and find out if anyone brings children to work from Belsize Park,' said Bimsley.

'Don't you want to wait for Dan to match the footage from Coram's Fields?' asked May.

Bryant shook his head. 'We wasted time before and a man died.'

'Wait, does this mean we're going on to shifts, because I'm supposed to be going out tonight,' said Colin.

'Where are you going?'

'I've got tickets for Coldplay at the Emirates Stadium. They're on at half nine.'

'I assume that's something to do with young people singing?'

Meera snorted. 'No, they're really old and rubbish.'

'Well, I'm sorry to disturb your plans for a musical evening, Colin, but this is rather more important. And get Dan back. And find out what the hell's going on with Raymond. Then call the Cedar Tree to check on Mrs Kasavian. And somebody make me some bloody tea!' Bryant stormed from the common room.

'Blimey,' exclaimed Bimsley, 'he's cheered up all of a sudden.'

'Of course he has,' said May. 'It's more than one murder now. He's got a conspiracy on his hands.'

'Are you all right?' May came back into the office and sat on the edge of Bryant's desk.

'No, I'm not all right. I'm very upset.'

'Who with?'

'With myself, obviously. I should have seen this coming. Waters went into St Bride's. What was he looking for? Did Sabira Kasavian send him there? And why did Amy O'Connor have to die in a church, of all places, when we know Sabira is a Muslim?'

'Why won't Sabira talk to us?' May wondered. 'If she really believes she's the subject of a witch-hunt, why won't she try to convince us it's true?'

'You know, back on Monday, when we first heard about O'Connor, I thought I knew what had killed her.'

'Go on, enlighten me.'

'It looked like there was a wasps' nest in the bottom of the Roman excavation, just under the tree in the corner of Salisbury Court. I was pretty sure we wouldn't be given the case, but I did a little checking anyway. After I saw Ben Fenchurch at St Bart's, I talked to the admitting officer who emptied O'Connor's pockets at the hospital. I thought she might have been carrying an epinephrine pen. If you're allergic to wasps you're supposed to keep one with you at all times, so you can give yourself an injection to stop an allergic reaction. Then I thought maybe she was stung and didn't notice – it can happen if the sting occurs in an area where there are few nerve endings – or perhaps she didn't know she was allergic. But Ben didn't report any broken skin, and something would have showed up in the PM. So we're back where we started. I kept thinking about St Bride's being the reporters' church, that her death was a message of some kind, but that can't be it.'

'Don't worry, we'll find the little girl,' May assured him. 'We're closing in.'

Just after 10.00 p.m. on Thursday night the fine weather broke, and the effulgent skies dropped across North London. The streets cleared and Hampstead Heath took on the sodden appearance of a beaten-down rainforest. At the Cedar Tree Clinic, water bent the trees, sluicing over the garden slopes.

Standing at her bedroom window, Sabira Kasavian looked out and watched golden needles passing through the spotlight over the back porch. Even though the room was overheated, she found herself shivering. She prayed that Jeff Waters had done what he had promised, and that she would be saved. She listened to her mobile and counted the rings, seven, eight, nine, and then voicemail.

'Edona, please call me back when you get this. Please, it's very important.' She couldn't remember if her old school friend had gone home to see her parents yet. It didn't sound as if she had. She closed the phone and slipped it back in her jeans.

The clinic offered a safe harbour for now, but she also knew there was no way out. She was caught in a race between the forces of good and evil, darkness and light. It was hard being patient, not knowing what was going on in the outside world.

Not knowing what was going on inside her own head.

She could hear the rain drumming hard on the roof. The trees were moving beyond the window, and for a moment it seemed there was something dark inside them shifting back and forth. She came closer to the glass and tried to see what it was, but the light in the room was too bright, so she switched off the bedside table lamp.

Now she could see down into the garden. And there

he was, standing in the rain looking up *at* her window, watching, daring her to call the nurse, knowing that by the time she did so he would have slipped away into the dark wet greenery, and she would look even more disturbed.

Perhaps I am mad, she thought. *Perhaps this is their wish, to make me as mad as them. I'll show you some real madness before you get to me!*

17

DESTABILIZATION

At 8.00 a.m. on Friday morning, Meera Mangeshkar sat on a rain-sodden bench in the courtyard of St Bride's waiting for her partner to finish his second breakfast burger. In the branches of the tree above her head, people had tied coloured ribbons to commemorate the lives of journalists killed in the recent conflict in the Middle East.

'It's amazing how many of these little courtyards and alleys are still around,' Colin Bimsley said, sucking bits of bacon from his teeth. 'Mr Bryant lent me a book about them. You can still find old debtors' jails and the channels of underground rivers round here; they take a bit more digging out but they're there here all right. I walked past something called the Alienation Office on the way here; 1577 it said over the door, something to do with transferring feudal lands without a licence. And over in Fen Court there are loads of upright sugar canes covered in Old Testament quotes, something to do with the Stock Exchange and slavery. I was coming out of the Cock and Woolpack pub the other night and saw them. Amazing what you find when you get off the main roads.'

'It'd be even more amazing if you stuck to the job and

found a murder witness,' said Meera, taking the lid off her cardboard cup. 'Starbucks tea is horrible. Why hasn't the City got a decent chain of teashops? My gran still makes proper Delhi spiced tea.' She took a sip and grimaced. The flow of workers into offices was steady now.

'This is a chance to prove ourselves,' said Colin, balling up his fast-food bag and looking for somewhere to put it. 'Find the little girl and we'll come off bin duty and surveillance for good. Move up the ladder.'

Meera was doubtful. 'That's not going to happen. There is no ladder. We're at the bottom of the pile right now, and I can't see the unit hiring anyone else beneath us. They don't even have a teabag allowance. So, how do you want to do this?'

Colin looked up at the steel and glass buildings sandwiched between Georgian and Victorian brick houses. 'Looks like the offices go a long way back. We'll have to cover them all. We won't get much joy asking which employees bring kids in because it was a Saturday, so there would have been different receptionists working. It would be better to find out who was in each of the buildings at the weekend, then see if they brought children. And I think we need to keep it a bit vague. If we tell someone that their kid might have been a witness to two deaths, they're liable to prevent us from talking to them. City types can be dead arsey.'

'We're officers of the law, Colin, they can't "prevent" us from doing anything.'

'Maybe not, but we need a detailed account from a girl who looks like she's about eight years old.'

'OK – you take those two sides, I'll take these. And I'll find her first.'

Colin grinned. 'Why are you so sure?'

'Because if I was a parent who had to go into work on a Saturday and my kid wanted to play outside, I'd make sure I could see her from my office window, and those trees are in the way of the rooms on the south and east sides.'

'OK, you have a head start – do you want a little bet?'

'Not if it ends with me having to go on a date with you if I lose.'

'You read my mind. It's not like you haven't been out to dinner with me before.'

'Funnily enough, I don't count sitting on bins all night doing surveillance while you eat chicken jalfrezi out of a box. First one to finish calls the other.'

Meera headed into the nearest building on her side of the square.

Back at the unit on Caledonian Road, Dan Banbury went to see John May. 'We've got a match on the girl,' he said. 'It's definitely the same kid in both screen grabs, the shot from Salisbury Court and the Coram's Fields footage. It's just a piece of software that matches physical features and body shapes, so I don't have an ID for you, but we're working on it. I put a rush on the test fibres from the Waters apartment and we have a match on those, but it doesn't make sense to me.'

'You're talking about the hairs?'

'They're from Sabira Kasavian.'

'So they were lovers.'

'I didn't say that,' Banbury hedged. 'They just place her in his bed.'

'What, you think she came by to read the Sunday papers with him or something? If she isn't a murderess she at least cheated on her husband, which means she lied to us.'

'I don't know that she did.'

'What do you mean?'

'Something's not kosher. I tried to picture what happened. Waters invites her back to his flat, or Sabira calls on him. Women are more prone to shedding signifying evidence than men. Make-up and long hairs with traceable dyes, a wider variety of clothing materials; my missus leaves a trail of tissues and trash wherever she goes, and if she opens a handbag – well, it's like Vesuvius. It might have cost her a grand at a fancy store but basically it's a dustbin with a strap. God help her if she ever tries to have an affair, I'd be on her like—'

'Dan, get to the point.'

'Sorry, John. Kasavian often works late, but hasn't been out of the country in more than two months, which means she wouldn't have stayed over. Even so, I'd expect something else in Waters's flat – a visit to the bathroom, something from the lounge sofa or the kitchen, but there's nothing. And it's not just her, there's nothing from anyone else other than Waters in the whole place.'

'So you're saying someone planted the hairs.'

'More than that, I'm saying they cleaned the flat up so we'd only find the hairs. There's nothing else there to contaminate the evidence. It's like someone wants to guarantee that she gets the blame.'

'Is there any way of proving it?'

'That's the problem, John, it's just a feeling. Even in the cleanest flats you find alien matter and have to eliminate it piece by piece, but not this time. Two perfect long hairs, one placed where it couldn't be missed, on the pillow, the other conveniently left in the laundry basket between shirts so I can easily date it. I could buy the evidence as it stands, but in my experience it just doesn't feel right.'

'So you think someone planted the evidence to discredit her?'

'If there was someone in her husband's circle who was determined to frame her, this would be a bloody good way to do it.'

'With the motive of destabilizing Kasavian just as he's trying to push through the UK side of the borders initiative.'

'Exactly. The crazier the wife looks, the worse it reflects

on his judgement. I'd start looking into his department, and see who's got the most to gain by causing his downfall.'

'Home Office Security is notoriously secretive. I very much doubt there's any way of getting to their inner circle.'

'Then I think you need to find a way before something else happens,' Banbury warned.

18

LUCY

When she tried Royal Oak Recruitment Services, Meera Mangeshkar struck it lucky.

The receptionist immediately singled out one employee. 'Andrew Mansfield,' she said. 'A lovely man, but a real workaholic. He's here nearly every weekend, never takes his holiday allowance. His ex-wife works nearby and they look after the children between them, two boys and a girl.' She tapped the blurry photograph. 'And that's definitely little Lucy. She's wearing her favourite yellow top. She's – well, she's quite a handful. Knows her own mind, that one does.'

'Do you know if Mr Mansfield was working last Saturday morning?'

'Oh, he'd have been here. We had a rush job on all last week. Lucy was probably with him.'

'Could I see him?'

'He won't like being interrupted, but let me try.' She rang Mansfield's office and persuaded him to grant Meera an audience.

Meera called Colin and told him to come over. Together they headed for the fourth floor. The sight of so many tightly arranged cubicles made Meera feel claustrophobic. *Fancy working in here every day*, she thought. *Give me the streets any time.*

Mansfield could not have been older than forty, but looked as if he was about to drop dead. His grey suit matched his skin and hung about him like a flag. His shirt collar was a size too big, and his dark eyes were sunken and lifeless. He seemed to find it an effort to speak, and had already forgotten who his visitors were.

'We're from a central London crimes unit investigating two incidents that your daughter may have witnessed,' Meera explained again. 'It may be the case that she didn't register seeing anything she considered to be out of the ordinary, so we need to talk to her in order to form a fuller picture of the events.'

'Where did these "events" happen?' asked Mansfield distractedly. His BlackBerry buzzed and he reached for it.

'You can leave that for a minute,' Meera warned. 'One was on Saturday last around lunchtime, out in the square.'

'She was here. She went downstairs to play with Tom Penry, one of my colleague's boys. If you want to interview him, I can probably arrange that. You said there were two incidents.'

'Yes, the other was in Coram's Fields in Bloomsbury yesterday afternoon at around four p.m.'

Mansfield shook his head blearily, as if trying to clear it of clouds. 'No, I don't think I know—'

'It's a park just opposite the Brunswick Centre.'

'Oh God, yes. She ran off to look at the animals. We'd stopped in the farmers' market. I was dying for a cigarette and trying to take my mind off it, so I stopped at a bookstall. It was very busy there. Lucy was watching a man making pancakes. When I turned back I couldn't find her.'

'How long was she gone?'

'I don't know – five or ten minutes, something like that. She's very independent, quite fearless, always going up to strangers and chatting. I try to stop her. She's smart, though, a good judge of character.'

'She's still a little girl, Mr Mansfield.'

'I searched the market, then remembered the park opposite. That joins on to Coram's Fields, doesn't it? There's a petting zoo there.'

'So you left the centre and headed there?'

'There's a crossing going all the way over from the Brunswick Centre to the edge of the park. I followed the railings down the side and then I saw her running towards me. She said something about seeing a friend. She makes stuff up all the time.'

'Did you see the friend?' asked Meera.

'No, I was a bit angry that she'd run off again, but we were late for her optician's appointment, so I let it go.'

'Would it be possible to talk to her later today?'

'I'm picking her up from school at four because she needs to go back to have her glasses adjusted. Could I leave you to take her and do the interview afterwards? I've got a lot of work on this afternoon.'

'No, Mr Mansfield,' said Meera firmly. 'You have to pick up your own daughter. I don't think she'd want to be met at the school by a complete stranger.'

'All right,' said Mansfield finally, 'but it's just adding to my problems today.'

Meera scowled back at the offices as they left. 'When Mansfield keels over and dies on the job, leaving his children without a father, I wonder if his bosses will show their appreciation for all the hard work he put in,' she said.

'Divorced working parents competing over the kids.' Colin gave a shrug. 'I bet little Lucy gets a lot of terrific presents.'

Colin checked the name he had written in his notepad.

The boy, Tom Penry, attended the same school as Lucy Mansfield, but was in a lower year.

'Don't take this the wrong way, Meera, but I don't think you should interview the girl.'

'What, you think I'm going to scare her or something?'

'Sometimes you scare me. Get Janice to do it – she's brilliant with kids.'

'But I'm younger, I'm closer to the kid's age than her. Plus I'm a lot shorter, which kids like.'

'Yeah, but she's . . . you know, more patient.'

Meera finally agreed to the idea, but Colin could tell he had hurt her feelings.

'There was somebody out there in the garden, I swear to you,' said Sabira Kasavian. 'He was staring up at me, watching my room, but when I looked again he was gone. Believe me, I know exactly how that sounds but he was there. Go and look at the ground if you don't believe me. It was wet; he must have left footprints.'

Longbright had looked in on Sabira because the Cedar Tree Clinic was just up the road from Lucy Mansfield's school, where she had arranged to meet the girl with her father. She had been instructed to break the news about Jeff Waters before Sabira had a chance to find out accidentally. It was rare for someone to be killed in one of London's public parks, and reports of the murder had started to hit the press, although details were vague.

Sabira was seated in one of the clinic's empty afternoon lounges. Her mood had changed to one of tetchy anxiety. It was as if she was coming down from a night of drug-bingeing. Longbright sensed she would have to go easy with her.

'Why would someone come here just to watch you?' she asked gently.

'They want me to know that I'm always being watched, that I'll always be watched until . . .'

'Until what?'

'Until I kill myself.'

'What makes you think these people want you to kill yourself?'

'They leave notes telling me to.' There was something sinister about the way in which Sabira seemed resigned to her persecution.

'Do you have any of these notes that I could look at?'

'No, I threw them all away.'

'Where? At home, in the kitchen bin?'

'Oh, in the street somewhere.'

'Do you ever have suicidal thoughts, Sabira?'

'Suicide is for people who can't see a way out of their situation. Even when we had terrible problems at home, I would always try to find a solution.'

'And now?'

'Now there really is no way out. I suppose I could run away, go back to Albania, but I would not even be safe there.'

Longbright rubbed her arms. The room had grown suddenly cold. 'Sabira, I'm trying to think about this logically. What do you know that could make someone reach out and try to kill you in another country?'

'It's too big to talk about. As big as the world itself. A global conspiracy. They have people everywhere. They will track me down and kill me, then make it look as if I killed myself.' The sudden clatter of teacups in the next room made them both start. 'I had proof – I swear I did – but it disappeared without ever leaving my hands, just as if it never existed. They came and took it from me while I slept.'

This was textbook paranoid delusion, Longbright realized. Sabira didn't think one particular person was out to kill her, she thought the world meant her harm. 'If you don't tell us why people want to hurt you, it's very hard for us to help you,' she pointed out.

'There's no one I can trust. And if I could trust them, I wouldn't be able to protect them.'

'You can trust me.'

Sabira shook her head violently. 'No, you're the last person I can confide in – surely you must see that?'

Longbright had serious doubts about introducing the subject of Waters's death, but there were computers and televisions scattered throughout the clinic, and the last thing Sabira needed to see right now was a sensationalistic report on the murder of an acquaintance in broad daylight.

'Sabira, we know you befriended the photographer assigned to cover your public appearances. We spoke to him.'

'He has nothing to do with this.'

'You know what? If I had a problem and needed someone to help me out, he's the sort of man I would have picked to confide in. That's why I wanted to talk to you, before you heard it from anyone else. Someone attacked him yesterday.'

'Is he injured?'

'I'm afraid he's dead.'

Sabira said nothing. For a moment Longbright thought she had failed to understand. Finally she looked up at the detective sergeant and said, 'His killer is outside the window right now.' The casualness of her tone was chilling.

Longbright looked out, but the garden was veiled in rain.

'It's the same man who was there last night. Can't you see him?' Her voice began to rise. 'He's right there, you must be able to see.' And then she was yelling in a thin, high voice, 'He's there! Right in front of you! He's there!'

Longbright ran to the French windows and unbolted them, running out into the downpour, but there was no one in sight. The rain had beaded on the grass, giving it a silvered sheen that held no other footprints but her own. She searched inside the bushes and under the trees, but it was clear no one had been standing there. She headed back to the house, soaked.

Sabira had turned away from her, expecting failure. 'I knew he'd vanish before you got there. You must go now,' she said. 'Go and never come back.'

'Sabira, if there's anything I can do—'

'Just go. You can see how mad I am. Even I don't know what I'm saying any more. Go fast. It is safer for you.'

There was no point in remaining any longer. Longbright slipped her business card into Sabira's hand. 'This has my home contact details on the back. Please use them at any time.'

She said goodbye and went to speak to Amelia Medway, the centre's senior nurse.

'It may be more than just mental exhaustion,' Medway told her. 'Sabira is free to come and go as she pleases, but she's exhibiting quite serious symptoms, and may require more specialized health care. We're not a psychiatric unit, Miss Longbright. We're not secure, and don't provide long-term pharmacological solutions.'

'How long do you think she'll be here?'

'Certainly until Monday, when she'll be assessed by a King's College psychiatrist who'll decide the next step. If he thinks there's a genuine risk of self-endangerment, he'll refer her to the private ward of the Bethlem Royal Hospital in Bromley. She'll be well cared for there. If he thinks she's out of danger he may allow her to return home, providing she remains under local supervision.'

'There's one last thing,' Longbright said. 'I know your concern is for Sabira's mental wellbeing, but we need to make sure that she's not physically at risk from anyone else.'

'The doors of the centre are locked at night, but we're

not legally allowed to restrict her movements. If she wants to go out, we have no way of stopping her.'

'Then perhaps you could keep me informed of her whereabouts.' Longbright gave her a card with the unit's number, and then took her leave.

Lucy Mansfield's school was just off England's Lane in Belsize Park. It was privately run and so smart that it looked like an upmarket restaurant from outside. It was popular with executive couples, who placed their children's names on its waiting list years in advance.

Longbright caught up with Lucy's father by the main gates. The girl who came running up was slightly built and small for her years, but clearly filled with confidence and energy. Lucy had reached the age when she had just discovered the power of her opinions, and was already used to being heard.

They went to the Caffè Nero on Haverstock Hill. Andrew Mansfield bought his daughter low-calorie chocolate cake and explained why Longbright was here.

'I was playing with Tom,' Lucy explained between greedy mouthfuls. 'I didn't do anything wrong. It's a real game, with a rulebook and everything. It's called *Witch Hunter* and you have to ride across the countryside and find witches to kill.'

'She used to play the game with her brothers,' Andrew explained. 'It's an RPG. We vetted it, of course. My wife doesn't approve, but I don't see the harm in it. It's historically accurate, rather like those books, the *Horrible Histories*, so the kids learn about the English Civil War. I didn't know you still played it, darling.'

'I don't, but Tom had the cards on him, and we were waiting for you and Tom's dad so we played.'

'How do you play the game?' asked Longbright.

'You pick if you're going to be a witch or a hunter. Hunters ride to a town in a place called Suffork and listen to accusations from the villagers, and then they find the person who's a witch. It doesn't have to be Suffork. It can be wherever you like.'

'How do players know who's a witch?'

'There are lots of questions you have to ask, but you can tell 'cause of the way they look. We couldn't ask a lot of the player questions because we didn't have a witch, because Tom wanted to be a witch hunter as well as me.'

'So you found someone you thought was a witch? Why did you pick her?'

'Because she was pretty and witches can change their skin, and she was reading a book about eating babies.'

Longbright remembered that a bookmarked paperback of Ira Levin's *Rosemary's Baby* had been found in Amy O'Connor's handbag. 'Then what did you do?' she asked.

'We pretended to be playing ball so we could get up very close, and I had a good look at her, but I still couldn't tell if she was a witch. And then we killed her. Can I go now?'

Longbright frowned. Lucy had deliberately skipped the part she needed to hear. 'Why would you kill her if you couldn't tell whether she was a witch?'

'I talked to Tom and he thought she was.'

'How did you kill her?'

'We put a curse on her.'

'How do you do that? Did you talk to her?'

Lucy shook her head. 'No. We did this.' She rubbed her fingers together. 'And we said the thing on the card called an incant . . . an incant—'

'An incantation.'

'Yes, but I can't remember what it was. Tom has the cards.'

'I think you did talk to her, Lucy,' said Longbright.

'No – she just told us off. That's not talking, is it? Ask Tom, he'll tell you about her.'

'Then what happened?'

'She got up from the seat, put her sandwich box in the

bin and went into the church. And then me and Tom went back into the office.'

'All right, what about yesterday, when you went to Coram's Fields? Why were you in the park?'

'I was bored of waiting. Dad was being a grump and wanted to look at the books on the stall, so I walked away.'

Longbright addressed Lucy's father. 'Had you taken your daughter to Coram's Fields before?'

'No. I knew there was a garden square opposite but I only had a vague idea there was a children's park there.'

'Lucy, how did you find the park?'

'I crossed the road and there it was.'

'Have you seen it before?'

'I can see it when we drive to Waitrose. We never have time to stop.'

'There's a camera by the road that photographed you with a man. Who was he? How did you come to meet him?'

Lucy thought for a moment, but there was something too pantomimic about her performance. A finger on the chin, a roll of the eye. 'Oh, I remember now. He asked me where the farmers' market was so I told him.'

'Are you sure? How did he get into the park?'

'I don't know.'

'Because the camera shows the two of you entering through the gates at the same time.'

'He must have been walking close to me or something.'

'And that's all that happened? He didn't say anything else to you?'

'No. He asked me the way and I told him.'

'Why would he think you knew the way?'

'Don't know.'

'And then what happened?'

'I heard Dad calling and ran back to the road, but I had

to go all the way around because there's no gate on that bit.'

'Did you see what happened to the man?'

'No. Daddy, we have to go now or I'll be late for the optician.'

She's lying by omission, thought Longbright. *She spoke to Waters, but maybe he told her not to tell anyone. She might even know why he was killed. Something else must have happened at St Bride's when the children were putting a curse on Amy O'Connor.*

Lucy glanced back at her with wide innocent eyes, then picked up her pink rucksack and got up from her seat. 'Daddy, come *on*,' she commanded.

All right, Longbright thought, *we'll see what your playmate Tom has to say.* She checked the number for Tom's mother, Jennifer Penry, and called her. 'I'm afraid you've missed him,' said Mrs Penry. 'Tom's with his grandparents.'

'Do you have a number for them?' Longbright asked. 'It's important that I speak with Tom.'

'I'm afraid not. At the beginning of the week they flew to Bodrum and boarded a gulet – one of those traditional Turkish boats? They're going along the coast and will be returning from Göcek, I'm not sure when. My in-laws fancy themselves as free spirits. All very annoying. I don't suppose there's any way of contacting them until they stop in Rhodes, and I'm not sure when that is.'

'Don't you have a number for Tom's grandparents?'

'They don't use mobiles. I thought we had one for the skipper but it doesn't seem to work.'

That's convenient, thought Longbright before checking herself. *Now you're starting to get paranoid as well. Maybe it's catching.*

'What about the travel company who arranged the trip?'

'I have no idea who they used. You'd have to ask them.'

'But I can't do that.'

'They may call in at some point. If they do, I'll get a contact number for them.'

As Longbright headed back towards Belsize Park Tube station, she found herself checking the glistening pavement behind her.

19

METHOD IN MADNESS

Arthur Bryant stepped into the narthex of the baroque Wren church and slowly made his way up the nave towards the altar.

It was early on Saturday morning and the place was empty. Sunlight shone through the modern design of the stained glass, dividing the marble floor into patterns as richly coloured as Tetris blocks.

Bryant consulted the church pamphlet and read:

In 1375 Edward III issued a writ in the Tower of London confirming the Charter of the Guild of St Bride. Its first purpose was to maintain a light to burn before the statue of St Brigide the Virgin. The Guild continued until 1545, when it was swept away by Henry VIII.

He folded out another section.

St Bride's is known as 'the cathedral of Fleet Street'. After its devastation in the Blitz the parish rose again, as it had so many times before. Little of importance that has happened in England's story has not been echoed here in St Bride's. From Celts and Romans to Angles, Saxons and Normans, the church has acted as a parish pump to the world.

As the journalists' church, it facilitated the spread of information. Was that why Amy O'Connor had chosen it, to make a point?

It was certainly not her local parish. O'Connor had lived in Spitalfields. Before that she had resided in Wiltshire from the age of seven. Banbury had been up to her apartment, but the City of London officers had already conducted a search and submitted a report, and he had not uncovered anything new.

Bryant eased himself on to a wooden chair and looked up at the great stained-glass window. The entire ground floor of the church had been searched inch by inch. If O'Connor's death had been planned somehow, why did it occur here? O'Connor's family might have originally come from Ireland but they were Protestant, not Catholic. Amy may have visited St Bride's before the day of her death, but no one recalled seeing her. The question rose again: Why this particular church? *A parish pump to the world.* A message of some kind?

There was an answer drifting like a raincloud at the back of his brain, but every time he tried to focus on it the damned thing dissipated. *The perils of age,* he thought bitterly, *you have to think twice as hard as you did when you were younger, and it will just keep getting worse unless you force yourself to make connections. Everything is connected. Step back and see how it all fits.*

Jeff Waters had come here after talking to Sabira Kasavian. It could only be because she had told him something about the O'Connor death. That had to be the link. It was nothing to do with the little girl.

But now Oskar Kasavian's wife was stashed away in a

clinic and seeing ghosts emerging from the walls. After psychiatric evaluation she would probably be on her way to a more secure unit. She refused to co-operate with the State or the police because she was scared, but her claims made no sense.

Bryant shifted on the hard seat, his old bones aching. He was used to dealing with the aftermath of death, not the problems of the living. *All right,* he decided, *do what you do best: approach it instinctively. What do you naturally feel about Sabira Kasavian?*

The answer came back: *That she's innocent, and that she's telling the truth.* No matter how crazy she sounded, no matter how ridiculous her claims were, what if he assumed they were real? Then the key question became: What did she discover that could possibly make her so fearful?

He asked himself, *What would you do if you discovered something unthinkable?* You might run away, bury your head, pretend you didn't know anything. But if you were brave enough you would try to get proof, to stop people from implying you were mad. And that was exactly what Sabira said she did, stealing a file from Edgar Lang's office, but somehow her evidence turned into a bundle of taxi receipts. Either she had made a mistake, or somebody switched the folder to make her look crazy.

If you decide to go down this route, old boy, Bryant told himself, *you had better be damned sure of your facts, or you'll end up looking as loopy as she does.*

'Are you ready for more bad news?' asked John May a little later. 'Sabira smashed up her room at the clinic in the early hours of this morning. When the duty nurse arrived she attacked her with a knife, then tried to cut her own wrists.'

'Both wrists or just one?'

'One, the left – what does it matter? She tried to kill

herself, Arthur. They've been forced to sedate her. It looks like she'll definitely be moved on Monday, no matter what the psychiatric report has to say about her condition. And don't bother calling the clinic, because the staff are under strict instruction not to talk to us or anyone else until they've conducted their own internal investigation.'

'Who allowed that?'

'Who do you think?'

Bryant picked up the phone and called Oskar Kasavian's direct line. 'Why have we had our access to the clinic cancelled?' he asked.

At the Home Office Department of Security, the call was recorded.

KASAVIAN: Do you understand the serious condition my wife is now in, Mr Bryant? It's quite clear to me that your detective sergeant upset her with news of the photographer's death.

BRYANT: It would have been worse if she had discovered the news for herself. It's all over the papers. If we can't talk to her directly, can you answer some questions?

KASAVIAN: I don't suppose you'll desist until I do.

BRYANT: How did you find out what had happened?

KASAVIAN: I received a call to say that my wife had attempted suicide, and had tried to stab her carer. The poor woman was trying to calm her down—

BRYANT: Do you remember the exact time this happened?

KASAVIAN: No, and I hardly think it's relevant.

BRYANT: Do you have any idea what sparked off her attack?

KASAVIAN: She was apparently difficult at dinner, refusing to eat and so on. I looked in on her in the day lounge at around eight thirty p.m. and found her asleep in her chair. She went up to her room at half past ten, and the nurse was disturbed by the sound of furniture being thrown about a couple of hours after that.

BRYANT: So the nurse went up to the first floor and found your wife in a state of distress.

KASAVIAN: That's right. She had overturned her dressing table, smashed a wall mirror and torn her clothes. It appeared she had also started to cut her wrist with a knife.

BRYANT: Where did she get the knife from?

KASAVIAN: She had taken it from the dining room. She warned the nurse not to come any closer or she would kill herself. When Miss Medway took a step forward, Sabira lunged at her. Luckily, Medway managed to disarm her. She called a doctor, and one of the other nurses stitched and bandaged my wife's wrist.

BRYANT: Why didn't Medway see to her wound?

KASAVIAN: My wife wouldn't let her near.

BRYANT: Did she give any reason for her actions? Make any demands?

KASAVIAN: Nothing. The nurses agreed there were no warning signs. There's no rationality behind my wife's actions. Mental instability is unfathomable by its very nature, although I understand it can be inherited. I assume you are cognisant of the fact that both her aunt and grandmother suffered mental breakdowns and were institutionalized for periods?

BRYANT: So you informed me. I ran some checks but was unable to verify the details.

KASAVIAN: Then I suggest you check the files more thoroughly; it's all on record. There seems to be little more I can do for my wife now, so if you have no more questions—

BRYANT: Can you tell me: the cut on her wrist, was it transverse or vertical?

KASAVIAN: Across, I believe, although I don't see what that has—

BRYANT: And you saw the knife?

KASAVIAN: Yes.

BRYANT: How sharp was it? Was it serrated, very sharp, a bit blunt?

KASAVIAN: They use them in the dining room so I suppose it's rather blunt, but it clearly served its purpose.

BRYANT: That will be all for now. Thank you.

'Are you aware of just how much Oskar Kasavian has to lose?' May asked after Bryant had rung off. 'It's not just his wife, although that would be enough. His government career, his entire future is at stake, and you're asking him for details of how his wife cut her wrists.'

'The details are important,' said Bryant. 'I fear she won't survive much longer. There may be nothing we can do about that, but we may be able to save others.'

'I don't understand what you're talking about. There is no one else to save.'

'You're wrong. I want Lucy Mansfield watched night and day. Get Fraternity on it, he's a smart lad. And get in touch with the Turkish authorities, see if you can find a way of reaching her playmate, Tom Penry.'

'You still don't believe Sabira Kasavian is mad, do you?'

'Oh, I always thought there might be madness involved, but let's just say there's method in it.'

'We're supposed to be a team, Arthur. There aren't meant to be any secrets between us, and yet here you are holding cards to your chest.'

'I know I can be a little proprietorial about evidence,' Bryant admitted, 'but this time I have good reason.'

'Well, I'd love to hear it.'

'What I may be about to confirm about this case could place me at risk. I'm not speaking metaphorically or talking about risk to the unit, I mean it could physically harm me. And if I share that knowledge with you, it would place you in the same position.'

'So what am I supposed to do, let you go your own way until something happens to you?'

Bryant thought for a moment. 'There may come a time when I have no choice but to confide in you. Right now, though, it's better that you form your own theory.'

'I'm sure you have a good idea about what that might be.'

'Yes, as it happens. I need you to reach the conclusion that Sabira Kasavian has undergone a mental collapse due to the exorbitant pressures of her social life. Write it up in a report this weekend and submit it to the Home Office on Monday, ahead of the psychiatric report. Send copies to four people: Oskar Kasavian, Edgar Lang, Stuart Almon and Charles Hereward.'

'To do that, I'll need something more than anecdotal evidence,' said May.

'That's why you need to talk to Edona Lescowitz. Someone at *Hard News* had her surname on file.'

'The Albanian girlfriend? I thought she'd returned home.'

'So did I, until Jack Renfield ran a passport check and found that she never left the country. While you're at it, check out the records of Kasavian and the others' company, Pegasus. Give me something on the directors' background. I have to go to the British Library.'

May recognized the furtive look on his partner's wrinkled face. He looked like a Shar Pei tricking its owner out of dinner. 'And of course you can't possibly tell me what you're up to.'

'I'm conducting some research. It's rather esoteric, but I

think it will turn out to have a bearing on the case. I have some books on the subject, of course, but they don't cover the particular time period I want.'

'Which is when?'

'The early thirteenth century. I won't bore you with the details.'

May was mystified. What could his partner possibly have found that would be relevant to a murder case over seven centuries later?

'Fine,' said May. 'We'll conduct all the interviews while you go and poke about in a cobwebby old library.'

'It's not cobwebby,' said Bryant, nettled. 'It's a new building. Although I'll also be in the old archival annexe in Clerkenwell, which is not only cobwebby but partially flooded.'

20

A FATAL FLAW

Walthamstow is a north-eastern district of London that has lately become home to a large Polish community, spilling over from neighbouring Leytonstone. Bombing raids and development projects have replaced many of the terraced Edwardian houses with grey concrete blocks of flats. On a warm summer evening the local lads hang out in the scruffy high street beneath a riot of plastic signs offering cheap booze and easy ways to send money abroad. In this sense it is like any other working-class London borough.

John May found the flat easily enough, but was surprised to find Sabira's friend living in such straitened circumstances. Edona Lescowitz lived above an Indian shop that sold tinned vegetables, mobile-phone covers and alcohol from distilleries with unpronounceable names. May figured that calling first would only alarm her and decided to take his chances, but there was no answer from the door buzzer.

'She's gone to the laundromat,' said the tiny Indian boy behind the counter of the Am-La Late Nite Groceries Store.

'Where's that?' asked May.

'Turn left out of here, next corner.' He returned his vacant stare to a Bollywood rock video.

May followed the street to the laundry. The printouts taped to the window informed him that the proprietor was also available to unlock phones and could cater for hen parties.

He recognized Edona from her photograph, although she was without make-up now, dressed in torn jeans and a blue Dodgers top. Her auburn hair was tied back, exposing pale skin and high cheekbones. When she finished unloading her washing from the tumble-dryer and stood up, it seemed to May as if she had been expecting him and fearing the worst.

'It's all right,' he said, approaching, 'it's nothing bad, I just wondered if I could talk to you about Sabira Kasavian.'

'You are police, yes?' She closed her arms over her sweatshirt protectively.

'Yes, but we're an independent unit. We're doing all we can to help her. I thought, as you're her friend, you might be able to shed some light on Sabira's recent behaviour.'

Edona sat down on the bench with a defeated look on her face. 'I knew you would come eventually. She called me to say they had locked her away.'

'She's been sent to a clinic to recover, but I'm afraid she's not getting any better. I think she's in good hands, but I'm not a doctor. I want to understand what's happening to her. I'd like to ask you a few questions.'

'She's not crazy, if that's what you're thinking. But she is . . .' Edona chose her words carefully. '. . . vulnerable.'

'How long have you known her?'

'All my life. Our fathers grew up together. My uncle married her cousin. We used to be very close. Weddings, funerals . . . despite everything, we managed to be together for family occasions.'

'But you're not so close now?'

'A lot of things happened. She was determined to move to England and find herself a rich husband, and that's exactly what she did. Meanwhile, I met an Albanian man who said he was a TV producer and I moved to Tirana to be with him. We laughed about it, said we had both met the men we were going to marry.'

May moved back to let a woman through with a plastic tub of washing. 'But you came here,' he said.

'It turned out my future husband didn't have a job but he did have a wife and three small children, so I left. And before you ask, no, I don't see much of Sabira any more, because of her husband. When she married Oskar, he turned her into someone I don't recognize. Me, I'm a reminder of her past, so I'm not welcome. I am too low class.'

'But surely he can't choose his wife's friends?'

'All the government husbands do it,' said Edona wearily. 'They can't afford to make mistakes. The wives aren't invisible women who sit in the background any more, they help to lift up their husbands' careers. It's the same in my country.'

'Do you think Sabira believes in her husband?'

'I don't think she believes in the power of the British government.'

'Why do you say that?'

'She's seen it from the inside now, the way it works. When we were small, we dreamed of living in England, the lovely gardens, the friendly police, *please* and *thank you* and everyone saying *sorry* all the time. So proper, so well behaved. The well-spoken county ladies, confident and sure of themselves. We thought that to be English was to be fair, to be decent. *Reasonable.* Now she knows this is a lie. To survive in the English government, it is to hide, to cheat, to bury the truth. This is what she told me.'

'Was there some specific incident that changed her mind?'

'Something – yes. I don't know what it was. Something I think she discovered.'

'From her husband?'

'No, from the wife of one of his colleagues. Her name was Russian – Anastasia.'

'Edgar Lang's wife.'

'I think so, yes.'

'Do you remember when this was?'

'Maybe two months ago.'

'That was when her behaviour started to change?'

'Yes. She had a relapse.'

'A relapse? What do you mean?'

'Do you not know this? How Oskar and Sabira met?'

'She was here on holiday, wasn't she?'

'She came here for treatment,' said Edona. 'Sabira was . . . well, she had an addiction problem. Nobody knew. She had never broken the law. But there had been bad episodes. Sabira is the one person who should never take drugs. She's very emotional, very highly strung.'

'What was she taking?'

'At first, prescription medicines, painkillers, anti-depressants. She trained as a figure skater but hurt her back when she was sixteen. Then later, I think it was cocaine and ecstasy.'

'You think.'

'I don't know for sure. But I have seen such behaviour before.'

'How did these "episodes" of hers show themselves?'

'She became paranoid, thought everyone was out to get her. Oskar cleaned her up, kept it out of the papers, looked after her. She's one of those people who has everything going for her: brains, drive, beauty; but she has this one fatal flaw. So you see, Mr May, Sabira is not having a breakdown. She's having a relapse.'

*

Janice Longbright was arranging interviews for Kasavian's colleagues. While Jack Renfield went off to meet Charlie Hereward, she met up with Edgar Lang at the Athenaeum Hotel on Piccadilly.

Seated in a wing-backed green leather armchair sipping coffee, Lang looked like a dissipated film star. He was an advertisement for old-school grooming, wrong for television but right for government. His eyes were hooded and half-shut, useful for hiding secrets, his skin too frequently buffed by red wine. Longbright sensed his type and involuntarily recoiled as she shook his hand. Lang seemed like the kind of man who would humiliate waitresses.

'Thank you for taking the time to see me,' she said, seating herself. 'I'll only be a few minutes.'

'Good – I'm due at the House of Commons for a meeting in about half an hour.' He checked his watch, an unobtrusive but vintage silver Cartier.

'So it's not a Monday to Friday job.'

'The government part is – you won't catch civil servants working late unless they have to. At the weekend I operate in a private capacity.'

'That's right, Pegasus. Security arrangements?'

'You make it sound as if we repair locks. We act as the link between scientific institutions and the press.'

'Like a PR company?'

'It's more to do with the government-sponsored prevention of negative publicity.'

'I'm sorry, I'm being a bit thick . . .'

'Scientific breakthroughs are achieved by the global sharing of information, or at least they used to be. Now that information is in the hands of private companies, it has to be safeguarded rather than published.'

'We were discussing a similar problem with the privatization of forensic laboratories.'

'Then you'll understand the issues at stake. But I thought you wanted to talk about Sabira Kasavian.'

'We were unofficially appointed by Mr Kasavian to uncover the reasons behind his wife's behaviour. Excuse me.' She felt her phone vibrate and saw a message from John May: 'SK former addict may have relapsed'.

As she talked, she texted back: 'Find out what Ana Lang said to SK that upset her'.

'I understand she's had issues with substance abuse in the past. The obvious conclusion is that she's using drugs again.'

'There are stringent precautions taken to protect everyone in the department against coming into contact with dangerous substances, Miss Longbright. The country is on a high alert from the threat of terrorism. All packages are opened; all rooms are checked.'

'Does that include home premises?'

'On occasion, yes.'

'But there are a great many social events to attend. It is feasible—'

'If someone is desperate enough, they'll always find a way. If you were to ask me, is there drug-taking in the House of Commons, I would have to answer yes. One has to be realistic about such things.'

'How often does Mrs Kasavian see you or your wife?'

'Once a week or so, sometimes more often. We cover the same ground.'

'Do you discuss your work in front of her? Does she know what you do?'

'You mean, is she a security risk? No, we don't discuss the Home Office or our private work with spouses. I doubt they'd understand it if we did. To outsiders, it would seem abstruse and tedious. Now, if that's all . . .'

'Just one more thing. Do you like her?'

'What do you mean?'

'I mean as a person? Do you like her?'

'I neither like nor dislike her. My business colleague chose to marry her, so my opinion hardly matters, does it?'

'I'm sorry to have taken up your time.' Longbright rose. 'I believe Mr Kasavian is putting up the UK's plans to revise Europe's border-control initiative some time in the coming week?'

'Indeed, on Friday morning, at the conference in Paris.'

'I read that you oppose his views. I assume there are others in the department who share your opinions?'

'I rather think that's outside of your jurisdiction,' said Lang curtly, closing his briefcase and rising. 'You shouldn't listen to gossip.'

'My point is that it might be in his opponents' interests to discredit Mr Kasavian through his wife.'

'If you're planning to make a case of that, you'd better be very sure of your facts.' Lang abruptly turned away from her.

'Well, I'll make sure you're the first to know if we do,' Longbright couldn't resist adding.

Renfield was having a little more luck with Charles Hereward. They met late on Saturday afternoon in a cramped coffee bar used by barristers behind the Inns of Court.

Hereward was a blunt, broad-beamed Yorkshireman with a spectacular comb-over, an old-school former Labour politician who still went for a beer with his pals on Saturday night and a kickabout with his kids on Sunday morning.

'She's a smart lass,' he told Renfield. 'Gives as good as she gets. But she's not the sort of woman the other wives take to. They support their men unreservedly but they can be a bunch of vituperative bitches when they get together. I tend to stay out of their way.'

Renfield didn't really understand why he had been asked to interview Hereward. There were two unsolved deaths to investigate, and he thought they should be concentrating on Waters's friends and relatives first. Taking the obvious route of inquiry had never been the way at the PCU.

O'Connor hadn't made any close friends in London and even her workmates barely seemed to remember her, so they had drawn a blank there. The photographer, though, he must have made plenty of enemies. Renfield didn't think the deaths were related, and certainly didn't believe that Sabira Kasavian's breakdown was connected in any way, but who was he to argue with his bosses?

'She nicked a file from your office,' said Renfield. 'It was full of taxi receipts. How did she get hold of it?'

'The registered headquarters of Pegasus is not at the Home Office, obviously,' said Hereward, looking suspiciously at his minuscule coffee cup. 'It's in Whitehall Place. Sabira knows the place well enough. She sometimes comes there to meet Oskar. I can only assume she picked it up when she was last in.'

'Security a bit lax, is it?'

'Well, I don't pat down my wife's pockets each time she leaves, but then I don't leave stuff lying about.' He sipped his coffee and was shocked to find the cup now empty. 'I doubt an outsider would be able to make much sense of anything unless they knew exactly what they were looking for.'

'I suppose you must have a lot of sensitive documents locked away,' Renfield pressed.

'Enough to bring down this and the last three governments,' Hereward admitted. 'I probably shouldn't say this but we'll all feel a bit safer with Sabira locked up, away from her junkie friends. Lean over and ask the waitress for two more, bigger cups this time.'

'You mean people gave her drugs?'

'That photographer who got stabbed in the park, he was supplying her with cocaine.'

'You know that for a fact?'

'It's common knowledge he was knocking her off and

paying her with coke. It's no skin off my nose what people get up to in their own time, but she should have remembered who she was married to.'

'You think it's damaging Mr Kasavian's career?'

'Not if he acts fast and puts her out of harm's way. I'm on record as being in favour of Oskar's border-control proposals. It's in my interest to make sure that he pushes it through.'

'Bit of a dilemma for him, though, isn't it? Choosing between his wife and his work?'

'You don't know the Home Office.' Hereward gave a graveyard chuckle. 'If you saw your spouse or your career falling out of a window, you wouldn't think twice about which one to save. The wife's replaceable, the job's not. Everyone thinks Oskar's a hard case, but I can tell you he's done everything he can to help her out. Depending on his performance in Paris, Her Majesty's Government will decide whether he'll take complete control of the initiative. That's why his wife has been taken off the streets. She can't do any more damage while she's banged up. The doctors will bring her to her senses, but it'll be too late by then.'

'What do you mean?' asked Renfield.

'If you ask me, one of the government's conditions to Oskar will be to keep his wife far away from the corridors of power.'

'You mean . . .'

'He'll have to file for divorce. Mental illness is good enough grounds.'

21

BREAKING FREE

Having planned her escape, she waited until they had finished serving afternoon tea.

There was no point in trying to get away at night because the front door was locked and alarmed. The shift ended in ten minutes' time, and the staff nurses would go to change out of their uniforms in five minutes. Sabira's screaming fit had not singled her out as someone to watch more carefully. The clinic's brochures didn't advertise the fact, but such behaviour was hardly out of the ordinary; much worse happened in the solitary first-floor bedrooms of the east wing at night. She had heard the nurses telling stories and laughing behind the patients' backs when they thought no one was listening.

And now they were heading off duty, down to the back of the house.

She listened for their tread and conversation on the stairs. The danger was that the first of them – Sheryl Cooper was the most gimlet-eyed of the clock-watchers – would reach the front door in under five minutes. Sabira searched about her room, trying to think what she might need, but her head was still full of clouds. She knew there was something she was supposed to take with her, but whenever she tried to think what it was the object of her attention slipped away.

She knew she was not at all well.

Being well meant being able to exert control over your actions, but it grew more difficult to do so with each passing day. It was much worse now than it ever had been before. If she remained here, in a place that encouraged so much introspection, she would never find her way back to the normal world. It was better to get out now and worry later about the consequences.

Taking a bag would only slow her down and make her more visible. It was a warm evening and she needed to travel light. She wished she had written the plan down somewhere, just so she could remember it, but at some point she had decided not to leave evidence.

Take it one step at a time, she told herself, fighting down panic. *The first thing you must do is get out of the building without being seen – if you fail to do that, there will be no plan.*

She had gathered the few things she needed in the top drawer of her dresser, and stuffed them in her pockets. Outside, she could hear footsteps in the corridor – not a nurse because they wore trainers, but one of the other patients.

Opening her door a crack, she peered out and saw the door on the other side of the hall close. That room belonged to Spike, an American musician with a shock of dyed black hair and a body so thin that he could surely feel his bones rubbing when he walked. He looked seventy but Sheryl had told her he was just forty-two.

Stepping into the silent corridor she ran quickly to the head of the stairs and looked down. At the moment there was a clear path to the front door, but the ground-floor hall was fed by four corridors. Any number of people could appear within seconds.

I could be going to the newspaper stand, she told her-self, remembering the stack of journals that stood beside the front door. *That's what I'll say if anyone stops me. If I get caught this time I mustn't make them suspicious enough to report me. There won't be any second chances.*

She headed downstairs as if it was the most casual thing in the world. It was just five or six metres to the front door and the brass latch that could be popped smoothly and silently. She peered into the side corridors as she passed. Someone was laughing in the dining room but there was nobody in sight.

Dinner was being prepared in the kitchen – she could smell the usual stale aroma of warm potatoes and boiled vegetables. The clinic offered a full international menu including vegetarian and gluten-free options, but everyone seemed to opt for mash and pastry and gravy. Denied drugs and alcohol, they comforted themselves with carbohydrates.

She tried to imagine what would happen beyond the door, the path, the gate. She would make her way to Hampstead High Street and the Tube station, head south on the Northern Line. It was too risky to take a taxi. Taxis had talkative drivers.

She needed to do more than just get out of the clinic. She had to break free in her mind and start thinking clearly again, but try as she might she could not find a shape to her thoughts. Something grey and cotton-woolly had soaked them up and wiped them away.

Her hand was on the lock, pressing down on the trigger that would spring it, when an image appeared before her. Less a man than a devil in black, watching and waiting for her to make an attempt at escape, knowing that she would try and fail.

She faltered, heart speeding, hand dropping, suddenly sure that if she stepped across the divide from captivity to freedom she would be playing into his hands. She saw a thin crimson line extending and dripping, staining the floor, and realized that the wound on her wrist was bleeding through the bandage.

She pushed at the lock and heard it pop, felt the cool evening draught come in around the lintel. But there was nothing she could do to make the black figure step aside. He was smiling benignly at her, amazed by her capacity for self-delusion, blithely coming to this country and marrying into the upper echelons, and then happily assuming she could destroy reputations and wreck the status quo without any risk to herself.

You are dead to us now, he told her.

I am dead, she repeated, losing her resolve and lowering her hand from the lock as the door to the outside world began to swing open.

'Where are you going?' asked smiling Amelia Medway, the senior nurse who was just arriving to start her shift. 'You know you're confined to the clinic now.'

She gripped Sabira's arm firmly and led her away to the dining hall, passing her over to the annoyed Sheryl, who was now out of uniform, thinking that her duties were over for the night.

'Take Sabira back upstairs and keep her there until the dinner bell,' said Nurse Medway, adding softly, behind Sabira's back, 'I don't want her left alone this evening, do you understand? Not after what happened last night.'

'I'm off duty now,' said Sheryl, who was going on a date, and didn't get them very often. 'I made sure she took her medication at three. I can get someone else to take over.'

'No, I'll do it,' said Medway, raising her voice to the patient. 'You can have dinner in your room tonight, Sabira. Would you like that? We usually have a special treat on a Saturday.'

Sabira stared dumbly back, barely seeing the face before her. Nurse Medway didn't like the lack of focus in her patient's eyes, and made a mental note to check her prescription dose. Paroxetine was an anti-depressant that helped treat panic disorder and social anxiety, just part of the cocktail of drugs her doctor had insisted on administering, but she wondered if it was cross-reacting to cause somnolence.

'Let's not take the stairs,' said Nurse Medway cheerfully. 'We'll use the lift for a change.' Looking down, she noticed the freshly stained bandage. Sabira felt her hand being raised, but offered no resistance. 'And we'll get that nasty old dressing changed for you while we're at it.'

22

AT HOME

Nobody answered the doorbell.

As usual, it was a war of nerves to see who would last out the longest. Alma Sorrowbridge was elbow-deep in kitchen soapsuds and Arthur Bryant was on his hands and knees, trying to reach an eyeball that had rolled under the bed.

The doorbell rang a third time, staccato and impatient. 'Can you get it?' called Bryant. 'I've lost Rothschild's eye.'

The outcome of any battle with Bryant was pre-ordained. With a sigh, Alma dried her hands and headed for the front door. She opened it to find a bull-necked man in a stained white wifebeater vest. He was staring angrily past her. 'Where the bloody hell is he?' he demanded to know.

'Mr Bryant, it's for you,' said Alma, heading back to the kitchen.

Bryant stuck his head around the door and raised an eyebrow. 'Who are you?'

'I'm Brad Pitt,' replied their next-door neighbour, 'and I want a bloody word with you.'

'I'm rather busy right now,' said Bryant, still appearing

as little more than a disembodied head. 'Do you want to come back when you've got dressed?'

'Are you trying to be funny?'

Bryant's brow wrinkled in puzzlement. Why did people always ask him that? 'Brad Pitt the actor? You don't look anything like him, not unless he's been in prison. Very well, I suppose you'd better come in for a moment.'

The neighbour, whose overdeveloped body clearly kept him from walking with his legs together or his arms by his sides, made his way down the narrow hall. When he looked into Bryant's study, his mouth fell open. 'What the bloody hell are you up to?'

On the desk in front of him was half a cat. To be precise, the bottom half of a stuffed, sandstone-coloured Abyssinian cat. The top half lay on its side, filled with straw and old newspapers. 'Oh, that's Rothschild, my old friend Edna Wagstaff's spirit medium. I'm afraid he's past his peak condition. I was trying to restitch him, but this popped out.' He held up Rothschild's glass eye. 'Edna sometimes used him to summon the ghost of Dan Leno.'

The neighbour gawped at Bryant as if he had started speaking Chamicuro.[2]

'Of course, Leno was a music-hall comic, so his advice wasn't very useful. But he did give us the details of his clog-dancing routine. You can sit down if you don't get dirt on anything. Who are you again?'

'The poor sod who lives next door to you.'

'Do you have a name?'

'Just Joe to you.'

'Well, Just Joe, what seems to be the trouble?'

'Do you want a list?' Joe scrubbed a hand through his stubbled hair, then ticked off his fingers. 'One, there was a smell like burning rubber and rotten fish coming from

---

[2] Endangered Peruvian language now spoken only by seven elderly people in the world.

in here at three o'clock yesterday morning. Two, there was a noise that sounded like someone rupturing a duck just as I was trying to get my kid to sleep last night. And I'm guessing it was you who left the pig outside my front door yesterday.'

'Ah. Well, they tried to deliver it while we were out, you see.' Bryant had arranged for his old butcher to drop off a pig carcass that was past its sell-by date, but the butcher had been forced to dump the meat outside Bryant's new flat after a traffic warden threatened to have his van towed away. Pigskin was genetically close to human flesh and ideal for experimentation. Bryant attempted to push a trotter under his desk as he was talking but the pig rolled into view. It had a dozen pub darts sticking out of its flank. 'I'm trying to prove something,' he explained lamely.

'I think you've already proved enough, mate,' said Joe. 'When they took the last bloke out of your flat I told the council to warn us if they were thinking of renting to another nutcase, but I must have missed the call.'

'I think we've got off on the wrong foot,' said Bryant, forcing Joe to shake hands with him. 'The smell caught me by just as much surprise as you. I needed to reach the boiling point of an ammonia-potassium compound and accidentally burned through one of Alma's saucepans, so I set it down on the hall table, but the varnish melted and reset, so I had to unglue it with a blowlamp. And the noise you heard was actually atonal avant-garde German music from the school of Schoenberg.'

'It sounded like it was from the ironmongers down Chapel Street market, and if you do it again after lights out I shall come round and fetch you a punch up the bracket,' warned Joe. 'Are we seeing eye to eye on that? What are you doing with a dead pig, anyway?'

'That's something I can't tell you while the case is still open,' Bryant admitted. 'I'm a police detective.'

The news alarmed Joe, who in his time had been no stranger to the world of stolen goods. 'How long ago did you retire?'

'I haven't.'

'Well, you might want to think about it before we come to blows.'

'There's no need for that, we'll be fine from now, I assure you. It's just that I'm new to all of this.' Bryant waved his hands around the wall with vague distaste.

'All of what?'

'You know – tenement living. Here. Slum dwellings. The stews. The rookeries.'

Joe looked at Bryant as if trying to tune in a broadcast on a broken radio. 'What do you mean, rookeries?'

Bryant sighed impatiently, struggling with the effort of communication. 'Poorly constructed low-quality housing with third-rate sanitation constructed in overcrowded, impoverished areas for the poor, often occupied by criminals and prostitutes,' he recited, 'named after the nesting habits of the rook, a bird that constructs large, rowdy colonies consisting of multiple nests—'

He got no further, because Joe had lifted him two inches off the floor by his shirt collar. 'Are you saying I'm a criminal who lives in a dirty house?'

'No, I'm referring to traditional Victorian definitions of working-class accommodation, although Albion House is technically an Edwardian construction dating from 1909, I believe. Could you put me down? I have a weak bladder.'

Joe quickly lowered the elderly detective to the floor. 'So why are you staking the place out? Who are you after?'

'Nobody. I live here. Mrs Sorrowbridge and I were left destitute after the council placed a compulsory purchase order on our old property in Chalk Farm, so they rehoused us here. We're your new neighbours.'

'I'm pleased to meet you,' said Alma, shyly coming

forward with an outstretched hand. 'I'm just making cabinet pudding with ginger tea, if you'd like some.'

Joe was flummoxed. 'Er, no thank you, ma'am,' he said politely. 'I'd best be getting back.' He glanced uncertainly at Bryant. 'As for you, try and keep the weird noises and funny smells down after midnight, will you?'

'My fault entirely,' Bryant called back. 'I'm not used to community living, but I'll soon get the hang of it. Do drop by again some time. You never know, you might be able to help me in my investigations.'

This last remark only served to quicken Joe's pace to his front door.

23

THE FOURTH SOLUTION

As Bryant watched his neighbour go, his attention was caught by a man standing below the balcony in the street, looking up in puzzlement. 'Raymondo, is that you? What are you doing here?'

It being a Saturday, Raymond Land was attired in civvy clothes, which consisted of the kind of trousers one saw advertised in the backs of local newspapers, and a Marks & Spencer's shirt that would have looked unfashionable on Denis Thatcher. It was as if, after his brief flirtation with youthful fashion, he had thrown up his hands in defeat and fallen into a sartorial black hole.

'I couldn't remember your door number,' he called up. 'I needed to see you.'

That was Alma's cue to head for the kitchen, leaving Bryant to welcome the acting chief of the Peculiar Crimes Unit. 'If any more of your little pals are going to drop in,' she called, 'could you let me know?'

'Is it all right to leave my car down there?' said Land, arriving up the staircase. 'It won't get broken into, will it?'

'Certainly not,' said Bryant, ushering him in. 'This is a

semi-respectable neighbourhood. Don't tell me you were passing by and thought you'd drop in.'

'No, I had an appointment with Edgar Lang.' He looked around. 'This is very nice, better than your last place. Very homely. Lang asked to see me. Not about Kasavian's wife – there's nothing much that can be done on that score until after the psychiatric assessment. About his boss's appearance before the European Commission.'

Bryant was amazed. 'Since when did important people start asking your advice?'

'My opinion is valued in some circles,' sniffed Land. 'He wanted to know how Oskar was bearing up.'

'Why would he want to know that?'

'I suppose he's protecting his assets.'

'What happens if Kasavian messes up and the Deputy PM fails to award him a new senior position?'

Land thought for a moment. 'He won't be able to remain as chief security supervisor. He'll have nailed his colours to the mast, trying for a new department outside of the Home Office. I imagine he'll be reshuffled.'

'And who's next in line for the position if he goes?'

'Well, I imagine it's Edgar, as he's technically Kasavian's number two. After that, it would be Charles Hereward.'

'So you could say it's in Lang's interests to see that Oskar's initiative fails. And if the wife is sectioned and Kasavian is forced to take time off to look after her, there'll be a no-confidence vote and Lang can step in again. In fact, every way you play it, first Lang and then Hereward and the others win, because even if the doctor decides that Kasavian's wife is fit enough to go home, she'll still require special attention.'

'Not everyone's as Machiavellian as you,' said Land, exasperated. 'Lang just wanted to know how the investigation was progressing.'

'I hope you didn't tell him,' said Bryant. 'I'd love to bottle your innocence and sell it to old streetwalkers.

You do realize this initiative has been in various planning stages since the 7/7 bombings? It was pushed on to the last Labour agenda by the US, then redesigned by the Tories. It's a total overhaul of the UK's terrorism security system, and one man will be in charge of it. Except that now maybe he won't be, because seven weeks ago his wife started going mad.'

'I hadn't thought of it like that,' said Land in a very small voice.

'No, of course you hadn't,' said Bryant, giving him the sort of valedictory pat one would give a horse just before having it shot.

'I have to admit, I'm at a total loss with this one.' Land accepted a consolatory plate of cabinet pudding from Alma, who had decided he needed feeding. 'Death, madness and political suicide, all knotted together in one unfathomable bloody case. I mean, I know Oskar Kasavian is a hatchet man but he's losing the love of his life. Believe me, I know how that feels.' He chewed ruminatively. 'She's left me, Arthur. I don't suppose you know how to use a microwave oven.'

'I thought you and Leanne were getting on better after you confronted her about her affair?'

'So did I. But now she's told me she wants a divorce, and she'll be moving the rest of her stuff out next week. I've been married for twenty years. I always assumed we'd be together for ever. Before she met me, she'd never even been on a caravan holiday. She had absolutely no idea how to change the rotor blade on a lawn mower. She didn't even know that you could save a fortune by shopping with coupons. I taught her everything.'

'Perhaps she wants to go places you can't take her.'

'What – Barbados?'

'I don't mean literally. I mean in her head. Maybe she needs more mental nourishment.'

'She won't get it from a flamenco instructor in Cardiff.'

'Perhaps we'd better stick to the case for now,' said Bryant, exhibiting rare tact.

'Good idea. It'll take my mind off the thought of having to do my own washing. What are your new neighbours like?'

'Brad Pitt lives next door.'

'He's come down in the world. So, what's John's take on all this?'

'My partner thinks more logically than I do. He reckons there are only three possible solutions. One: Sabira Kasavian is genuinely undergoing mental problems brought on by a return to drug dependency, which would mean it's nothing more than unfortunate timing that she should become ill during the most important month of her husband's career. But we can rule out that possibility.'

'How so?'

'First, cocaine stays in the system for three days, and the clinic ran a urine test as part of her admittance procedure. She was clean, so the rumours about her are false. And if this was all just coincidence, it would mean that Amy O'Connor died a natural death and Jeff Waters was attacked in the park by chance. Now, there are fifty-five churches in the Square Mile alone, and that's not counting the non-English ones, yet O'Connor used the same one that Waters visited.'

'OK, unlikely, I agree.'

'Ben Fenchurch at St Bart's recorded an open verdict on O'Connor, but Waters was hit by a professional. The attacker didn't touch his wallet and mobile, and kept him out of the line of the CCTV cameras around the park. We're still checking Waters's electronic diary and address book, trying to find out if anyone hated him enough to have him killed, but it's not looking good. Which means that it's unlikely these events were entirely coincidental.

'This brings us to solution number two: Sabira Kasavian is faking her madness in order to destroy her husband's career. Now this is an interesting one. First, it fits with what we know about her. She boasted to her best friend that she would come to England and find herself a rich husband. She grew dissatisfied with him, and came to hate all of his friends and the world in which they moved. She felt trapped and shut out of society, and decided to take revenge for the way in which she had been treated. She could do this by wrecking her husband's career and paving the way for a favourable divorce settlement. The next step requires a leap of the imagination.'

'Go on.' Land sat wide-eyed with a forkful of pudding halfway to his mouth.

'One person knows too much about her. The man in whose apartment we found two of her hairs. So she has to ensure he won't ruin her plans. She pays to have him killed. This leads to two problems. First, how did she pay Waters's killer? We have no evidence of money being transferred from her account. Second, where does that leave Amy O'Connor? Is she entirely coincidental to the whole sordid business, or did Sabira meet her and possibly fall out with her at some point in the past?'

'Well?' asked Land. 'What's the answer?'

'Janice and Jack are still working on that one. It's not easy going through personal histories when, of the two people involved, one is dead and the other appears to be in a state of nervous collapse. Which brings us to the third solution.'

'That Sabira Kasavian is being framed,' said Land.

'Well done, you. Now the pieces slot together much more easily. This time the motive is power. Someone wants Kasavian's position as the man with the ear of the Deputy Prime Minister, and sees a way to achieve it. We know from our own dealings with the security supervisor that he's not a man to be trifled with. But he has one weakness: his beautiful, fragile young wife. If she can be taken out he'll be weakened, and if this can be done publicly so that everyone knows of his difficulties, there can be no cover-up. Oskar Kasavian used to date Janet Ramsey, the editor of *Hard News*, but even she refused to hold off the damaging stories – like all dyed-in-the-wool journalists, she's a scorpion who can't resist the chance to sting.'

Alma popped back in and left a fresh pot of tea from sheer force of habit.

'Where does this solution leave the deaths of Waters and O'Connor?' asked Land.

'We already have a connection between publisher and photographer; she employed him. It might be that Waters had a big mouth and too much inside information. The same goes for O'Connor, who is connected to Waters by the child who was at both death sites.'

'I'm getting confused,' Land admitted. 'How does all this connect to Mrs Kasavian?'

'It seems likely that Ana Lang talked to Sabira about work matters, although she doesn't recall any specific conversation that might have set her off. So we have two deaths and a case of mental instability, and it's possible there are threads between them all. Unfortunately, the most feasible solution is also the hardest to prove, because in this scenario the killer appears to have thought of everything. We've been unable to form a proper link between Amy O'Connor and Jeff Waters, and the evidence has been carefully stacked against Sabira Kasavian. It appears she slept in her photographer's bed, and will be responsible for destroying her husband's career.'

'But there's a flaw to the killer's plan,' Land pointed out. 'All we'd have to do to catch him is wait until he achieves his ends.'

'Ah, *cui bono*, precisely. Once it's possible to identify the victor with the spoils, the case solves itself. Unless that's part of the plan too.'

'What do you mean?' asked Land, deflated.

'We're dealing with politicians,' said Bryant. 'People

who are capable of formulating schemes over a long period of time and studying them from every angle. The most successful crime is the one that nobody knows has been committed, yes? Let's imagine that next week, Kasavian's wife is sectioned and Oskar fails to reassure the European committee, so he's reshuffled and Edgar Lang is appointed in his place. This doesn't necessarily prove that Lang is the culprit.'

'It doesn't?'

'Oh, Raymond, do try to think. Who benefits?'

A light went on above Land's head. Quite literally, as it happened, because at that moment Alma came in and switched on the standard lamp behind him. 'Hereward, Lang or Almon would be poised to take over,' he said. 'They might even have hatched it between them. Or they could have another partner we've yet to meet, someone who has successfully kept themselves out of the limelight.'

'You see, now you're using the parts of your brain you don't usually use, which in your case is pretty much all of it,' said Bryant cheerily. 'I mentioned that it was John who came up with the Three Solutions theory.'

'It doesn't matter whose idea it was—'

'In this case it does, because I intend to propose a fourth solution. The Arthur Bryant solution.'

Land groaned silently, but even so a little bit of disappointed noise came out.

'And this solution involves witchcraft, madness, secret codes and an ancient London myth – perhaps its greatest and most terrible secret.'

'Oh no . . .' Land buried his face in his hands.

'But I don't want you to worry,' Bryant reassured him. 'I'm on the case. Although in the coming week, you may find me doing some very strange things.'

Land winced. 'How strange?'

'Oh, strange even by my standards.'

'But why?'

'Because I think we have to chase down the culprit in order to prevent the deaths of others,' said Bryant. 'I knew we were wrong to try and do this by the book. There's no more time left for strategies and suppositions. We need to start fighting back.'

Land looked faintly unwell. 'Is this likely to get us into trouble?'

'I should think so, yes.'

'But you think it's the only way?'

'I'm afraid so.'

'What should I do?'

'Get ready to take cover,' said Bryant. 'There'll be a lot of flying debris.'

PART TWO

The Chase

24

THE ESCAPE

Sabira Kasavian was beset by devils.

She opened her eyes to find them crawling up the bedspread towards her face. Tiny blood-blackened imps with scratchy claws and seed-like eyes, they pulled at the counterpane and tried to slip under the sheets with her.

Screaming, she awoke.

Amelia Medway appeared at the door within seconds. 'Were you having a bad dream?' she asked solicitously.

Sabira decided not to answer. Speaking to the enemy would only place her in danger. She checked under the sheets and scratched furiously at her hair.

'Come along, then, sleepybones, it's nearly nine o'clock.' Medway parted the curtains on the kind of glorious summer morning London sometimes springs on the unwary, momentarily fooling them into believing the weekend will be fine. 'You can have breakfast in your room today. We have some new arrivals. Everyone's rushed off their feet downstairs. Get yourself into the shower and I'll bring you up something nice in a few minutes. I think it's time you washed your hair, don't you?'

She looked around the room, checking that everything

was in order. A pair of nail scissors had appeared in the top of Sabira's make-up bag. 'I think I'll have to take these for the time being,' she said, pocketing them. 'Just in case of accidents.' She wasn't going to be caught out twice.

Medway was not prepared to take any nonsense today. Spike, the burned-out musician across the hall, had already thrown his toys out of the pram this morning after having his tattooing needles confiscated. It always surprised her how many of them were like children, squabbling over places at the dining tables, accusing others of taking their seats and stealing their magazines. Much of it was connected to the physical symptoms of withdrawal, of course, a rising and falling ache that contracted the muscles and burned the stomach lining.

Sabira Kasavian had dabbled with drugs, but wasn't an addict. She was clever and bored and fearful, but there was no craving in her eyes. Medway knew all the signs and found few in her new patient. So much rubbish was talked about addiction, now that everyone was an expert. Just last night some television doctor had been telling his audience that there was no way of stopping someone who wanted to do drugs, that addicts had to hit rock bottom before they could start healing, that they had to want help; all stupid, wrong advice in her opinion.

Sabira puzzled her, though. In one way she seemed exactly the type who might end up inadvertently escalating her use of recreational drugs, but there were glimpses of a steelier woman inside. It didn't add up. And it seemed there was a pattern to her behaviour, an erratic but calculable change of mood as each day progressed.

She took one last look at the wild-haired young woman sitting lost on the edge of the bed and resisted the urge to reach out to her. Some of the patients were devious, and if you became involved in their power plays they would make sure you'd suffer. She went to get the breakfast.

Sabira walked to the window and looked out at the

day. The milky mist was burning off, and would unveil a pitiless blue morning too grand to last. Such mornings were dangerous; everything could be seen. She pinched the skin on her right arm hard, rousing herself from her torpor.

It was time to move fast.

There was no point in heading for the front door in daylight. The manager's office was right beside it, and she always kept her door propped open. Sabira pulled on jeans, a sweatshirt and trainers, added a scarf, and then went back to the unbarred window. The room beneath hers had a bay, which meant she could climb down on to its slated roof. It belonged to a woman called Francina who was in thrall to some vague pervasive sorrow and never got out of bed before noon. With any luck she would still be asleep.

Sabira opened the window as quietly as she could and climbed out. The drop seemed greater from the outside and she had second thoughts. But even falling and breaking her ankles would be preferable to another night spent with the tiny crawling creatures that invaded her dreams.

She stretched as far as she could and let her trainers find purchase on the angled roof of the window. She could see that the room below still had its blinds closed, so she was able to take her time, gripping the guttering and carefully lowering herself on to the sill before dropping to the wet grass.

She felt sure she would be seen if she went up the garden, so she ran down the alley at the side of the house and opened the gate into the front, quickly swinging behind the box hedges, where she could not be seen from the manager's window.

Then she set off down the hill towards the Tube station.

London starts slowly on Sunday mornings, but the walkers were up and out in Hampstead, reining in their straining dogs until they were within sight of the woodland, then slipping their leads and yelling out names intended to show that their owners were also from good breeds: 'Jasper!', 'Montmorency!', 'Florinda!'

Sabira kept her head lowered and her pace steady. She had her mobile and purse, but no credit cards and just a small amount of cash. She needed to formulate the next part of the plan. If she had stayed at the clinic they would have found a way to kill her before the psychiatric hearing, just to be on the safe side. She had to make the detectives understand. Mr Bryant seemed the most sensitive to her situation. But she had to let him know without alerting his superiors. What could she do?

Her enemies had occult powers, of course, ways of watching her from afar, ways of turning her purpose and changing her direction. They could cloud her brain and render her dumb, or invisible, or simply mad. They could reach her anywhere. But this morning something had gone wrong. They were still asleep, perhaps, for she could see things clearly for the first time in days and act upon her instincts.

She had a chance.

Behind Hampstead's white pillars and grand porticos were council estates, flats for the porters and nurses of the Royal Free Hospital, lives less celebrated but more essential to the running of what residents still termed 'the village'. Sabira had to be suspicious of them all. No turning head could be trusted. Trotting briskly to the Tube, she descended in the lift and found a southbound Northern Line train waiting at the platform with its doors open. She avoided the most obvious carriage and moved further down, watching to check that no one else boarded, but her view was blocked by an immense woman festooned with shopping bags.

The train sat in the station with its doors wide while the driver apologized for the delay. She had no clear destination in mind, and badly needed to order her thoughts for the day ahead. She needed to be somewhere calm and safe.

But that's what poor, scared Amy O'Connor had told herself. She had gone into her church and died. Even there, they had got to her.

Well, it wouldn't happen again. She could think properly now, although who knew how long this clarity would last, and she wasn't scared but filled with righteous anger.

Alighting at King's Cross she changed lines, heading for Great Portland Street. There was one place where she would be safe long enough to come up with a stratagem.

The unit was usually closed on a Sunday, but today Raymond Land had asked for all staff to be on hand.

At Bryant's request, the emphasis was on May's third solution: that someone in power was framing Sabira Kasavian to destroy her husband's career and advance themselves. But everyone could sense that even as they made calls and conducted interviews, the traces of guilt were being covered over. It was a race they could feel themselves losing.

Raymond Land usually enjoyed being in the office when it was like this: quietly humming with purposeful activity. If he was honest with himself, he was no longer upset that the case hadn't been solved after a week; being here stopped him from thinking about his suddenly empty home. People only really cared about themselves. It wasn't their fault – Banbury always said humans were hardwired that way. If you applied that to the case, for example, a real survivor would do everything necessary to keep a secret, and keep on keeping it until it was truly buried again . . .

He hurried to Bryant's office.

'Arthur, I've been thinking. If we accept your idea that everything is linked, this has to have started in O'Connor's past, and it occurs to me that we still know nothing about her.'

'I'm way ahead of you, old sausage,' said Bryant. 'That's why I'm going to Bletchley this afternoon.'

'That's somewhere near Milton Keynes, isn't it? I've no details of her ever living there.'

'She didn't. I'm looking up an old friend of mine. Didn't I tell you? I've received a most suggestive communication from Sabira Kasavian.' He held up a Hallmark greetings card featuring a cartoon duck and the words 'Thank You!'

Land was puzzled. 'Did Banbury find that in O'Connor's flat?'

'No, it's from Sabira to me. Posted to the unit yesterday. I'm afraid I only just got to it.'

'It looks like a thank-you card.'

'That's exactly what it is.'

'I don't understand.'

'I do. She's sending me a message.'

'Yes, it says thank you.'

'No, it says something else. I'll be back with the answer in a couple of hours.'

The London Central Mosque stood at the edge of Regent's Park near the top of Baker Street. It had been founded during the Second World War in recognition of the British Empire's substantial Muslim population and their support for the Allies during the war. The Churchill War Cabinet requisitioned the site and King George VI opened an Islamic cultural centre. A mosque was the crowning touch, adding a golden dome to the London skyline.

It was the first mosque Sabira had visited in London, and still the place where she felt most at home. The minaret and dome provided a change from the surrounding 1970s architecture, and on a Sunday morning the park opposite was filled with walkers.

Slipping off her shoes and placing them on the rack,

Sabira covered her head with her scarf and slipped inside. The thick red carpets and great gold chandelier above her reminded her of the mosque in Tirana; unlike the churches she had visited in England, most mosques were fundamentally the same, the better to concentrate one's attention upon prayer and reflection, but this one had immense windows filled with peaceful greenery.

She made her way up to the women's balcony, thinking, *This is how she died, ending her fears within the sight of her God. If I am to be taken, do it now and cleanse me of my fears.*

But she did not die. After an hour of reflection, Sabira rose and left the mosque. Now that she was thinking clearly once more, she considered calling the PCU to check that they had received the card – but what if her calls were being monitored?

While she had been in the mosque an idea had formed in her head; she had no proof of any kind to offer, and yet there might still be some gesture she could make that Mr Bryant would pick up. It had to be something her tormentors would never understand. Assuming the detectives were smart enough to appreciate her motives, she would give them as much as she dared, then head for Victoria Station to lose herself somewhere in its rural branch lines.

She set off in the direction of London's law courts.

At Holborn Tube station she alighted and tried to see if anyone was following her, but the platform was crowded with Sunday tourists now, and she had no idea whom she might be looking for.

Slipping behind the station, she headed into the green square of Lincoln's Inn Fields. Here, in the dappled shadows of the great plane trees, she felt protected. It seemed that fate had drawn them all in this direction. It had brought Amy O'Connor to St Bride's Church just up the road, and Jeff Waters to Coram's Fields, within five minutes' walk of here. Now she too had returned, to leave a trail that only someone with a very particular turn of mind might think of following.

Across the road was number 13, Lincoln's Inn Fields, one of the strangest buildings in London. Without looking back she climbed the steps and pushed open the front door.

There was nobody at the reception desk. The great room to her right appeared to be empty as well. Ahead was the narrow hallway that led to the stairs and the secret gallery. But now that she placed her foot on the first stone step, she started to feel a familiar dizziness.

She could not allow the imps to return. Swallowing hard, she rubbed at her arms and climbed the gloomy stairway, but she could see them suddenly swarming in from every corner of the dark, divorcing themselves from the dusty shadows, the heavy velour curtains, the bookcases and skulls, the busts and statues.

You don't exist, she shouted silently. But the house was designed for ghosts, and she realized she had fallen into a trap. There was no worse place to be in the whole of London for having tricks played on the mind.

Fighting blindly on, she made her way to the first floor, but the imps were batting at her face and arms, scratching at her legs, and now the voices began, urging them on.

I know what I must do, she told herself, *and all the demons of hell can't stop me.*

25

DEATH'S PUZZLE

Longbright broke the cordon of yellow plastic tape and rolled it up. 'Let's not draw attention to this,' she told Renfield. 'Keep the door shut. If anyone wants to come in, they can knock.'

She stepped into the hallway of number 13.

'There's supposed to be somebody on the door all the time,' said Renfield. 'She's a voluntary helper. Things were a bit slow because the listings magazines had the place down as closed. Usually it's open from Tuesday to Saturday, but they just added Sunday mornings. She went off to get a cup of tea. Said she wasn't gone longer than a couple of minutes. She's very upset. The EMT couldn't hang around; they had an emergency call to Holborn tube station so I let all but one of them go. He's upstairs.'

'Wait until Kasavian finds out what's happened,' said Longbright. 'Have you managed to get hold of Arthur?'

'His phone's off, or maybe his train's in a tunnel. He didn't drive to Bletchley, did he?'

'God, I hope not. John's on his way over with Dan. He's trying to find a place to park. Security's tight around here.'

London's legal centre consisted of two Inns and two Temples, each complex with its own great hall, chapel, libraries and barristers' chambers, built around a garden covering several acres, like an Oxford college. They operated as their own local authorities, a miniature city within a city.

May was not far behind them. 'Arthur's going to blame himself for this,' he said, looking around as he entered the hall. 'What the hell was she doing in here? Are there any other visitors?'

'A few of them on the lower ground floor,' said Longbright. 'Chinese architectural students.'

'Nobody on the first floor or above?'

'No, just downstairs.'

'You'd better go and take statements. Does everyone have to sign in when they arrive?'

'It's not compulsory.' Longbright pointed to an old-fashioned leather-bound visitors' book lying open on the hall desk.

'I'm thinking the killer probably wouldn't have signed in,' said Renfield sarcastically.

'No, but I want to see if the victim did.'

May checked down the brief list of the morning's names and found Sabira Kasavian's signature clearly written. It was the sort of thing Bryant would have checked at once. 'Five past eleven,' he noted. 'She left the clinic at nine. Why did she take two hours to get here? Where else did she go? No sign of the City of London Police, then?'

'I think as soon as they got the victim's ID they stepped back. Home Office turf.'

'I presume someone's spoken to her husband?'

Renfield looked blank. 'We figured you would do it. Not sure anyone else is up to the job.'

'Great, thanks for that. I can't wait to tell our client that his wife is dead.' He checked his watch. 'Where's Dan? I need to get some idea of the situation before I talk to him.'

Banbury was coming up the stairs with an aluminium box and a large clear plastic object that looked like a giant breadboard.

'What's that?' May asked.

'It's an anti-contamination stepping plate,' said Banbury. 'It means you can walk within the crime scene without messing it up. I thought Mr Bryant might be here. You know what he's like.'

'Go on, then, lead the way. We need to get started as quickly as possible. What else did you bring?'

'I didn't know what we were dealing with so it's a bit of a grab bag for now.' He slapped the side of the box containing a swab dryer, biohazard precautions, chain-of-custody labels, a haemostat, a trace-evidence collector and presumptive blood IDs. 'I meant to bring a laser measurer – the wife bought me one for my birthday, but my nipper flattened the battery trying to confuse the cat with it.'

Everyone followed Banbury. 'Is it possible to get some more light in here?' he asked.

'There is, but not in the normal way,' said May. 'What do you know about this place?'

'I've walked down this street loads of times but I've never noticed it. Some kind of museum?'

'Sir John Soane was an architect. This is his house and his shrine. There are classical, medieval and renaissance antiquities here, a lot of paintings and drawings, death masks, statuary, artefacts, glassware . . . you name it. You won't be able to take anything away without a lot of paperwork.'

'What did you mean about the light?' Banbury asked.

May pointed up. 'You know how dark terraced Regency houses can be. When Soane used to teach students here he wanted them to be able to see their work properly, so he arranged a system for reflecting light down through the atrium from the skylight via a series of tiltable mirrors hidden throughout the floors. I think there's one in the pedestal of a statue.'

The house was an extraordinary clutter of alabaster, marble, gilt, ironwork, stained glass and wood. Every wall, every pillar, every arch, every ledge and surface was covered with arcane pieces. The corridors and open-sided rooms all appeared to be interconnected, turning the house into an indoor maze. Once it had been entered, first-time visitors were rarely able to find their way out easily.

'I'm glad Arthur isn't here,' said May. 'We'd never have been able to pay for the breakages.'

They were met on the landing by Catherine Porter, one of the current custodians. London's more personal museums were guarded by an army of middle-aged ladies who knew how to get chewing gum off plasterwork and orange juice out of tapestries.

Porter clearly took what had happened as a mark of personal failure. 'I feel dreadful about the poor young woman,' she said. 'I never leave the front door but it was so quiet this morning. Usually there's a queue outside before we open, but people haven't got used to the Sunday hours yet. She came in and must have gone straight upstairs to the picture gallery.'

'You don't have cameras?'

'No. We're old-fashioned; we have attendants. Visitors were always allowed to open the gallery by themselves, but we were worried about wear and tear so Terry, our chap on the first floor, operates the display himself every fifteen minutes. But he called in sick this morning.'

She led the way through the claustrophobic maze of art-works and unearthed treasures until they reached a tall, perfectly square windowless room of unnatural height, lined with wooden panels. On each of the walls, paintings were displayed beneath each other in pairs.

The body of Sabira Kasavian lay crumpled in one corner

of the room, her legs folded beneath her, right arm tucked beneath her torso, the left raised above her head.

'Has anybody touched her apart from the EMT?' asked Banbury, kneeling.

'You mean the ambulance man?' said Porter. 'No, he checked her for signs of life and immediately called you. I just touched her sweatshirt. I thought she'd fainted and fallen, but I could see she wasn't breathing. I thought it best not to try and move her.'

'You did the right thing. You're sure there was nobody else on this floor or above it?'

'Positive.'

'Are there any other exits up here?'

'No,' said Porter. 'There's only one way in and out, through the front hall. The back door leads into a small courtyard, but you can't get out from there.'

'So she died alone. Poor kid. She came to London looking for a better life, and all she got was ridicule – and this. What an end.' Banbury turned to the young Indian medic who waited at the edge of the room. 'Does that look like cyanosis to you? Her skin colouring's wrong.'

'We checked for blockages,' said the boy. 'There aren't any outward signs of trauma. The wound on the wrist—'

'We know about that. Anything else?'

'You'll have to get a toxicology report.'

'And this is the exact position you found her?'

'We put her back after checking her status. She was like that, the left hand extended.'

'Seems wrong to me,' said May. 'She was right-handed. And why would she have taken the dressing off?' He looked back at the livid red slash on her wrist, then knelt down. 'Shoes.'

Banbury looked at the black leather trainers. 'What about them?'

May pointed to the left. 'Looks like she tried to kick one off.'

'It might have come loose when she fell.'

'No, Dan.' May slipped his hand into a baggie and wiggled the toe of the shoe. 'Look how tight they are. She pulled it off.'

'I don't see what you're getting at.'

'Neither do I, but let's take them anyway. Bag them up for me, would you?'

Banbury finished documenting the body position and allowed the technician to step in and shift the body into a slender white bag with carry-handles. Together they carried it down to the hall passage to await the return of the ambulance.

May checked Banbury's pictures. 'It couldn't be clearer if she'd written in blood,' he said. 'She fell but still had the presence of mind to point at something. That wasn't a natural position.' He thought about the line of her extended hand and found himself looking at a sixteenth-century engraving of Covent Garden and a drab painting of Bristol's Avon Gorge Bridge. 'But that's not a lot of help.'

'Sorry, there was one other thing,' said Porter. 'We closed the walls because I thought it wasn't a good idea to have visitors looking in.'

'What do you mean?'

The curator pointed to the polished brass handles that were set in the walls. She reached over and pulled the first one.

The wall was hinged in the middle and folded back, revealing a second wall of paintings. She proceeded to walk around the little gallery opening each of the walls in turn until she had revealed a completely different set of pictures. May was amazed to see another set of handles on these inner walls. Porter pulled at these and the second set of walls folded back on three sides. The fourth pair revealed that the room was, in fact, open to the atrium on one side. From here you could see into the rest of the house, up to the skylight and down to an empty Egyptian sarcophagus in the basement.

'Sir John Soane had trouble storing all of his treasures,' she explained, 'so he arranged for the paintings to be hidden behind each other.'

'Wait – which one of these sets of walls was open when you found the body?'

'The middle ones.' She closed the set of doors above the spot where the body had lain.

May found himself looking at William Hogarth's paintings of *The Rake's Progress*. 'Tell me she wasn't leaving us a clue,' he groaned. 'Nobody does that for real.'

'It's either the paintings, the ceiling or heaven,' said Banbury, opening his kit and setting to work. 'Ask Mr Bryant. If it's something to do with one of the pictures, he's bound to know.'

26

THE CARDANO GRILLE

Arthur Bryant's taxi brought him from the station to Bletchley Park at 11.45 a.m. The mansion had been built by a wealthy city financier in 1883 and finished in a bizarre mix of architectural styles, crenellated and gabled, with Tudor beams, decorative cream stonework, turrets, arches and redbrick octagonal rooms, one with a green copper dome.

In the Second World War it had been rechristened Station X and filled with MI6 codebreakers who masqueraded as 'Captain Ridley's Shooting Party' to disguise their true identities. Here Alan Turing and his colleagues had cracked the secret of the German Enigma machine, following a path of discovery that would eventually lead to the birth of the modern computer.

Angela Lacie was a former MI6 cryptography expert who had retired to help raise sponsorship for the restoration of Bletchley Park. As she swung down the stairs Bryant saw her as he had first seen her not long after the war: an elegant no-nonsense scientist in a boxy suit, with her blonde hair in slides made from wire clips tied with beads; there had been no money for fashion items, and the MI6 girls had made their own.

Her hair was ash-coloured now but she had kept her figure and her smile. 'I wonder what it's taken to get you to visit me,' she asked, giving him a hug. 'You smell of disinfectant.'

'Wintergreen mixture.' Bryant bared his false teeth to reveal a bright green boiled sweet. 'I need your help.'

'I rather thought you might. You know I'm still under GCHQ employee rulings, so there won't be much I can tell you.'

'This won't compromise you, I promise,' Bryant assured her.

'Let's go to the day room. There's less chance of appearing in the background of tourist photographs in there.'

They walked through the ground-floor visitors' centre to the rear, where Angela pushed open a door leading to a small sunlit conservatory fitted with empty refreshment tables. She poured tea from an immense urn.

'I can't be long,' she warned. 'I've got a big Polish delegation in shortly. Polish cryptographers were fundamental to the breaking of the Enigma Code, and we're seeking to honour their contribution.'

'You mean you're trying to get money out of them,' said Bryant. 'I wonder what poor old Alan Turing would say if he could see how ubiquitous computers have become.'

'I imagine he'd be thrilled.'

'Funny how it's always society's outsiders who create the things that bind society together. Other countries venerate their national heroes. What did we do to one of our best minds? We chemically castrated him. Any biscuits going?'

'Here.' She passed him some Garibaldis. 'I suppose

you're still smoking that disgusting pipe. And I'm sure you were wearing that scarf the last time I saw you.'

'Remind me when that was?'

'The winter of 1985. Something to do with a woman found strangled in a Carnaby Street shop window. How are you?'

'Well, it never gets any easier. Sometimes I doubt myself. I'm probably having a mid-life crisis.'

'You've left it a bit late.'

'I didn't say which life.'

Angela laughed. 'Come on, then, what have you got for me?'

Bryant poked about in his overcoat and pulled out the thank-you card and envelope. 'This was sent to me in yesterday's mail. Take a look at the inside.' He passed it over.

Angela carefully opened the envelope, and holding up the card between her thumb and forefinger, raised it to the light. The back half was striated with two sets of narrow slits, the top four vertical, the bottom four horizontal. 'Oh Arthur, you didn't need to come to me with this, you could have gone to a toyshop and figured it out.'

'I'm assuming you haven't been near a toyshop in a very long time, Angela. Children only like flying robots that fire lasers now.'

'When I was a little girl I spent my days with bits of paper, making up codes.'

'And I built Spitfires, but children aren't allowed knives and glue any more. This also requires the use of a sharp knife – or a pair of nail scissors, which I believe the sender has.'

'Then I guess you already know what it is.'

'I wanted confirmation.'

Angela turned the card over and examined it carefully. 'Well, it certainly looks like a Cardano grille. A four-hundred-year-old cryptographic technique. They were still in use during the last war. I often made them with my schoolmates. I read somewhere that kids still play with them in parts of Eastern Europe. Not having money for toys makes you inventive.'

'So how does it work?'

'Easy. You cut some slots the height of a line of text in a piece of card, write in a ciphertext, remove the card and fill the rest in with anything you like. To decode the message the receiver places a similar card over the text. Where's the other part? Your friend, what did she put these over?'

'I don't know. There wasn't anything else. What's the difference between a code and a cipher?'

'A cipher's an algorithm, a series of steps for encrypting a message, and a code usually uses a key. There are hundreds of ways of doing it, but the simplest is to use a key like the Caesar shift.'

Angela took out a pen and began lettering her napkin. 'You write out the alphabet, then underneath it you do the same again, but change the order of the letters. Julius Caesar kept his messages secret by shifting the second line three places to the left. An additional trick is to start with a short keyword, then follow it with your text. If that eludes you, there are still various ways to decipher the message. You could "garden" for it, which means you contact the other party and ask them questions which encourage the use of certain words in their reply, or you can look for common repeats – words like "can", "you" and "today". Or you can overlay a second encryption. It was a great Victorian pastime, coming up with ways to hide messages. The personal columns of *The Times* were filled with cryptic messages sent between couples, sometimes over decades. One of Edward Elgar's best-known compositions became known as the Enigma Variations because he wrote part of his programme notes in a code that has never been broken.' She handed the card back to Bryant. 'You need the message page that goes with it.'

'I don't know what she did with that.'

'Then why bother to send this if she didn't tell you where you could find the other part?'

'She wasn't thinking clearly. She's been unwell for a while.'

'I'm sorry I couldn't be more helpful.'

'This message – what would it look like?'

'Well, the lettering would be of a point-size that could fit into those card slots. An old trick is to retype it and bury it in a bunch of other throwaway stuff – notes, receipts and so on.'

'In that case I think I might know where to find it. Angela, you should take off the "a" – you're an angel.'

'Well, you'll never know, will you? You had your chance once, long ago, but you didn't take it.'

'Didn't I?'

'No, you arranged to meet me and never showed up.'

'No, I distinctly remember showing up – you didn't.'

'I most certainly did, Arthur. I waited for you on Waterloo Bridge in the rain for a whole hour.'

'You should have been able to see me, then. I was on Blackfriars Bridge. I waited for ages. You must have failed to understand my message. Not very impressive for a codebreaker.'

'Oh, *Arthur.*'

'We could always pick up where we left off.'

'I think my husband and my five children might object to that.'

'Ah. No doubt.' Bryant rose and planted his hat on his head. 'Well, you were very helpful.'

'Give my regards to your lovely partner. And, Arthur . . .'

'Yes?' Bryant's watery blue eyes swam up at her so sweetly that she wanted to give him a cuddle.

'Please' – she picked a piece of lint from the frayed collar of his coat – 'do try to look after yourself.'

Bryant smiled vaguely and set off back to London, still thinking about being on the wrong bridge.

27

THE WARNING

Bryant was amazed to see the unit's common room in a hurricane of activity. 'Why the hell is your phone turned off?' May asked. 'We've been trying to get hold of you for ages.'

'It's not turned off,' said Bryant. 'I even recharged it before I left. Oh.' The object he removed from his pocket bore very little resemblance to a mobile. It was covered in what appeared to be sticky black tar.

'What on earth is that?'

'Liquorice Allsorts. I had a bag of them in my pocket. I left my coat lying on a tea urn at Bletchley.'

'Sabira is dead. I've just come off the phone from talking to Kasavian. He's – well, you can imagine. They're taking the body to Giles at St Pancras.'

'Lumme, that's horrible,' said Bryant, not sounding unduly surprised.

'Obviously, there are going to be horrendous repercussions. Take a look at these.' May passed Banbury's laptop to his partner and thumbed through the photographs of the crime scene.

'What killed her?' Bryant asked.

'We've no idea yet. There are no obvious marks on the body, but the EMT think she had trouble breathing. We've already managed to track her movements through her Oyster card and her phone. She slipped out of the clinic first thing, caught a Tube to Baker Street and went to the Regent's Park mosque, then headed down into Holborn and was found dead on the first floor of the Sir John Soane Museum.'

'The Soane? Why would she go there?'

'We've no idea yet. The duty nurse at the Cedar Tree said she'd been behaving more strangely than ever, talking to herself, exhibiting signs of paranoia.'

'That can't kill you, John. Something else must have happened. Who else was in the museum?'

'We've no idea. There was just one custodian on duty. The other was off sick. Four Chinese architectural students, a young German couple and a very old lady who restores frescoes in Italian churches. They're all in the clear.'

'Nobody else got in or out?'

'It doesn't appear so. There are no cameras inside the house but there's one on the wall next door and it didn't pick up anyone entering or leaving the front door.'

'No other exits?'

'The back door leads into a walled courtyard three storeys high.'

Bryant studied the photograph, tracing the livid crimson mark on Sabira's wrist. 'Did someone remove her dressing?'

'Not at the clinic. They think she must have taken it off after she left this morning.'

'Interesting. The arm. She's pointing. What's above her hand?'

May flicked to the next shot showing the muddy sepia paintings.

'Hogarth. *The Rake's Progress*,' said Bryant. 'She's been trying to leave evidence for us.'

'What do you mean?'

'A few weeks ago Sabira was as sane as you or I. Do you remember how vivacious she was when we first met her? I'm an idiot, I should have been able to prevent this.'

'I don't see how you could have done. Why didn't she confide in us?'

'She couldn't tell anyone connected with her husband's office. That's why she talked to Jeff Waters, and possibly to Edona Lescowitz. They were both outsiders.'

'No, Waters had a connection. He was employed by Janet Ramsey, Kasavian's former girlfriend.'

'I don't think that has a bearing on it. Those kinds of connections sometimes just happen, even in a city the size of London.'

'When you say she was trying to leave evidence, what do you mean?'

'Sabira broke out of the clinic prior to her psychiatric examination and tried to show us what was happening to her.' Bryant tapped the screen. 'She'd studied art history. She told us so herself. She headed here deliberately. I'll show you.'

He went to his room and returned with an immense, sooty, leather-bound volume, throwing it open and thumbing carelessly through the pages.

'Look. Hogarth's pictures tell stories. All you have to do is study the allusions in each one of the series, and you have a tale as complex as any novel. Look at them: each picture peppered with tiny details, signs and maps, globes and scrolls and what-have-you, all these bits and bobs mean something. There are classical references, of course, but any first-year art history student could figure out what's going on. It doesn't take a genius to tell you that when there's a black dog in the painting it means the subject is suffering from depression. *The Rake's Progress* tells the story of a spendthrift son wasting his father's fortune in boozers, marrying the wrong woman for her fortune, going to debtors' jail and ending up here, in scene eight, the painting Sabira was indicating.' His index finger stabbed at the page in question.

'In Bedlam,' said May, studying the picture.

'Indeed. The Bethlehem Royal Hospital where the mad were committed, and where Sabira was also to be sent. I think that's why she tore her dressing off. Look at the red ring around her wrist. The mark of madness. She wasn't trying to kill herself; she was sending us a message. No, more than a message – a warning. But I didn't pick up on it the first time, so she gave us another chance. She was far too frightened to talk, but if we reached the same conclusion by ourselves that was a different matter.'

'I don't understand,' said May. 'What was the message?'

'Amy O'Connor was found in St Bride's with a red cotton thread around her left wrist. I knew that it was once considered a treatment for madness, although in olden times you were supposed to attach a clovewort to the cord. Clovewort was a herb that supposedly cured lunacy. Sabira may not have known what the cord signified, but was leaving a mark on her wrist to link herself to O'Connor and the condition of madness.'

'That makes absolutely no sense. She didn't know she was going to drop dead in the museum, did she? So what was it – suicide? How did she think this "message" was going to reach us?'

'Er, I don't know.'

Bryant turned back to Banbury's photographs of the body. 'We don't know why her left shoe was half off?'

'We think it happened when she fell,' said Banbury, 'but I'm going to examine the shoes as soon as I have a chance.'

'Kasavian has demanded to see us first thing tomorrow morning,' said May. 'I assume he's going to throw the book at us.'

'Then we have to stall him while we look into this.'

'I don't see that we can do much until Giles has obtained permission to carry out an autopsy.'

'Sabira sent me a clue, a very simple code-reader, but I don't have the document it was intended to read. I think it might be in the folder of taxi receipts she brought back from the Home Office. Do we still have it?'

'It's on my desk,' said Renfield. 'Hang on.'

He brought it in and emptied it out on the common-room table.

Bryant pulled on his spectacles, withdrew the Cardano grille and started running it over folded pages of figures. 'There's something you can do,' he said while he checked the slots for codes. 'Go back into Amy O'Connor's history. Get inside her life. We need to know how the pair of them came to meet. That's the key to this. Whether she was mad or sane, Sabira left evidence that would lead us to the truth. She can still help us even though she's gone.'

The door opened behind them with a bang.

'Well done, lads, seems like you've done a grand job.' The tone was sarcastic. Charles Hereward studied each of them in turn, as if he was seeking to memorize their faces. He appeared to have been drinking. 'You couldn't manage to keep her alive, though.'

'We didn't exacerbate the problem by shutting her away in a private clinic,' said Bryant angrily. 'Who let you in?'

'We own you, pal. Oskar was under pressure from the Deputy PM to do something about his bloody wife before the Paris summit. All you had to do was keep an eye on her until she could be sectioned, you realize that, don't you? You weren't *supposed* to do anything.'

'Look here, you can't blame my team,' said Raymond Land. 'They were acting under Mr Kasavian's instructions.'

'Yes, to keep his wife under lock and key for a while, not to let her die and let the Paris talks die with her. I'm sorry, does that seem callous? Putting our work ahead of a dim, unfaithful cokehead who decided she wanted a bigger wardrobe than her Albanian leks could buy her?'

'We've done nothing that should affect Mr Kasavian's meeting,' said Land, alarmed.

'No? I put six years of my life into building the case for stronger border controls.' Hereward drifted between them like an untethered blimp. 'I started work on it three years before Kasavian even joined the department, and now all that work will be wiped away, and the clock will be reset to zero because Oskar will be in no fit state to present his case, and there's no one who can deputize for him.'

'I thought you were a supporter,' said Renfield. 'Why can't you do it?'

'Because it's more than just a change in the law. He's the one who's up for the job of co-ordinating the entire initiative, not me. He's the one who's meant to keep every tax-dodging immigrant and would-be terrorist out of this country. And now he's in no fit state to do it. Instead of studying the protocols of twenty-seven member states this week, he'll be burying his wife. And you know what? He blames himself for not being there for her when she needed him, instead of blaming you.'

'It's not our fault that—'

'I want you to know that we're going to take you down, no matter whose fault this was.' Hereward looked like an unpricked sausage about to explode. 'It's time you learned something about the department that pays your salaries. When it comes to getting rid of its enemies, it's a vindictive, irrational son of a bitch. We'll grind you into the dust, then blow the dust away. It'll be as if you never existed.'

He turned to leave, but then stopped. 'Don't think I dropped by just to threaten you. I happened to be in the neighbourhood.'

'I didn't know the currency of Albania was the lek,'

said Bryant, stacking his books in the stunned silence that followed Hereward's departure. 'Where were we? Ah yes, I think we need to conduct a search of the Sir John Soane Museum.'

Raymond Land was not a violent man, but he shot Bryant a look that could have blasted his desk into splinters and stripped off the wallpaper behind his head.

'What?' asked Bryant innocently. 'Have you never seen a civil servant explode before? You'd better get used to the sight. This is just the start.'

28

THE STRANGENESS OF CHURCHES

Bryant sent the rest of the demoralized team home at 10.00 p.m. on Sunday night. John May stayed behind, settling into his partner's huge leather armchair to keep him company, trying to stay awake while Bryant attempted to break encrypted messages in the taxi-receipt folder.

Two problems stood in the way of his success. If Sabira had been leaving a message for them, she would have written the encryption herself; he knew he wouldn't find it in a receipt or a company document unless she had forged it. Second, if it was written with the commonplace Caesar shift, he would need the keyword that preceded the coding.

'It's no good,' he said finally, poking May awake and handing him a Calvados. 'It's far too elaborate. If Sabira hid anything at all, surely she would have kept it simple. It only needed to be something her tormentor wouldn't spot. Until I can understand the nature of the algorithm, I'll have to pursue another tack. Would you see Kasavian by yourself tomorrow morning? I need you to buy me some time.'

May was exhausted and depressed. He had never felt

closer to taking retirement. 'What for?' he asked. 'You heard Hereward. Whatever happens now makes no difference. We're done for.'

'That may be, but we have to right a wrong. Sabira Kasavian was hounded to her death.'

'If you believe that, you're accusing everyone of complicity,' said May. 'We're talking about conspiracy to murder, cover-ups, perverting the course of justice and God knows what else. Don't you see what's happening? We've finally come up against something that's utterly impervious to investigation. All we can do is react to each new disaster.'

Bryant tapped his teeth, thinking. 'I need to understand the nature of madness.'

'No, Arthur, we need basic evidence, cause of death, witness IDs, not some fanciful research into the history of bloody Bedlam.'

'You're wrong,' said Bryant simply. 'No traditional approach can possibly work against something like this. I'll prove it to you in the next twenty-four hours. I'm so close, I can feel the answer shifting beneath my hands. The threads – those red threads that cured madness – they're coming together but I can't see *why*.' He held out his hands and examined their backs. 'It's right in front of me. Whoever did this has done it all before. It's second nature to him. I just need to find a way in. I know I'm probably not making any sense right now, but—'

'You're right,' said May wearily. 'You're not making any sense at all. It's late; let's go home.'

'No, you go. I'll stay.'

'Then I'll stay too.'

The light in the old warehouse at 231, Caledonian Road, burned on through the lonely night.

May felt lousy. He washed in the unit's leaky, cantankerous bathroom and had a shave, but still felt ill prepared to face Kasavian's wrath first thing on a Monday morning.

To his surprise the security chief's assistant called to tell him that Kasavian was walking on Primrose Hill and would meet him on the bench at the top in half an hour, so that was where he headed.

The day was cool and pinkly misted. From the height of the emerald mound, London was softened in a silky haze that smoothed out its mean-spirited edges. Some of the city's buildings appeared to have been speared with cranes in order to keep them from floating away.

Oskar Kasavian was slumped on the bench in his long black raincoat, looking like an abandoned umbrella. May sat some distance from him in case he decided to lash out. There was a coiled power in the security supervisor that almost everyone found threatening, but today his mood caught May by surprise.

'I used to come up here with her,' he said, looking down towards Westminster. 'I never saw the beauty in it myself. I asked her, why this park? It's so small. She told me that when she was a little girl she saw a Disney cartoon at her local cinema, *One Hundred And One Dalmatians*, and this park and the surrounding houses were in it. She liked it because it looked so neatly organized. She arrived here expecting it to look different, but it was exactly as she'd seen it in the film. That's the difference. I looked and only ever saw the exits, the railings, the inadequate lighting, the signs and the bins. She saw what this place represented. An image of Englishness, something she craved.'

May had no response. Words of comfort would sound false, so he waited.

'I don't hold you responsible for what happened. I should have spotted the signs and done something about it. I should have been there for her. One gets focused on work to the exclusion of all else. It becomes very hard to live a normal life.'

'You'll have to sign the autopsy consent today,' said May. 'There's no obvious cause of death.'

Kasavian sighed, an exhalation of air that sounded like defeat. 'I'll be at the office. The Belgians and the French are planning a final attempt to derail the border-control process.'

'Nobody needs to know about your wife just yet. We'll keep it out of the press for as long as we can.'

'John – may I call you that? There's no point in you continuing the investigation. My wife is dead. I'm withdrawing you from the case. I'll make sure there's no reflection on your abilities.'

'It's a murder investigation, Mr Kasavian. I'm afraid even you don't have the power to stop it now.'

'But your unit is under the jurisdiction of the Home Office.'

'You have the ability to direct prosecutions and investigations, but not once a murder inquiry is under way. We have a responsibility to the public. If we call a halt it means that somebody out there is tempted to kill again. Each time we catch a criminal the desire is lessened in others. Prevention of public disorder; it's a fundamental part of our remit.'

'Then you must do what you have to do. I'm going ahead with the presentation of the initiative. It won't be easy to get the work done in time, but I can't lose this as well. I'll need your findings before I leave. I can't go into the chamber of representatives without knowing what happened. I have to start putting it behind me as soon as possible.'

'Then I'll personally provide you with the report before you set off,' said May.

Maggie Armitage acted as a PCU contact point for crimes containing elements of mental and spiritual abnormality. In addition, she offered advice on anything from ghostwriting to rhinoplasty. What she lacked in logic she made up for in a kind of deranged effervescence that sometimes shed light into penumbral corners.

Today Bryant had arranged to meet her at Liverpool Street Station. The white witch and self-proclaimed leader of the Coven of St James the Elder turned up in a purple woollen tea-cosy hat, a green velvet overcoat and orange leggings. Her glasses, winged and yellow-tinted, hung on a plastic daisy chain around her throat. She looked like a small seaside town celebrating a centenary.

'The colour of vitality and endurance,' she said, pointing to her tights. 'I thought we might need it today, judging by the tone of your call.'

Bryant explained the case as they passed through the diaspora of commuters. 'The remains of Bedlam were uncovered here, right beneath Liverpool Street Station in 1911,' he explained. 'It had been on this site since the thirteenth century. The workmen found dozens of layers of human skulls. The patients had died of sweating sickness. I thought you might pick up some useful vibrations.'

'At the moment all I can feel is the Tube trains through my trainers, but I'm pretty insensitive until I've had my first cup of coffee,' she replied, taking his arm. 'I would have worn blue had I known – it was Bedlam's trademark colour.'

Bryant barely heard her. 'Madness, melancholy and distraction, that's what they attempted to cure. The so-called "moon-sick" had a red thread tied on them. Kasavian's wife didn't have any red thread – Janice looked through the clothes in her wardrobe – so she scratched a red line around her wrist and died pointing to the picture of Bedlam. She was telling me that her madness held the key to this.'

'I don't know why you thought I'd know more about being barmy than one of your textbooks,' said Maggie. 'OK, I've lit the teapot instead of the kettle occasionally and I once used Strangeways in a revivification ritual by mistake.'

'How could you revive a cat by mistake?'

'I thought he was dead but it turned out he was asleep. The ritual had a reverse effect and put him in a trance for the entire winter. I just stuck him in a box with the tortoise and he woke up in the spring. I'm not really the person to ask about madness. You'd be better off with Dame Maud Hackshaw. She's been inside, you know. Bethlem, the real one, now in Bromley. That was back when she was still getting visits from Joan of Arc. The hospital's still going strong, although they've got rid of people poking the patients with sticks for a shilling a time.'

'Tell me anything you know about madness.'

'I know a few bits and pieces. The hospital was called "Bethlehem", from the Hebrew meaning "house of bread", because its founder got lost behind enemy lines during the crusades and followed the Star of Bethlehem back to camp. Let's see, what else? Madness is known as the English Disease. Wasn't Hamlet sent here because it was thought his behaviour would go unnoticed? Bedlam was used to incarcerate political prisoners, which sounds apt to your case.'

'My thinking precisely. They considered her a political danger.'

Maggie gripped his arm more tightly. 'I'm feeling something now. Yes, a definite sensation. We must be over the site. Something's pulling me back. I'm rooted to the spot.'

Bryant looked down. 'You've trodden in chewing gum.'

'Oh.' She looked for somewhere to scrape it off. 'Did you know there were once two great statues over the entrance of Bedlam, *Acute Mania* and *Dementia*? Can you imagine how that made arriving patients feel? Half of the problem with madness is its definition. Tell someone they're crazy and they soon start acting crazy. Look at the way they dose children up these days for merely exhibiting normal healthy high spirits. What signal does that send to them? We've always thought that the human body has to be balanced in order to work properly. It was said to be made up of four humours that matched the seasons and elements: yellow bile and fire for summer, black bile and earth for autumn, phlegm and water for winter, blood and air for spring. A lot of alternative therapies still conform to those rules.'

'I've been trying to understand why Amy O'Connor died in St Bride's Church,' said Bryant. 'It has to mean something.'

'You know, there's hardly a church in the whole of London that doesn't have something unusual about it. St Bartholomew the Great in West Smithfield has the ghost of a monk who's said to haunt the church looking for a sandal stolen from his tomb.'

'They must have been very good quality sandals.'

'And there are wonderful puzzles in churches. St Martin-within-Ludgate has a seventeenth-century font with a Greek palindrome inscribed on it. *Nipson anomemata me monan opsin.*[3] "Cleanse my sin and not just my face." And I suppose you know about the devils of St Peter-upon-Cornhill? In the nineteenth century its vicar noticed that plans for the building next door extended one foot on to church territory. He bullied the architect into adding three leering devils to frighten his neighbours. You can still see them.'

They threaded their way past WH Smith and Accessorize, two senior citizens discussing esoterica in the most mundane of settings.

'Churches have become almost invisible in London,' Maggie said with a sigh, 'but they hide their own secret

---

[3] As the Greek 'ps' is represented by a single letter, this is a correct palindrome.

codes. There are compositions of hymns hidden in stone-work and all sorts of runic curses, but it all comes down to man's clumsy attempts at balancing good and evil. Madness is always seen as evil. So perhaps someone was just trying to blacken your victim's name. Did you search the church?'

'Yes. There was nothing,' said Bryant. 'Sabira's husband is about to transform Europe's attitude to policing its borders. But after a week of psychological torture, culminating in the death of his wife, it's possible he won't have the strength to succeed in pushing through our government's demands. I think Sabira knew a hawk from a handsaw. Behaving strangely allowed her to say what she felt more easily, but that isn't why she died.'

'Perhaps she was taking revenge for the way she'd been treated, and went mad in the process, like Hamlet.'

'I wish I knew. I followed the red thread and it led me to Bedlam.'

'But madness has no logic, Arthur, and you don't have a cause of death,' said Maggie.

'If anyone can find one, Giles Kershaw can,' said Bryant. 'And when it comes to being irrational you're the perfect person to talk to, so keep talking.'

'I suppose I'll have to take that as a compliment,' said Maggie, holding on to his arm. 'Let's walk a little while and go a little mad.'

29

CAUSE OF DEATH

Bryant met his partner at the St Pancras Mortuary and Coroner's Office, housed in the diseased gingerbread cottage that lurked behind the cemetery of St Pancras Old Church. He was always cheered by its connection to both Frankenstein and Dracula; in the adjoining graveyard were the tombs of Dr Polidori and Mary Wollstonecraft.

Rosa Lysandrou, Giles Kershaw's dour housekeeper, admitted them. Rosa was a natural Greek mourner. Her features appeared to have spent at least two-thirds of their life streaked with tears. She eyed the detectives warily.

'Is that a new perfume you're wearing, Rosa?' asked Bryant, sniffing the air.

'It's incense from the chapel,' Rosa replied. 'For the dead. Their spirits are all around us.'

'That's nice. I suppose it's company for you. All in black again, I see. Did somebody just die?'

'It's a morgue,' said Rosa. 'Somebody has always just died.'

'I thought perhaps you were dressed out of respect for your country's economy. Is Giles in?'

She led them along the hall to the main autopsy room and pushed open the door.

Giles was excited to see them. 'Just in time,' he said, pulling off his hairnet and releasing a mop-head of glossy blond locks. An unnerving array of body-opening tools had been rolled out behind him like a car-repair kit. 'I started as soon as Mr Kasavian emailed back his consent form, and I have a result for you. I can explain why no one else was found near her in the Soane Museum, and can probably account for the fact that Amy O'Connor died alone in St Bride's Church.'

He was standing beside a silver Mylar sheet ominously concealing a human shape on one of the steel tables. 'You heard the EMT thought she might have had difficulty breathing?'

'That suggested a poison to me,' said Bryant. 'Nobody came near her in the building, but if she'd ingested something harmful it would have taken time to work, which means Sabira took it before reaching the museum.'

'Exactly so. The problem was administration. She didn't eat breakfast yesterday morning. Her stomach was empty, so how could she have ingested anything?'

'A tablet and water,' said May. 'That would indicate suicide.'

'Unless a doctored tablet was disguised as something harmless, say an aspirin,' said Bryant.

'Come on, who knows when they're going to need an aspirin? Are you suggesting the killer gave her a headache first? I examined the stomach lining for residue,' said Kershaw. 'There was nothing. Obviously poisons have other ways of entering the body: gas, spray, liquid administered via an injection, so I looked for a break in the epidermis but found nothing. Until I looked here.'

He folded back the bottom of the sheet to reveal Sabira's bare feet.

'Look at the sole of the left foot. It's easy to miss

because there's some hard skin, but there is a very tiny puncture mark here, near the heel.' Kershaw indicated the spot with the collapsible antenna he had inherited from the PCU's last coroner.

'I can't see anything,' said Bryant.

'No, but that's because you can't see anything,' said Giles. 'You need to put your glasses on. Trust me, it's an entry mark. Something entering the body here would have taken longer to feed into the main circulatory system, so by timing it back—'

'The mosque,' said Bryant. 'She took her shoes off to enter it and left them by the door.'

Giles removed a bag from beneath the table and carried it over to his desk. Opening it, he lifted out Sabira's shoes. 'Take a look at this.' He slipped on plastic gloves and carefully removed the inner sole of the left shoe. 'Don't touch, but if you look very closely you should be able to see it.'

They could make out a tiny indentation in the rear of the inner sole. 'This is what caused the mark,' said Kershaw, tweezering a crystalline sliver of glass from a microscope plate. 'She tried to kick off the left shoe because by the time the toxin started to take effect her foot would have been itching.'

'So Sabira was right,' said May. 'Someone was watching her at the clinic. They followed her to the mosque. She put her shoes in the rack, and while she was inside, they inserted a sliver of glass into the inner sole. She came out, put the shoe on—'

'Probably didn't notice anything beyond a faint irritation, because there are very few nerve endings there,' Kershaw added, 'and she continued to the museum, where she would have started to feel sick and disoriented, collapsing in the room upstairs.'

'Do you have any idea about the kind of poison that could do such a thing?' asked May.

'It was obvious to me that we weren't dealing with some

street amateur,' Kershaw explained, 'so I started thinking about professional toxins, and immediately came up with TTX. Tetrodotoxin. It's an incredibly lethal neurotoxin that blocks the nerves by binding to the sodium channels in the cell membranes. There's currently no known antidote and, more interestingly, there are no biological markers to indicate that a sufferer has been exposed to it, which is why it proved so difficult to diagnose a cause of death. It occurs naturally in nature, and can be found in newts, octopi and those Japanese puffer fish people insist on trying to eat. Fugu.'

'I beg your pardon?' said Bryant.

'It means "river pig". The poisonous fish.'

'How much would you need?'

'If it's injected, you'd only have to use half a milligram to kill a normal-sized person.'

'What are the symptoms?'

'The victims experience numbness, shortness of breath, then complete respiratory collapse.'

'And how long would that take?'

'Could be anything between fifteen minutes and three hours, depending on various health factors and conditions. It's diagnosable in blood and urine, but that would involve mass spectrometric detection after liquid or gas chromatographic separation. In other words, not your average high-street cause of death. It's not easy to get hold of, but it's very much the sort of trick the Russians pull on foreign agents these days, using toxins that go undetected or remain misdiagnosed. In the last few years they've pioneered this method of assassination.'

'But if the same murder method was used on Amy O'Connor, why weren't there any puncture marks on her body?' asked Bryant.

'Are you sure there weren't?'

'I'll go back and talk to Ben Fenchurch again.'

'Toxins,' May repeated. 'Sabira Kasavian's mood swings

had an irregular pattern, but it was a pattern all the same. She seemed perfectly fine some days, confused and angry on others. You think she was being slowly poisoned? Is there something that could have made her display symptoms commonly associated with mental imbalance?'

'You're talking about the administration of a drug cocktail on different days,' said Kershaw. 'She was in a variety of locations, at home, in restaurants, in the clinic . . .'

'Any number of people could have got close enough to her to do it, but for one person to administer a toxin on a regular basis – wouldn't she have noticed?' asked May.

'That depends on the method of administration,' said Kershaw. 'There are some highly sophisticated ways to deliver drugs to the system.'

Bryant was following the thought. 'Jeff Waters was killed by someone dressed as a motorcycle courier. If his killer was hired to do the job, why couldn't others have been bribed or blackmailed into giving Sabira medication?'

'That would imply a sizeable conspiracy,' said May.

'But that's exactly what she told us she was afraid of, isn't it? Something so big that she couldn't trust anyone, least of all us because we were connected to the Home Office.'

'Well, you said you wanted to go up against the government,' said May. 'It looks like you've got your wish. What do we do now?'

'Giles, perform whatever tests you have to do to find out if Sabira Kasavian was being slowly poisoned. John, can you get Janice to try Tom Penry, the boy who was with the little girl at St Bride's Church? He should be back from his holiday shortly. I'm going to find out why Fenchurch failed to find a cause of death for O'Connor.'

As the trio set off on their tasks, Rosa Lysandrou looked at them and shook her head sadly, wondering why they couldn't leave the dead in peace.

30

THE WITCH TEST

Jennifer Penry knelt beside her son and smoothed his sun-blond hair in place. 'I don't see why you can't ask his friend about this,' she said. 'Lucy is older, and seems to have led him on.'

'Lucy won't tell us the truth,' Longbright explained. 'It's difficult with children.'

'Do you have children of your own?' Mrs Penry asked defensively.

'No,' said Longbright. 'My hours would never have suited motherhood.'

'That's your choice, of course.'

Longbright decided not to argue. She had never felt it was a matter of choice. Her mother had been a police sergeant, and so had her former partner. Public service was in her blood; there had never been a question of giving it up. 'Tom, how well do you know your friend Lucy?' she asked, kneeling down to bring herself to his level.

'We have to wait for our dads,' said Tom, 'so we play.'

'*Witch Hunter* looks like a very exciting game. Did you make up your own rules?'

'No, course not. It's an online game and a card game. There are rules you have to obey.'

Longbright had spent a fruitless hour attempting to play the game with other online players, but had given up after realizing that she had no enthusiasm for absorbing hundreds of arcane rules and updates. It had been like spending the afternoon with Arthur and his books.

'Do you always follow the rules? Are you sure you didn't make some up?'

'Lucy makes things up 'cause she says she knows them all and I don't. But I think she cheats.'

'That's not very fair, is it? Did she cheat before you went on holiday, when you were waiting for your fathers?'

'Yes, she said the lady was a witch but I knew she wasn't.'

'How did you know that?'

'Because there are ways of finding out.'

'What ways did you use to try and find out if she was a witch?'

'One way is to tie them up and drop them in a pond and if they sink they're not a witch and if they float they are a witch, but we didn't have a pond.'

'So what did you do?' Longbright kept her tone chatty and light. It was important not to put the boy on his guard.

Thomas squirmed a little. 'There's another test. You can stick a pin in them and if they don't feel anything or make any noise they're a witch.'

'And Lucy knew about this?'

'Yes, but she didn't remember it until the man said it.'

'What man, Tom?'

'The one who came up to us. He asked us what we were playing and we said witches, and he said you can tell who's a witch by sticking a pin in them. And Lucy said she didn't have a pin but the man did, and he gave it to us.'

'What did it look like?'

'It had a bit of red plastic over the end. He took the plastic off and handed it to her.'

'Then what happened?'

'Lucy ran around behind the bench and crawled underneath it, and she stuck the pin in the back of the lady's leg.'

Longbright glanced at Tom's mother. 'The lady must have been very angry with her.'

'No, she was angry with the wasps. She thought it was a wasp. But she felt it so I knew she wasn't a witch. We ran away.'

'What happened to the man?'

'I don't know. He went away.'

'And what about the pin? What happened to that?'

'Lucy dropped it. And then the lady went off towards the church, and we waited a bit longer to see if she was a witch because Lucy still thought she was, but then I went upstairs to see my dad.'

'This man – can you remember what he looked like?'

'He was a motorbike man with a helmet.'

'Can I show you a picture?' She unfolded a printed frame from the CCTV camera that had caught Jeff Waters's attacker at Coram's Fields, and showed it to him.

'Yes,' said the boy without any shade of uncertainty. 'There's supposed to be a bike badge. Triumph. There.' He tapped the courier's right shoulder.

'How did he give Lucy the pin? Was it in his hand?'

'He took it from his pocket. It was in the plastic thing.'

'What is this all about?' asked Tom's mother, concerned.

'The children had a lucky escape, Mrs Penry,' said Longbright. 'You should be very thankful.'

On the way back to the unit, she took a detour to the courtyard of St Bride's and searched the pavement, but a receptionist in one of the offices told her it was swept every evening, and there was no sign of the toxic needle or its plastic cover.

'I had an idea,' said May, seeking out Bryant at his desk. 'The algorithm – you've got the means of solving the code but not the code itself, right?'

'I don't know where else to look,' Bryant admitted.

'I think Sabira posted you the Cardano grille as a precaution. Have you looked in her belongings? It would have appeared innocuous without the grille to complete it.'

'She had a case of clothes at the clinic,' said Bryant. 'Dan will need to search her flat. It would help if we knew what we were looking for.'

'Maybe she left it with someone she could trust. How about Edona Lescowitz? She might not even know she has it.'

'I'll call Colin and Meera and get them to check with her,' said May. 'They're still keeping a watch on her flat.'

'I feel a fool,' said Bryant suddenly. 'I think I was wrong. I jumped in as usual.'

'What do you mean?'

'Sabira and the Hogarth painting. She died under a depiction of a madhouse. I assumed she was trying to tell me something about her mental state, that somebody drove her to it. But it was you who figured out that the madness might have been chemically induced, which means that Sabira wasn't aware of the cause. So that wasn't the message she was trying to leave.'

'Then why did she go to the museum? It makes no sense.'

'I don't know. Maybe she really was having some kind of relapse. I'm not sure I can rely on my instincts any more.'

'No,' said May. 'Arthur, if you think she was leaving you a message, then we failed to understand its meaning in time to save her. And that means we must keep looking.'

31

TUNNEL RUN

Colin Bimsley pocketed his phone. 'They want us to make contact.'

'With Lescowitz? Why?'

'They need to know if she was given anything by Kasavian's wife. Something no bigger than your hand, they said.'

'Why do I always get the feeling that we're being left out of the loop?' Meera complained. 'It wouldn't kill them to brief us properly for once. Or give us something with a bit more responsibility than babysitting someone who's not even involved in the case.'

'You don't know she's not involved.'

'Exactly. If we had more facts . . . Can you see her?'

'She must still be inside.'

Lescowitz was studying film design at Central St Martins. The pair had followed her to the South Bank, where she attended a screening of *The Parallax View* at the National Film Theatre. Now she was in the BFI bookstore looking at DVDs.

'Come on then, let's make ourselves known.' Meera led the way across the concourse towards the shop. The cinema was disgorging its audience, and they found themselves in a rapidly thickening crowd. Suddenly the area beneath Waterloo Bridge had become as busy as Piccadilly Circus. The bookstall owners were packing up their trestle tables and people were pouring into the bar at the front of the cinema. Colin pushed forward and managed to reach the shop, but there was no sign of Lescowitz.

Edona walked quickly away from the theatre complex, heading in the direction of Tower Bridge. She had tried calling Sabira at five to see how she was, but the call had gone straight to voicemail, and she couldn't help wondering if something was wrong.

Her friend had changed so much in the last few weeks that it was hard to believe she was the same person. The last time they had spoken, Sabira had deliberately distanced herself, warning Edona that it would be better to stay away from her. Her free-spirited friend had become haunted and fearful.

It was a warm evening, and the walkways of the South Bank were crowded with strollers. The river was a default destination for Londoners, as if it had a pacifying effect on them. They were such strange people that Edona doubted she would ever fully understand them. Why, for example, when they lived in such nice houses, did they leave their rubbish bins right by their front doors where everyone could see? And their language was so dense with allusions and references that it was often impossible to understand what they were really talking about.

She had no real destination in mind, but walking cleared her head. After a while she found herself in a quieter reach of the river, past the Design Museum, where roads cut down to the water only to double back on themselves or suddenly come to an end. She passed a housing estate, its bright lamps emphasizing the emptiness of the streets. A forlorn pub appeared in the distance, standing by itself at the edge of the foreshore, its rear veranda overlooking the river. She decided to have a drink there, and see if they had something to eat.

Walking over to the low wall, she looked down into the pebble-strewn mud. Her grandparents had told her about coming to London and swimming in the Thames on a hot summer's day, but surely they could not have swum here, where the brown water raced between moored barges at such a speed that the river's detritus became trapped between them?

She heard the motorcycle before she saw it.

The Triumph was on the riverside pavement heading directly towards her, its rider leaning out at such an angle that it seemed he would overbalance.

There was something in his hand – something that gleamed brightly—

With the river wall at her back she had nowhere to retreat, so she dropped down instead, and his arm passed above her head, lightly catching at her hair. Braking hard, he swung the bike around for another pass.

But now there was another motorcycle, a Kawasaki ridden by a slender Indian girl with a crop-headed man on the pillion. As it slowed, the man jumped off, fell hard and righted himself, running between her and her attacker.

The Triumph tried to get close but was driven back as the Indian girl, who looked far too slight to be in control of such a powerful machine, blocked its path. The two bikes circled in an awkward display of attack and defence before the Triumph took off.

'Stay with her,' Meera called to Colin, heading off after the leather-clad rider who had just fishtailed around the corner. His engine was more powerful than hers, but she had grown up in these streets and knew every one-way system and cul-de-sac.

Throttling hard, she thought that if she could get alongside him, she might be able to force him into one of the roads that dead-ended at the raised river wall.

She and Colin had picked up Edona's trail again by covering the route along the embankment wall. They had just spotted her when the other bike had appeared. The rider's physique matched that of the courier in the CCTV shot from Coram's Fields, but he was wearing a different crash helmet.

Meera was in jeans and a nylon jacket, and didn't fancy her chances coming off at high speed. She needed back-up, but there was no way of radioing in the suspect without losing her concentration. The rider ahead left Cherry Garden Street and hit the busy dual carriageway of Jamaica Road, turning hard left into the traffic. Meera followed and was almost fendered by an immense refrigerated truck.

He hit the roundabout and came off at the first exit before hitting another hard left and doubling back. She knew he would try to head for the Rotherhithe Tunnel, passing under the Thames. If he made that, he would be able to reach the chaotic traffic on Cable Street and the Limehouse Link, and there would be a good chance that she would lose him.

He was forced to turn on to Bermondsey Wall, which right-angled into Cathay Street, heading back to Jamaica Road. If he missed the tunnel approach he would only be able to take the painfully misnamed Paradise Street, which she knew dead-ended at St Peter's Church.

The road was narrowed by parked vehicles and braced with speed bumps, but she needed to pull alongside. Accelerating as much as she dared, she raised herself and jockeyed over the bumps as he tried to go around them. The time she gained brought them neck and neck. The sound of their engines reverberated from the passing house-fronts. The junction for Paradise Street came up faster than she had been expecting. He needed to cut straight across on to Jamaica Road to catch the tunnel.

Meera saw that there was no possibility of pulling up beside him without killing herself, and was forced to fall back. Checking his rear-view mirror, he roared ahead and crossed the junction.

Or at least, he would have done, if the refrigerated truck she had veered around earlier had not ploughed into him.

The Triumph rolled under the lorry's front wheels and was mangled to scrap. She did not see what happened to its rider. Braking hard and skidding to a stop, she stood the bike down and ran to the junction. The truck could not brake fast without shifting its load, and came to a halt on the far side of Paradise Street in a blast of air brakes.

When she found the rider lying ten metres further on, she saw that he had been thrown into a wall and had compacted the vertebrae in his neck, severing his spinal cord. He had died instantly. With a hiss of anger, she sat down beside him as sirens surrounded her.

32

METHOD AND MADNESS

It was almost midnight by the time Meera and Colin brought Edona Lescowitz into the unit. The three of them had only been released by Bermondsey Police after direct intervention from Kasavian. Edona had been informed of her friend's death, and had agreed to come in on the condition that somebody drove her back to Walthamstow. She appeared accepting and unruffled by what had happened, as if she had been expecting the worst, although she was taken aback by the shabbiness of the police unit.

Longbright made her comfortable in Bryant's old armchair. May had warned his partner not to upset their witness, but he need not have bothered; for the moment she seemed surprisingly composed.

'We appreciate your help,' May said. 'You're taking this very well.'

'Mr May, I was raised in an Albanian orphanage. We were taught to expect the worst.'

'Well, I'm afraid it gets worse. We think your friend Sabira was being poisoned, which would have accounted for her changes of mood. But even before that process began, she was unhappy.'

'Of course she was unhappy,' Edona replied. 'She was losing her identity. You don't know what a shock it is to come here and build a life in a strange country, as she did. And then to be treated so badly by those conspirators.'

'Why do you call them that?'

'She told me how they hatch their plans. Always trying to destroy their rivals. And the wives, always looking for excuses to meet in fancy restaurants. Sabira was always trying to get out of the lunches.'

'Did she tell you anything about them?'

'She said the wives met in order to agree on certain things.'

'What kind of things?'

'How to protect their men from bad press and keep a united front. They could apply pressure on any wife who failed to conform.'

'You think that's why Sabira had the fight in Fortnum's? She was goaded into reacting?'

'I'm sure of it. She couldn't be controlled, so she had to go.'

'So the wives might know more than they've told us?'

'If they have any suspicions, I doubt they'll share them with you.'

'I wonder if we could get them to open up,' said Bryant. 'We'd have to field someone they would trust. That rules out anyone from here. We're all a bit too rough around the edges.'

May took exception to this. He had always had great success with women. 'What about Janet Ramsey?' he suggested. 'She used to go out with Kasavian.'

'They close ranks against former partners, divorcees and journalists,' said Edona.

'It would need to be someone they don't know,' May agreed. 'Miss Lescowitz, what do you think caused your friend's breakdown?'

Edona hesitated for a moment. Her natural instinct was

to distrust the authorities, but it was hard to refuse May's friendly, open face. 'I think her behaviour was deliberate at first. It allowed her to say things that had no voice. But at some point it stopped being a method and became a madness. Whether this was real or induced is for you to find out, no?'

'Nobody ever figured out Hamlet,' muttered Bryant.

'Did Sabira say anything about the photographer who always followed her?' asked May.

'I saw him a couple of times when she and I went out together. He told her to be careful, and not to trust anyone.'

'Why would he tell her that?'

'I don't know. I got the feeling it was part of another conversation they had had earlier.'

'Do you know if Sabira had any feelings towards Waters? Were they intimate?'

'No, certainly not. She was loyal to her husband. It was in her nature. Waters was a handsome man. He spent his life calling out to attractive women to get their attention, so I guess he had to be charming, and Sabira enjoyed being paid a little attention, nothing more than this.'

'Come on, let's get you home,' said May. 'We know how to contact you.'

'One last thing,' said Bryant. 'Did Sabira send you anything when she was in the clinic? An envelope, perhaps?'

Lescowitz thought for a moment. 'No, nothing.'

'I'm sending Colin back with you. There will be someone posted outside your apartment until this is over. Colin, you can take my car.'

The detectives arranged for Fraternity DuCaine to share shifts with Colin and Meera outside the flat in Walthamstow until they could come to an arrangement with local officers.

Bryant stood at the window watching Bimsley and Lescowitz crossing the Caledonian Road, heading for the car spaces the PCU rented from the aged Russian extortionist who had been smart enough to buy up empty lots in the seventies, when the area had been a violent no-go zone. 'We're running out of time to uncover something that will probably kill us off for good,' he said. 'Not much of a deal, is it?'

'It's your call, Arthur,' said May. 'We could take Kasavian's advice and drop the whole thing right now.'

'Amy O'Connor's killer is dead, but other lives may be at risk. Someone's cleaning house to stop information from getting out. They're in the habit of hiring thugs, so they won't think twice about hiring another.' He rubbed his eyes wearily. 'I have to go home. Tomorrow's going to be a tough day.'

33

CONSPIRACY THEORY

'I don't need to be coached on how to behave like a lady,' said Longbright, straightening her jacket in the full-length mirror she had installed in her office. Looking over her shoulder, she caught Meera trying not to laugh. 'What? This jacket's a Biba classic.'

'It's not the jacket,' said Meera, 'it's how you wear it.'

'And how's that?'

'With your shoulders hunched up. You look like a boxer, or a tranny. Try to relax a bit.'

'That's rich coming from you. I've never seen you in anything but jeans and workboots. You don't exactly exude femininity.'

'It's not about being feminine, it's about looking classy. Like you were born to the style. Don't you ever watch any makeover shows?'

'No, of course not.'

'Drop your shoulders. What have you got on your hair?' Meera reached up and touched Longbright's blond mane. 'God, there's enough lacquer on that to create a new hole over the North Pole.'

'It's Silvikrin Twelve-Hour Hard Hair with Highlights.

They stopped making it in 1968. I found some in a warehouse off the Edgware Road. Lady Anastasia has already agreed to introduce me to the rest of her coven today. She said she's happy to gather them at short notice.'

'Why would she agree to do that?'

'Because I represent an organization that's setting out to limit the freedom of the press in the wake of the recent phone-hacking scandals, and need to canvass opinion from women in the public eye. John came up with that one. She was very impressed by the CV he wrote for me.'

'What's the organization called?'

'Dunno, I'll come up with something. We're meeting in a Mayfair restaurant called La Cuisine des Gourmets. It's the closest I'll ever get to being taken out to a posh dinner, I can tell you.'

'You'd better eat something first. They'll order green salads. You'll be the only one asking the chef if he can knock up a cheeseburger.'

Longbright's hands went to her hips. 'Is that what you think of me? I admit I was born into a family of public-service-industry employees, but I do actually know how to eat with a knife and fork, thank you.' She checked herself in the mirror. 'Maybe I'll ditch the Bowanga Jungle Jaguar lip gloss, though. It's a bit too Ruth Ellis.'

'Who's she?'

'The last woman to be hanged in Britain. Call yourself a copper? Blimey. I mean, goodness gracious.'

Arthur Bryant found his way back through the corridors of the Robin Brook Centre at St Bart's Hospital, looking for Dr Benjamin Fenchurch. He spotted his old friend through the smoked-glass panel of the mortuary door, hunched over his desk as usual, and quietly entered. Creeping up behind the coroner, he tapped him on the shoulder.

Fenchurch jumped. 'God, you gave me a fright,' he said. 'It's a good job I wasn't holding a scalpel.'

'It's funny,' said Bryant, shaking some Dolly Mixtures out of a paper bag and offering them, 'I didn't startle you last time. But then you heard my shoes, didn't you? You made a comment about me still wearing Blakey's. You didn't see me reflected in the mirror above your desk. And then there was the mix-up with the cadaver drawers. You had trouble finding O'Connor. How long have you been having a problem with your eyes?'

Fenchurch looked devastated. 'Don't say anything, Arthur. Please. I've got eight months to go before retirement.'

'But it's affecting your work, Ben. You didn't spot the puncture mark on O'Connor's left calf, did you? She was stabbed with a needle-tip coated in Tetrodotoxin. You had no assistant but still went ahead with the post-mortem, and you missed it.'

'I started having trouble with my right eye a year ago. I knew it would exempt me from finishing my term if it got on to an official medical report, so I delayed my check-up. I'll lose my payout if they make me go early.'

'Is O'Connor still here?'

'The funeral parlour is due to pick her up tomorrow morning.'

'Let me see her. My eyesight's not much better than yours, but I know what I'm looking for.'

Fenchurch led the way to the cadaver drawers and they extracted the chilled bag containing O'Connor's remains. Bryant donned gloves and turned her left leg without waiting for Fenchurch's permission. He was always surprised by the way in which the absence of life left bodies looking smaller and less substantial, as if the soul could be weighed.

Leaning forward, he saw that a tiny but definite lump could be found now that the skin had lost its elasticity and started to retract.

'OK,' said Bryant, 'I won't say anything about your

eyesight on two conditions. You need to change the report, and promise me that any further post-mortems you handle in your remaining time are conducted in the presence of a qualified medic.'

'Of course,' said Fenchurch gratefully.

'And I'm still expecting you on my bowling team next Saturday,' said Bryant, tightening his scarf in anticipation of the sunlit streets.

Meanwhile, Dan Banbury and Giles Kershaw were at the St Pancras Mortuary with the remains of a black Triumph Thunderbird and its rider, bagged and labelled with scanner codes.

'You know how many of these we see a month?' said Kershaw. 'Trucks still have blind spots that let them take out bikes at corners.'

'He was really travelling,' said Banbury. 'It must have been like driving straight into a concrete wall. No tags in his clothes. The bike was bought through an online site, and the seller was given an alias. Think it was the same guy who went after Waters?'

'Well, he's right-handed, and his build is consistent with Waters's attacker. He recently took anabolic steroids and cocaine.' He held up the rider's left wrist and turned it over to reveal a dense, smudged square of deep blue ink. 'That's a Russian prison tattoo, made by repeatedly jabbing yourself with a darning needle dipped in household ink. He tried to remove it with a razor blade but it looks like it went septic. If you're thinking of a criminal career, it's not a good idea to trademark yourself with something traceable.'

'What is it?'

Kershaw drew closer and shone a penlight over the patch. 'A church or a monastery. The crucifix at the centre indicates that the wearer is the prince of thieves. Four spires on the church. That's either the number of years or prison terms he's served. And that looks like a spider at the base. The joints on the legs are irregular, see? That gives us his admittance date. Russian prisoners take great pride in tattoo codes, and they're evolving all the time. I've seen these coming out of St Petersburg.' He snapped his gloves off. 'You should be able to fill in the rest, shouldn't you?'

'An ID won't be enough to link him to anyone at the Home Office. They're covering their tracks.'

'My dear fellow, you give up far too easily,' said Kershaw. 'This chap wasn't supposed to go under a lorry. That was a mistake, and mistakes leave a trail you can follow.'

'You say that, but I can't even check to see if he ever contacted Kasavian's department because we can't access their phone records, and this guy wasn't carrying a mobile. We retraced his route on the off chance that he dropped it somewhere, but no luck.'

'There'll be a mistake,' Kershaw insisted. 'I can't be doing with conspiracy theories; they really are the province of crackpots. Just how many could be in on this? I mean, really?'

'I don't see that it would have to be that many. We're talking about, I don't know, five, maybe six people who are already signed up to the Official Secrets Act.'

'So what's the big international secret they're all protecting?'

'You tell me. Amy O'Connor worked in a bar. Anna Marquand, if it goes back that far, was a biographer. Jeff Waters was a *paparazzo*. And Sabira Kasavian, when it comes down to it, was an Eastern European bride coming to the realization that she'd made a bad marriage. They *must* have shared a piece of knowledge that got them all killed, but I can't begin to think what it might have been.'

'You've linked them,' said Kershaw. 'Good. Let's go from there.'

34

DOXIES AND RAKES

Longbright checked her reflection in the glass-covered menu outside La Cuisine des Gourmets and barely recognized herself.

She was dressed in a high-necked grey trouser suit she had only worn once before, to her aunt's funeral. Meera had tied her hair back with tortoiseshell clips. In gold earrings, a single strand of pearls, a tiny gold Cartier watch Bryant had borrowed from the 'unclaimed' drawer in the PCU's evidence room, low patent black heels and a matching bag, she looked like the Thames Valley wife of a professional golfer.

The restaurant was hung with gleaming copper pots, bunches of dried lavender and other pastoral French knick-knacks. And there they all were, seated at a long table by the largest window, the Home Office wives.

Longbright still had Sabira's notes about them. She could identify Cathy Almon, whose spouse headed the HO's Workforce Management Data System, and Lavinia Storton-Chester, whose husband Nigel was the security division's public relations manager. She also recognized a highborn woman named Daniella Asquith, wizened and birdlike, from staff files Dan had downloaded. At the head of the table sat Lady Anastasia Lang and Emma Hereward. The empty chair beside Ana Lang had clearly been left for Longbright.

Introductions were made, but the swirl of conversation was barely interrupted by Longbright's arrival. Apparently the chef had prepared a set menu for the group. Longbright's dietary requests – not that she had any – were obviously not to be taken into account. A first course arrived, something with asparagus tips and quails' eggs fussily laced with a crimson sauce that reminded her of blood-spatter patterns. Longbright trotted out her prepared mission statement about building evidence against the press.

'If you could do something about the *Guardian* journalists,' Emma Hereward piped up, 'we get a very rough ride from them.'

You picked the only left-wing newspaper out of nearly a dozen national dailies, thought Longbright, who had been in the public-service sector long enough to be able to spot a table of union-busters at a hundred paces.

'All our expenses have to be checked now,' Ana Lang agreed. 'All this rubbish about the taxpayer having to fork out to have MPs' moats cleaned. The lack of trust is appalling.'

Inflamed by the topic, Daniella Asquith became so animated that there seemed a chance she might be going into cardiac arrest. 'I think we behaved impeccably,' she said. 'You know what it's about? Jealousy. We have got a very, very large house. Some people say it looks like Balmoral, but it's a merchant's house from the nineteenth century. It was Labour who introduced the Freedom of Information Act and it is Labour who insisted on the things that caught us on the wrong foot.'

The wives chorused their dissatisfaction with the press reportage of the expenses scandal. The irony was that the story had been brilliantly uncovered by the *Daily Telegraph*, traditionally a bastion of right-wing journalism, but this appeared to have passed them by.

'They won't leave us alone,' Ana Lang told Longbright. 'It's because we've publicly voiced support for our husbands. They think we're fair game now.'

'I imagine Mr Kasavian's wife made the situation worse,' said Longbright.

'Well, of course it was terribly sad that she chose to kill herself, but she was unstable,' said Lang.

'I didn't know she killed herself.'

'They'll keep it out of the press, but it was an overdose. She was taking so many prescription drugs that she couldn't keep track of them all.'

'You know that for a fact?'

'No, but it was common knowledge.' All the wives nodded agreement.

'Did anyone actually see her take pills?'

Ana Lang spoke for the others. 'Not as such. But she told us she was taking them. She couldn't even remember what they were all for. She said something about not wanting her doctor to find out. In fact, I advised her to stick with her prescribed regime and avoid taking anything else.'

'When you have a situation where one wife doesn't . . . fit in,' Longbright said carefully, 'what usually happens?'

'We never exclude,' Ana Lang pointed out. 'We just don't go out of our way to *in*clude.'

As Longbright watched everyone playing with expensive meals that didn't really interest them, she suddenly felt sorry for the government wives. She could see their endangered lives laid out like forgotten biological displays: winter in Barbados, summer in Tuscany, kids at Eton and Harrow, lunches at the Delaunay and Hawksmoor, an endless round of preparations for cocktail parties and charity dinners that nobody actually wanted to be at, an atrophying existence of slow and steady strangulation until divorce or dotage beckoned.

'So you'll do what you can to help us limit press intrusion,' said Ana Lang, as if a decision had been reached by consensus. '*Hard News* has been especially tiresome about publishing pieces that don't have our approval. I suppose you know that the editor, Janet Ramsey, had an affair with Oskar? We simply had to put a stop to that.'

'How did you do it?'

'We leaked information about her unsuitability. She had an abortion a few years ago. We couldn't have him going out with a leftie.'

'Surely public servants have to undergo vetting and are required to be non-partisan?'

'Our spouses are required to make public displays of even-handedness, but that doesn't affect their beliefs,' said Emma Hereward. 'Whether they realize it or not, they do what we tell them. The public thinks we simply turn up at photo opportunities to support our husbands, but I assure you, we are the power behind their thrones.' She pointed at Ana Lang and laughed. 'And she is our Lady Macbeth.'

'Edona Lescowitz was right,' Longbright told her fellow detectives that afternoon. 'They are witches, all of them. They honestly think they can make or break their men, and I get the feeling that the fourth man in the Pegasus set-up, Stuart Almon, isn't long for this world. They want him out.'

'Why?' asked Bryant.

'Because he's weak. He's "not performing", whatever that means. The husbands all toe the line around their women, if the wives are to be believed. They do what they're told. Except when they go to their club, which they're allowed to do no more than once a week.'

'Where do they go, the Garrick, White's, the RAC Club? I'll bet most of them still don't allow women members.'

'Exactly. I can't blame the wives for forming their own society, so long as men want to play their little power games behind closed doors. Apparently the boys belong to a place in Westminster called the Rakes' Club.'

'You have got to be joking,' said Bryant. 'The Rakes'? The club founded around the time of Guy Fawkes – *that* Rakes' Club? How could I have missed it?'

'What?' asked May.

'The Hogarth painting Sabira died beneath. She wasn't trying to tell us something about the nature of madness, because she wasn't indicating the picture of Bedlam. I was misled by the red string around O'Connor's wrist and the red wound around her own. She was drawing attention to the series, not a specific painting. *The Rake's Progress.*'

'Now wait a minute,' said May, raising his hands, 'this sounds like one of your loopy—'

'It's obvious, isn't it? The directors of Pegasus Holdings all belong to the Rakes' Club.'

'So they're the kind of men who belong to clubs. What does that prove?'

'You don't understand the Rakes',' said Bryant. 'That's why she went to the Hogarth room. We have to go back there.'

With the second layer of doors closed, the high-ceilinged, wood-veneered room was too poorly lit to reveal its corners. May used his Valiant torch and shone its beam into the cracks and crevices. When he examined the base panels he saw the sliver of an envelope sticking out.

'This is why she came here,' said May, carefully extracting the paper.

Bryant seized the envelope and examined it in the torch beam. 'She must have slipped it behind the Bedlam painting.'

'Then when she fell against the doors it shook loose and slipped down between the lower layers. It looks like it matches the one that was posted to you. Sabira took a chance leaving it here. We might never have come across it.'

'Except that she knew we would follow her route,' said Bryant. 'I wonder where she was planning to go after this.'

'She didn't have her passport. I guess she could have hidden out somewhere, waited to get straight in her head. Maybe she suspected she was being poisoned.'

'And with any luck, this will give us the poisoner's identity.' Bryant pocketed the envelope with a flourish.

35

BRING IT DOWN

At 11 p.m. on Tuesday night, the staff of the PCU were still at their desks.

Banbury and Kershaw were collating notes when they received an ID confirmation for the Triumph rider. His name was Luka Terebenin, a 32-year-old Russian who had spent four years of a twelve-year sentence in the Kresty Prison in St Petersburg for armed robbery before having his sentence commuted. Emerging from the over-crowded jail when he was twenty-six, he disappeared from state records before turning up in the UK eighteen months ago.

'Sounds like somebody recognized his talents and recruited him,' said Kershaw. 'What have you got on his residence here?'

'Bugger all,' said Banbury, checking through the online file. 'No residential address, no employment stats, nothing except a visa entry. It looks like he had a sponsor to take care of his file for him. I'll try to contact his family in St Petersburg.'

'It seems unlikely that they'll know anything about his life here.'

'It's not that. If they want his remains shipped home, they have to pay for it. Not our responsibility.'

Everyone looked up as Bryant sauntered in looking more at peace with himself than he had for a while. He was definitely less wrinkly. 'This is what you're all waiting to see, I take it.' He removed Sabira's envelope from his pocket. 'The card. I need to compare it to the Cardano grille.' He held out his hand.

'I don't have it,' said May. 'You must have it.'

'Ah yes, I think you're right. I put it somewhere safe.' He stroked his stubbled cheek, thinking for a moment, and then wandered out to the kitchen, returning with the card. 'I hid it in the staff refrigerator.'

'Why would you do that?' asked May.

'Oh, my mother always used to keep the rent money in the larder, where my old man couldn't get at it.'

'Why can't you do things the normal way, just once?' asked Land, irritated.

'I find your emphasis on conformity hard to fathom,' Bryant replied as he donned his glasses and matched up the cards. 'Really, if you look at the average family there's no such thing. People behave in the most extraordinary fashion and think nothing of it. They have weird food habits and funny sleeping arrangements and rituals: "Oh, we always go to Grandma's the day before Guy Fawkes Night to lock the cat up" – that sort of thing. You know the butcher on the Cally Road, the big bloke with one eye? When he's not chopping up cows he plays gypsy accordion with three other fat butcher-musicians who call themselves the Gastric Band. Is that normal? Of course not. There's no such thing as normal.'

'Well,' said May, 'does it fit?'

'Perfectly. Take a look.' Bryant had placed the card with the cut-out sections over a pencilled panel of random letters on the inside of the second card. 'It clearly isn't just a Caesar's shift. We need a keyword.' He grabbed a pen and wrote out a list of names. 'Let's try everyone in the security department first.'

'Your notions of normality fascinate me,' said Land, unable to drop the subject. 'You formed your opinions in London in the second half of the twentieth century, you must have some idea that they go against the generally accepted flow of things.'

'*Au contraire*, old sausage, I've lived through Marilyn Monroe, the Suez crisis, the Kennedy assassination, Watergate, Edward Heath, the Jeremy Thorpe scandal—'

'Jeremy Thorpe?' said Land. 'That name rings a bell.'

'It should do. A fairly definitive example of political lunacy,' said Bryant without looking up. 'Thorpe was the Old Etonian leader of the Liberal Party in the early 1970s. He was alleged to have had an affair with a male model, Norman Scott, and to have hired a hitman to murder him. The hitman shot Scott's Great Dane, Rinka, but the gun jammed before he could kill his target. Thorpe was forced to resign and was eventually acquitted in a very peculiar trial, while the hitman married a woman who fell nine hundred feet off a mountain in the Alps. This was the chairman of the United Nations Committee, and this is British politics, not some tiny Mexican town where the mayor is sleeping with the chief of police's daughter. You think you know normal? You have no idea.' He slapped down his pen. 'I've tried everyone's names and none of them work. Anyone got any bright notions?'

'Try "Pegasus",' said May.

'Hang on.'

The room went quiet as it became clear that Bryant was transposing letters.

Bryant creaked back in his green leather armchair, looking at his partner in alarm. He showed none of the satisfaction that usually crossed his features when he knew something the others did not.

'I think we're in more trouble than we realized,' was all he said.

He refused to speak to any of them. When May approached, he shook his head and left the room.

'Tell me,' said May, stepping over Crippen as he tracked his partner along the corridor, 'what's wrong? What did it say?'

'I thought I might be putting you in danger before,' said Bryant, 'and now I know it. Sabira was killed by her knowledge. I don't want any of you to be targeted.'

'Arthur, how long have I known you? How many times have we been threatened? Have we ever backed down or given in? Of course not. What can you achieve by yourself?'

Bryant's silent glance chilled him. He had only seen his partner like this twice before in their long careers together.

'All right, I'll put it another way,' tried May. 'Nobody knows we have both the cards, do they? So long as they're not aware of their existence, we're safe.'

'Has Dan gone?'

'No, he's still at his desk.'

'I want the unit swept for bugs. The phones, the walls, under the floorboards, everywhere. He needs to check the cars as well.'

'With all this clutter? Aren't you being a little—'

'I know, John. Do you understand? I *know*.'

'You know who's behind all this?'

Bryant quietly folded the card away inside his jacket.

May held out his hand. 'Show me, Arthur. I'm in this with you. We agreed to see it through, didn't we?'

'I'll deal with it by myself,' said Bryant. 'There's no need for us all to go down.'

'Let me see what you've got,' May persisted, 'and I'll decide what to do. You have to trust me. I've always been the sensible one, haven't I?'

Bryant held his partner's gaze with a long, steady look,

and then removed the envelope from his jacket pocket and handed it over. May took out the card and read the decoded sentence that Bryant had highlighted.

It said: 'OSKARKASAVIANMURDERER'.

May went to the window and opened it wide, setting a flurry of pigeons to flight and allowing the late-night traffic noise to flood the room. 'That isn't possible,' he said. 'It can't be him. Why would he hire us to find out what was wrong with his wife if he was secretly destroying her?'

'Oh, you won't find his fingerprints on anything. He's made sure he's untouchable. Think about it. He hired a thug, Terebenin, and got hold of untraceable toxins. He has all the right connections. He was with his wife every night. He's convinced we can't solve the case, and he'll be ready to damn us when we fail. He's lied to everyone. Sabira discovered something that could destroy him, and was removed because she was too difficult to manage. You know it's true, John. In your heart, you've always known it couldn't be anyone else.'

'I don't know . . .'

'Think about it. Why didn't Sabira tell us outright? Because she was being watched, and everything was overheard and reported. Whatever she discovered about her husband was so big that she knew he'd do anything to keep it hidden. If she'd communicated with us directly, he would have found out about it and acted with even more celerity. What on earth could she do? When we met in Smith Square, I told her I was good at breaking codes, remember? I gave her the idea.'

'So she used the only code she remembered from her childhood and sent you a message.'

'But the two envelopes were separated in such a way that we might never have found the second one. '

'What you're saying is that we have nothing, and he knows it.'

'He knows how to cover his tracks, John. Even if we did get physical proof, he'd make sure that everyone closed ranks against us. His wife's meeting with the photographer made him realize she was on to him, so he had evidence planted in Waters's flat to damage her reputation. And when that didn't work, he knew he would have to get rid of her.'

'I suppose Sabira's message wouldn't hold water in a court of law?'

'Of course not. Even if we could prove she sent it, she's on record as having psychological problems. We have evidence but no proof. If Giles can show that she was being systematically poisoned, we would still need to trace the trail back to Kasavian.'

'All right,' said May. 'We'll put this to the vote.'

He walked back into the common room. 'Listen up, everyone, I need your attention. Arthur and I have new information, but if we share it with you it could put everyone at risk. The two of us are prepared to go it alone. If you choose to abstain from taking the investigation further, your decision will be treated without prejudice.'

Renfield was the first to speak. 'This is a police unit, not an insurance office. We didn't sign up for the health-and-safety aspect of the job.'

'Jack's right,' said Meera. 'I don't see why we should let you have all the glory when it's cracked.'

'Hear what we've got first.'

'No,' said Renfield. 'Does anybody want to opt out?'

No hands went up.

'Very well. Arthur will brief you on the latest turn of events.'

'Right,' said Bryant. 'I just want you to know that we'll all be going to hell for this. This is the sort of thing that can undermine an entire government.'

'I knew it,' said Renfield. 'Bang goes the knighthood.'

36

RUNAWAY

It was 2.30 a.m. when the briefing session finished, and the detectives sent everyone home. 'We can't protect them all,' said Bryant. 'It's spreading fast and it's going beyond our control. There are already too many people who know too much: Edona Lescowitz and the children, Lucy Mansfield and Tom Penry. Perhaps the kids are safe – no one would believe them.'

'Then we have to wrap it up fast, before Kasavian gets wind of anything,' said May. 'And every step of this case has to be solid. I suggest we keep the PCU operational as a safe house for anyone who needs it.'

Banbury had run a check on everything emitting an electronic pulse in the building, and had been able to account for all devices except Bryant's radio, which continued working even after it had been unplugged and had had the batteries removed. It was agreed that no unknown callers would be admitted until they had closed the investigation, which needed to be before Kasavian headed for Europe, when he would move beyond their reach.

'I suggest we hit the Rakes' Club tomorrow, as soon as

it opens its doors. Sequester their accounts and membership log. We're dead without evidence.'

They headed out to collect May's BMW. The light rain
was sheening the roads with yellow diamonds.

'What's so special about this particular club?' asked
May, unlocking the car door.

'The Rakes' was founded by a group of classically
educated young Catholics who agreed to meet together
on certain nights and kill the first man they met on the
city streets,' said Bryant. 'For three years they literally
got away with murder. London magistrates tried to find
legal precedents to curb them but failed to do so, and of
course there were no police. There's evidence linking the
ringleaders to the so-called Jesuit Treason, or Gunpowder
Plot, of November the fifth 1605, after which most of them
were eventually transported, but the Rakes' Club survived.
In fact, several other versions of so-called Hellfire clubs
appeared in Europe, mainly in the Netherlands. Secret
clubs demanded a level of loyalty that ranked above that
to the Crown and the State. They terrified governments
because of their ability to destabilize the population.'

'But that's the most fundamental tenet of the PCU's
remit, to prevent the disintegration of society.'

'Indeed. The Hellfire clubs in their original form
petered out at the end of the eighteenth century, when the
old libertine ways died. The Napoleonic wars distracted
everyone. But the clubs continued in new, more acceptable
forms. Their members became wealthy landowners.
In many ways they represented the very things that the
Peculiar Crimes Unit was created to oppose.'

May pulled out into Euston Road and turned on
the windscreen wipers. 'Hellfire – didn't they practise
Satanism and witchcraft?'

'They did by the time Sir Francis Dashwood was re-
modelling West Wycombe House in the style of the Temple
of Bacchus. They were the opposite of guilds and Masonic

lodges. I suppose you could say they sought to encourage disorder, anarchy, drinking, gambling and above all promiscuity, strictly for the elite – peers, military men, the gentry. The Wig Club in Edinburgh required its members to make a toast from a penis-shaped glass after donning a wig made from the pubic hair of the royal mistresses from Charles II to George IV. It was full of lice, so they all ended up infested.'

May laughed.

'The concept of anarchic clubs still continues in other forms. For example, Churchill's Special Operations Executive, the 'Ministry of Ungentlemanly Warfare', encouraged espionage and sabotage in Nazi-occupied Europe. I think a sense of anarchy exists deep within the English bloodline, and a good thing too.'

'How on earth do you know all this?' May asked.

'Remember all those years when you were out late at night in West End supper clubs, chatting up young ladies? I was at home reading.'

Ahead the lights were turning red, but when May applied the brakes his foot went straight to the floor.

'Do you want to slow down a little bit?' said Bryant. 'There are several things going through my head at the moment, but I don't want one of them to be the windscreen.'

'Can't,' said May. 'No brakes.'

The BMW shot past a truck, heading towards the red lights.

'It's the big pedal in the middle, if memory serves,' said Bryant.

'Not working. Get in the back seat.'

'Certainly not. I'm not going to leave you up here in the front. What if you take the key out of the ignition?'

'That would engage the steering lock.' The BMW raced through the lights, narrowly missing a Marks & Spencer truck and a very angry cyclist.

'There are at least seven sets of lights on this stretch of the road,' said Bryant. 'The cabbies reckon you can hit something called a "golden run" when all of them are aligned green.'

'I think we're going to be hitting something else,' May warned. 'The next ones are turning red. We're going to catch all of them.'

'Wait – look.' Bryant pointed at the ambulance in the next lane. 'It's heading for University College Hospital. Get behind it and stay as close as you can.'

May swung the BMW in behind the ambulance and followed it across the next intersection, racing through the red lights. A police patrol car swung in behind them and started its siren.

'Elderly detectives crushed in multiple pile-up,' said Bryant. 'I was planning to die in my sleep, not in the Euston Road. I don't want to be remembered by a Sellotaph.'[4]

'Will you shut up a minute and let me concentrate?' May warned. They rode another red light. Miraculously, the next cross street was deserted. The ambulance was starting to slow in preparation for its left turn into the hospital. The BMW bumped its rear fender as the medics looked out of the window in alarm.

The ambulance turned. May tried to turn with it, but felt his tyres start to slide.

'I read somewhere that you're supposed to turn into it.'

'Thank you,' said May tetchily, 'I'm trying to.'

The BMW slipped off the wet road and bumpily mounted the pavement.

'Go over there.' Bryant airily waved his hand at the windscreen. 'Between the lamp-post and the wall.'

'I don't think there's enough of a gap.'

The BMW shot into the space and was brought to a jarring, grinding stop as wall and lamp-post scraped the

---

[4] A bunch of flowers taped to a lamp-post.

sides of the vehicle, dragging it to a standstill. Behind them, the patrol car halted and two officers jumped out.

'I'll let you talk us out of this one,' said Bryant. 'Then I'm going to go home for a cup of strong tea and a change of trousers.'

37

THE SICKNESS OF THE MOON

'The next time you decide to go stock-car racing, you might not want to do it in the Euston Road,' said Banbury the next morning.

He had been called into the unit early, and still looked half-asleep. 'I've just been over your car, John,' he said. 'The brakes weren't cut. It looks like the shoes were oiled with some kind of industrial lubricant. Unusual molecular structure, like the sort of product NASA used to develop for the space program. Nice job, very professional. We'll run a trace. There can't be many people who could supply it. I think your repair bill is going to be more than the resale value. I hope you were insured.'

'I don't think my insurance covers assassination attempts,' said May. 'I need you to try and link it back to Kasavian.'

'I hope you're joking. There are no prints, nothing to suggest tampering.'

'But you said a lubricant—'

'There's a lot of high-intensity industrial work going on in the backstreets around here. It could have been picked up from a spill. That's the way these jobs are pulled off,

John. They're designed to leave room for interpretation.'

'But you could check around the area, and if no one was using—'

'How long have you got?'

'Fair point,' May conceded. 'Giles reckoned he'd get clear proof that Sabira Kasavian was being slowly poisoned. Anything on that?'

'Well, I spoke to him on the way in – didn't see why I should be the only one up – and I think he's run into trouble. So far all the isolated chemical components he's had back from the Institute of Tropical Medicine are commonly found ingredients. They need to run further tests, which is going to take time and be expensive. Knowledge and proof are different states.'

'But the only person who could consistently get close enough to her was Kasavian.'

'That's purely circumstantial.'

'What about Waters's death?'

'That trail's already gone.'

'Then what else do we have? It sounds like O'Connor is our best bet.'

'We could start at the other end, with the culprit.'

'Except that he knows how we work,' said May. 'He's been studying our methodology like a hawk for years, waiting for us to slip up.'

Bryant sat up with a start. Everyone had thought he was asleep. 'We still have a few tricks up our sleeve. There's someone we can use: Leslie Faraday.'

Faraday was the Home Office liaison officer, and the budget overseer of London's specialist police units. Answerable only to Kasavian, he had *carte blanche* to investigate anything that failed to meet his approval. In the hands of someone intelligent this power would have been absolute, but Faraday was the sort of man whose mental circuitry had been soldered into place.

'Faraday? He's an idiot. What can he do?'

'Oh, I absolutely agree. He's W. S. Gilbert's original "Disagreeable Man" – you know:

Each little fault of temper and each social defect
In my erring fellow-creatures I endeavour to
    correct.

'But that's the beauty of it; he'll never spot what we're up to. He can help us nail Kasavian. We need the Home Office agenda for the Paris presentation. I don't think he's just going to sign off on the initiative, I think he's going to bury something else there, something he doesn't ever want to surface here.'

'How can he do that?' asked May.

'By placing security information under one of the Europe-wide anti-terrorism secrecy laws. If it's neatly knotted with red tape and locked in an EU filing cabinet, we'll never be able to get our hands on it.'

'That gives us less than forty-eight hours.'

'Then we'd better get a move on. We need to get into the Rakes' Club. We're not the only ones being set up for a fall; Stuart Almon's card has also been marked. He'll be our ticket in.'

'Why would he agree to do it?'

'You heard what Janice said. Almon is being sidelined. He'll be there. Anyone who fears Kasavian is a potential ally. See if you can arrange to meet him when I get back. I want to be there.'

'Why, where are you going now?'

'I have to catch up with someone at the British Museum. I need some more specialist knowledge.'

As always, Bryant's thought processes were as mysterious as Mars and just as hard to reach. Gathering up his hat, stick and scarf, he set off for Museum Street.

*

Georgia Standing did not look like an archivist specializing in the study of Roman lunar symbolism. She looked like a Goth who had come to London for a Cure concert and, having accidentally got locked in the British Museum, had decided to make the most of it. Her jet mane was sewn with Egyptian beads that glittered darkly as she swung towards him on high rubber boots. 'Hey, Grandpa,' she called, 'you're looking good. Still wearing my favourite scarf. Long time no see.'

Bryant waved her hands away. 'Don't call me Grandpa and don't try to do any complicated young-people handshakes with me. How are you getting on?'

'Oh, you know. The female archivists try to trip me up with smart remarks and the married men have a tendency to hit on me. Meanwhile I haven't had a decent date since the Queen Mother died. How's the PCU?'

'Still going, although at this rate we may not make it to the end of the week.'

They strolled together across the gravelled forecourt, passing through a miasma of roasted frankfurters from the stall permanently moored at the museum gates.

'I went to visit Harold Masters in the Royal Bethlem Hospital last month. The doctors don't think he'll ever fully regain his sanity,' said Standing. Masters had been her predecessor at the museum before attempting to strangle someone. 'It amazes me how many academics have nervous breakdowns or start believing that God is speaking to them through the fireplace.'

'It comes with the territory,' said Bryant, taking her arm. 'They focus their attention on one area of study for so long that they lose their perspective. Poor Harold. He was always rather highly strung. Speaking of madness, I need your help.'

'I'll do my best. What's the problem?'

'Murder victims with crimson cords tied around their left wrists.'

'You're talking about the cure for lunacy, the sickness of the moon. A very clerical concept, the evil of the mind, just as illness was to the body. Was there clovewort tied on to the red string?'

'No. The cord wasn't there as a remedy. It was a warning to others.'

'Or a remembrance, perhaps. You know, madness has inspired some pretty irrational cures. Doctors tried to teach patients "therapeutic optimism" while attaching leeches to them, and when that failed they gave them blood transfusions from animals. The bloodline carried insanity, so I guess it made a warped kind of sense to try and bleed it out.'

'I thought this case must be about madness at first, particularly as the victim was found underneath Hogarth's painting of Bedlam from *The Rake's Progress*.'

'Was she wearing anything blue?'

'You mean the colour of Bedlam? No, she offered no other clues.'

'We were once the nation of the Mad Monarch, poor old King George. After that, Bedlam became a dumping ground for political prisoners.'

'Well, it appears to have become so once again,' Bryant explained. 'Which is why I'm rather more interested in the other meaning of the Scarlet Thread.'

'Oh no, you don't believe all that old claptrap, do you?' Standing led the way up the last flight of museum steps. 'My co-workers keep telling me Harold Masters believed it too. Wouldn't shut up about it, by all accounts.'

'He was interested in its mythology, just as am I. After all, the Scarlet Thread runs through the Bible as the blood of Jesus Christ, shed on the cross to wash away sin. But it also seems to run between a number of murder victims and the government.'

'I think you'd better tell me what you know.'

'To do that I have to take you to a part of the museum

with which even you may not be familiar.' Bryant pointed across the Great Courtyard with his walking stick. The vast arc of the glass roof glowed even on the dullest days. 'Far end, down the stairs at the rear, then turn left, right and left again. According to your predecessor, at the heart of the myth surrounding the Scarlet Thread is the idea that man can only be brought into a covenant with God through the spilling of blood.'

'Ah, the warrior Christians. Christ's own blood had magical properties. Stands to reason, if he could walk on water.'

'There's your first connection between madness and the blood of Christ, right there,' said Bryant as Standing clomped down the steps beside him. 'Goffredo de Prefetti was the Bishop of Bethlehem, and he supposedly brought Christ's blood to London. He put it on display at the opening of his asylum, and then placed it in the foundation stones of Bedlam at Bishopsgate. And somehow it ended up here in the British Museum.'

Standing gave a disbelieving laugh. 'No, that's not possible.'

'Oh, but it is. I'm about to show it to you now.'

As they headed back into the gloom between the interconnected rooms, they passed fewer and fewer visitors. 'I think people start getting Stendhal syndrome by the time they get down here,' she said. 'Too much choice; too much to try and understand. The permanent exhibitions in this section only appeal to academics.'

'It's probably what has kept this artefact safe for so long.' Bryant stopped before an illuminated glass case containing the six-inch-long reliquary. It was bottle-shaped, gilded and inset with precious stones, surrounded by a complex arrangement of enamelled angels, arches and sunbursts. There was a tiny inscription on the side reading '*Ista est una spinea corone Domini nostri ihesu xpisti.*' At the base of the case was a small plaque:

Holy Thorn reliquary belonging to Jean, duc de Berry, created between 1400 and 1410 to house Christ's crown of thorns from the Crucifixion. In the possession of the British Museum since 1898.

'There's a lot of theorizing that the Romans invented Christ by writing the New Testament,' said Standing, 'but if you dip into some of the online discussions you'll find yourself in a world of serious lunacy.' She walked around the crystal vial, studying it. 'I guess it's interesting from a mythological point of view. I can see a thorn of some sort, certainly. But the "crown of thorns" was never meant to be taken literally. It's a traditional metaphor indicating immortality. Which makes this bauble a nonsense. But it's a very attractive nonsense, a wonderful example of *émail en ronde bosse*. Pearls and rubies alternately arranged around the compartment that holds the relic.'

Surrounding the relic were trumpeting angels, a scene of the Last Judgement and cherubs raising the dead. At its centre, crimson and indigo jewels flanked a tiny dark brown sliver. 'Take a look at the edge of the so-called thorn,' said Bryant.

A faint line no thicker than spider silk ran through the crystal like a fine molten seam. 'It's oxidization. The air got in through a flaw in the crystal. Oxidized blood goes the colour of wood. What if the contents of the vial only oxidized on the outside?'

'Mr Bryant, I don't really see what you're getting at.'

'That's all right, nobody ever does.' He looked around for a bench and sat down with a sigh. 'My knowledge of the subject is only what Harold Masters told me. Hang on.' He dug out a crumpled sweet bag and his spectacles. 'I made a note. On October the third 1247, the Knights Templar presented Henry III with a lead-crystal pot that they marked with the symbols of the knights. They told him it contained the ultimate relic of the crucifixion: the blood of Christ. The gift came with provenance, confirmed by a scroll holding the seals of the Patriarch of Jerusalem, signed by the prelates of the Holy Land. They considered it the holiest and most important of all gifts.'

'That's understandable,' said Standing. 'Christ's blood consecrates and bestows eternal life. It's an elixir that leads to the gates of heaven.'

'There's a rather different story about the creation of the reliquary. Apparently there was once an imperial crown decorated with four of the supposed thorns from Christ's head, but when times grew harsh the crown was broken up and its parts were reused to make more treasures. The possession of such items wielded huge political influence, so four new reliquaries were constructed. But it turned out that three of them were created by forgers. You can tell them apart by looking for enamelling on the backs of their doors. The fake versions don't have that. This one does. So it turns out there's only one known thorn, and that isn't a thorn at all. Given that the crystal vial exactly matches the descriptions of the gift to Henry III, it would appear to contain Christ's blood.'

'R-ight,' said Standing slowly, clearly deciding that she was trapped in the museum's basement with a mad pensioner.

'So now I have a problem,' Bryant explained. 'What possible connection could a female bar manager, a photographer and the wife of a Home Office official have with the blood of Christ? Is there anything else that the red cord might signify?'

'If the blood was at the birthplace of Bedlam, I think the fact that your victim was found under such a painting confirms the connection,' said Standing.

'But what does it *mean*? You see my problem?'

'Maybe it stands for something else. You know, in the same way that the "thorn" does. To be honest, this is out of my league.'

'Did Masters ever talk to anyone else about his theories?'

'I have no idea.'

'So none of the other archivists have ever mentioned the contents of that case to you?'

'I'd never even noticed it before.'

'Then I'm sorry to have taken up so much of your time.' Bryant rose to leave. 'I should be off. I need to find a toilet, anyway. At my age, you always need to know how far you are from one, like petrol stations.'

'That's all right,' said Standing. 'If I think of anything, how can I contact you?'

Bryant gave her his card. 'If you have any ideas at all, no matter how strange, I'll listen to them.'

She watched as he jammed his trilby on to his head and set off by himself, as strange an exhibit as had ever graced the museum.

38

ROUGH MUSIC

The Rakes' Club existed behind a discreet ebony door with brass fittings at number 42, Dover Street, Mayfair. As Bryant had predicted, Stuart Almon had readily agreed to meet Bryant and May outside and show them around.

When the detectives arrived, they didn't spot the accountant at first. As thin as a chopstick, dressed in a loose grey Jermyn Street suit that matched the brickwork, he blended perfectly into the surrounding terrace.

Almon stepped forward and placed a cold, bloodless hand in each of theirs.

'I'll introduce you as prospective members,' he said. 'That way I'll be able to give you the tour. They can be spiky about who gets to see inside. We're a most venerable institution. It's said that to be accepted here you have to be a peer, a parliamentarian or a prick. Sadly we no longer have the ear of the government. It's an old-boys' network, and the boys are getting very old indeed. Many of London's landowners were members but they're dying off and being replaced by property developers.'

Almon led the way through a dun-coloured passageway

lined with sepia portraits of austere-looking lords and dusty busts of sour-faced duchesses.

'What else can I tell you? Still no women allowed, of course. The house drink is still saltheen – that's hot whisky and melted butter with spices, guaranteed to thicken the arteries. And of course we still have a resident black cat that gets served lunch before any of the members. That goes back to the days when the Devil was believed to adopt the form of a cat, making him the club's oldest member and therefore the first to be served.'

He signed the detectives into the visitors' book and led them to the library, a tall oval room buttressed with leather-bound volumes that appeared not to have been opened in two hundred years.

'It's said to be modelled on the library in the abbey at Melk,' Almon explained. 'Can't see it myself. Upstairs are the meeting rooms, snoozers and bar: everything you fear about such a place – boarding-school food, brandy snifters and smelly old geezers bottom-trumpeting in wing-backed armchairs. It's the sort of place where you'll still hear a coloured chap referred to as a golliwog, although at least these days all racist claptrap is reprimanded by the management.'

'Why does the Home Office still use the place, then?' asked Bryant.

'Habit, I suppose, and the fact that it's a network, as rickety as it is. Oskar seems to like it.' He led the way to the brass-fitted bar and called the barman over to plot drinks. Bryant looked around, wrinkling in complete disapproval. It seemed unlikely that anything nefarious had ever been planned by a handful of bibulous, bulbous-nosed aristos frittering away the last rents from their estates. 'Anything at all unusual about the place?' he asked.

'Well, there are a few arcane rules. One states that no club member may ever resign, living or dead. Some rot about the place collapsing if a woman ever sets foot inside the building, that sort of thing. But there was once a Hellfire club based here. The Duke of Wharton met here with his cronies. Young men gathered to discuss the existence of the Trinity. Questioning these things goes against the teachings of the Church, so it was said that the blasphemers were "raking the fires of hell", hence the Rakes' Club. In those days you could lead the most extraordinarily debauched private life and still command respect from your peers, who were, after all, peers of the realm. They weren't subjected to rough music.'

'What does that mean?' asked May.

'They exercised parliamentary privilege,' said Almon. 'Had they been commoners, acting as they did, they would have been tied to donkeys and driven through the streets to the noise of the public banging on saucepans.'

'So there was no rough music here for miscreants.'

'Not as such, but there *is* another meeting room.' Almon tapped the side of his nose conspiratorially. 'Sort of a club-within-a-club.'

'Does Mr Kasavian use it?' asked Bryant.

'Yes, from time to time. It has a name: the Damned Crew, a traditional title going back to 1602, all loosely connected to the Gunpowder Plot. It's actually a sub-club, with its own membership, initiation ceremonies and an annual subscription, operating separately inside the Rakes'.'

'How do you qualify to become a member?'

'Now that *is* a political network. You need to have worked for the government. It helps if you've displayed leadership qualities in the past and are committed to certain – ideologies. By that I mean it's full of barking fascists.'

'Is there any record kept of what goes on here? Meeting minutes, anything like that?'

'Absolutely not. There wouldn't be much point in

paying for an inner sanctum if it could be breached. You're looking to take Oskar down, aren't you?' A dark and eager light appeared in Almon's eyes.

'We're not at liberty to discuss—' May began.

'Yes,' said Bryant. 'Do you want to help?'

'I might be able to,' Almon replied. 'When men become arrogant they start to make poor choices.'

'Why would you help? Why now?'

'We always have to watch our backs, Mr May. My wife informs me of rumours when they become too loud to ignore. One currently circulating is that I will be scapegoated for certain failures of nerve within the department.'

'You mean there were things you didn't want to go along with?'

'The original members of the Damned Crew used their positions to get away with murder. I'm not saying we did anything quite that dramatic. I have no personal enmity towards Mr Kasavian. It would be a purely business arrangement between you and I.'

'We're not here to give your career a leg-up,' May replied. 'If you have something on your boss, it's your duty to inform us.'

Almon knocked back his whisky and clearly thought devious thoughts. Bryant instinctively disliked him; the civil servant was weighing up his options in order to maximize their advantages, but was nervous about crossing into territory from which he could not return. 'There is – a certain matter,' he said finally. 'It's something he wouldn't want uncovered. Something rather nasty. Let's not talk in here. I can show you the clubroom used by the Damned Crew.'

Almon led the way back to the stairs, turned on the darkened landing beneath a bust of Landseer and withdrew a key from his pocket. Bryant looked for a door, but saw none.

Almon stepped close to the wall and slid the brass key into what had appeared to be nothing more than a small stain on the wainscoting. The door had been painted to perfectly match the tobacco-coloured wall, and tipped inward to a narrow brick passage. 'This runs behind the walls of the bar,' said Almon softly. 'It was used by the servants to deliver the meals, so that no one would have to run into them on the main staircase.'

The passage smelled sharply of rising damp. Bryant and May followed the civil servant to its end, where a second door was unlocked to reveal an elegant smoking room lined with books. Six maroon leather armchairs stood on a sea-green carpet. There was a large globe, which Bryant suspected of being a bar, and a small walnut dining table. In one corner the conversation from the drinkers on the other side of the wall could be clearly overheard.

'I always assumed there were rooms like this in London, but I've hardly ever seen one,' whispered Bryant, barely able to contain his enthusiasm as he headed for the bookshelves.

'It's not as cloak and dagger as it appears,' said Almon. 'Enter any large building that once had plenty of servants and you'll always find rooms and passageways like these. The running of such houses depended on the efficiency and invisibility of the staff. We'd better be quick, Mr Bryant.'

Bryant forced himself to step away from the books. 'What do you have on Kasavian that's of practical use to us?'

'Before he came to head up our department, Oskar was employed as head of security for a biochemical company outsourced by the DSTL at Porton Down.'

'We know about that,' said May. 'It's old news.'

'How do you think he was able to step straight into a top position at the Home Office?'

'You tell us.'

'He knew how to keep a lid on things. He proved himself amply capable in his final months at Porton Down. There was a murder committed—'

'It was Kasavian?'

Behind them, the passage door opened. Oskar Kasavian stepped into the room. If he was surprised to find the detectives in his private sanctuary, he managed not to show it. Bryant, on the other hand, reacted as if Dracula had just appeared at a crypt entrance. He instinctively wanted to look around for a crucifix.

'Almon.' Kasavian smiled. 'I thought I might find you here. I do hope you're not planning to breach club rules. Gentlemen, if you would excuse us for a few minutes, my colleague and I need to have a private conversation.'

Unable to argue for a reason to stay, Bryant allowed his partner to steer him from the room.

39

BLOODLINE

The nondescript June weather had fractured into chill drizzle and darkness. Pustular clouds reached down to infect the top floors of buildings, sheening slates and blackening brickwork. Even the glass towers of the Square Mile were dimmed and streaked with condensation, as if they no longer wished to expose their interiors to the sinister streets.

Maggie Armitage furled her umbrella and huddled in the doorway of the Peculiar Crimes Unit, waiting to be admitted. Something had been bothering her ever since her meeting with Bryant at Liverpool Street Station. The sensation had grown with the passing hours, until she could bear it no longer.

Last night she had used her Ouija board to contact Starbuck, an unruly Edwardian child she occasionally used as her contact to the spirit world. He had proven impossible, throwing tantrums, tossing cups and vases, yanking at the tablecloth and tearing open the curtains, alternately angered and hurt by her questions. When she came to, she found herself shaking with cold.

The spirits were disturbed. *She* was disturbed. Death

held no terrors for her; she had brushed against it too many times, but evil . . . now that was a different matter altogether. Just as she believed there were forces for good in the world, it had to follow that there were forces for harm.

Janice Longbright opened the door with a look of surprise. 'We've never had you visit us before,' she said. 'Come on in before you get soaked through.'

'I had to come,' Maggie explained, bashing out the rain from her rainbow-wool hat. 'They're in danger, aren't they?'

'Someone sabotaged John's car last night,' said Longbright. 'We think it was intended as a warning.'

'I knew it. Are you brewing up? I'll think more clearly with a cup of tea inside me. Bags will be fine. Arthur will be on the first floor, to the right.'

'How did you know that?'

'He has to be able to look down into the street. Can you show me?'

'Of course.' Longbright led the way to the detectives' room. On the stairs they passed Jack Renfield. Maggie reeled, her spiritual sensitivity battered by unseen forces.

'Wow,' she said, staring after the broad-shouldered sergeant. 'So you're finally seeing somebody again.'

'Jack?' Beneath their Elizabeth Arden foundation, Longbright's cheeks coloured. 'We're colleagues. Well, friends. Well – I don't know, really.'

'There's a golden cord running between the two of you. I think he'll be good for you, although . . .'

'Although what?'

'There's something you have to discover about him first, a problem to solve. If you can surmount that, he'll be the one to save you.'

Longbright was becoming annoyed by the white witch's cryptic prognostications. Maggie reached Bryant's doorway and peered in, breathing deeply. 'Of course,' she said quietly. 'I knew it would be like this.'

Making herself comfortable in the green leather armchair while Longbright made tea, she took in her surroundings, studying the bindings of the books and folders, the chaotic spread of Eastern artefacts and Victorian bric-a-brac. She paid particular attention to a volume of Greek myths that lay on Bryant's desk.

'He knows more than he's telling you,' she told Longbright when the DS had returned with a tray. 'He doesn't want you to come to harm.'

'I'm sorry, I'm not with you,' Longbright admitted.

'The Greeks. Pegasus was the offspring of Medusa. The horse sprouted from her severed head. It was a half-brother to Theseus.'

'Nope, still nothing, I'm afraid.' Talking with Maggie was confusing at the best of times, but Longbright feared the white witch was finally dropping off the conversational bandwidth.

'I knew there was a connection, but couldn't quite recall having the conversation with Arthur about it. He already has everything he needs to solve the case.'

'I don't see how you could possibly know that, Maggie.'

'Because we talked about it long before any of this happened.' She accepted the hot tea and drank, but had to steady the cup with both hands. 'It's the Scarlet Thread, the line of Christ's blood. He's been misled. There's no such thing. It's just another myth. But the danger is very real, and he will die if he fails to remember the past. Your cat's about to have nine kittens, by the way.'

'Maggie, you know I love you dearly but I'm very busy and to be quite frank you're making as much sense as a box of rubber kippers, so if you don't mind, I think I'll just—'

'Did he go to the British Museum?' Maggie asked suddenly.

'Yes, I think he did. Why?'

'He went to look at the blood. But you see, it's not

Christ's blood at all. It's just a stupid old artefact manu-factured by the Church to encourage their devotees. And yet when you think of all the trouble it's caused . . .'

Longbright pointed along the corridor. 'Look, I'll just be in the next room, OK?'

'Your mother's with you, you know.'

Longbright froze. 'What do you mean?'

'I see Gladys standing behind you, almost as clearly as I see you. She never had the gift. It skips a generation, you see. She went to her death without knowing. But you have it, the power of second sight. You don't see it now but one day it will return to you.'

'Maggie, I'm a rationalist. I know you believe in . . . certain things, but I can't afford to accept that they're possible. My mother died in the course of her duty.'

'I know, dear. When I say I see her I'm speaking figuratively. I'll let you in on a little secret.' She leaned forward to impart the confidence. 'I don't believe in the supernatural either, not in its traditional form. I believe there are degrees of sensitivity that allow some of us to connect. I'm a connector. Think of me as a spiritual three-pin plug. And I've finally made the connection. Oskar Kasavian worked in Porton Down, didn't he?'

Longbright was taken aback. 'How did you know that?'

'Arthur once told me. You do know what they do there.'

'It's classified defence-of-the-realm stuff – biochemicals mainly, I think. They work quite closely with American military scientific units.'

'There's a reason why the Scarlet Thread keeps coming up.' Maggie searched the air, as if trying to catch the wavelength of a distant music. 'Madness. Loss. And blood, of course, always blood. A bright red through-line, if you will. I see it shimmering.' She raised a ringed hand before her eyes. 'Just there in front of me. Oh, how it shimmers! I must stay until Arthur returns. He needs to understand.'

Longbright could take no more. She left Maggie to her whispering spirits, returning to her office, but a feeling of dread had settled deep in her bones and would not be easily shaken off.

40

THE THREAD

Stuart Almon was a frightened man.

Kasavian paced slowly before him, seemingly studying the shine on his shoes. The packed bookcases and heavy tasselled curtains turned the room into something decadent and claustrophobic. They might have been in some overwrought Victorian bathysphere, miles below the surface of the Thames.

'What I fail to understand', Kasavian said carefully, 'is why you would bring the police here. Those poor senile flatfoots are merely marking time until their enforced retirement. Why do you think I appointed them to investigate my wife?'

'You wanted to be seen to be doing something,' said Almon sullenly.

'I didn't want the Met involved because they would have started throwing their weight around as they always do, ploughing through our company files out of sheer bloody-mindedness. There was never any likelihood of the Peculiar Crimes Unit doing the same, not with Bryant wandering off into old museums all the time. But you should know better. Or perhaps you thought you did.' He turned his hands, admiring their marmoreal sheen, and then wagged a finger at Almon. *Naughty boy*. 'You were pointing them towards something else, weren't you? Now, what could that be?'

'They knew about the club,' said Almon a trifle too hastily. Kasavian's sudden honesty made him nervous. 'I thought it would be better to bring them here, rather than just have them turn up. I only took them as far as this room. I wouldn't have taken them any further.'

'You know, Stuart, I've spent years listening to explanations about security breaches and I always start from the same position. I know people lie to themselves. They're hardwired to try and cover their mistakes.'

'This wasn't a mistake. It was damage limitation. I told you, they knew about the club. Your wife – she died beneath a painting of the bloody *Rake's Progress*, for Christ's sake. If that wasn't a clear enough pointer, I don't know what was. I don't know how or why it happened, but it did.'

Kasavian realized just how badly he had under-estimated his accountant. Concern momentarily flickered across his features. Then the mask came back into place. 'They'll get no further. You can undo the damage you've caused to the company.'

'How?'

'You're going to destroy the only remaining file for me.'

'You know I have no idea what's in it, Oskar, and I don't want to know.'

'I appreciate that. It's better for everyone that the contents remain a secret. What you don't know can't hurt you. If you want to open a Pandora's box, I won't be able to keep you alive.' Almon blanched. 'I mean, in terms of your career,' said Kasavian.

'I don't understand. Why don't you do it?'

'Good God, man, you led them here.' Kasavian's raised voice came as a shock in the deadened room. 'I can't be connected to it in any way. I'm about to present the government's case in Paris. The file is the reason why we were forced to tackle the border initiative in the first place. You talk about damage limitation – you have no idea of the damage that could be wrought.'

'Where is the file? Is it in here?'

'No, it's in a secure location. You need to go there tonight, remove the contents and burn them. Don't look at them, just burn them.' He took out a pen and scribbled down an address and some numerals, handing the paper over. 'Obviously it's imperative that nobody sees you. For your sake. You'll need this.' He removed a small brass key from his pocket.

The parting was awkward. As Almon moved back to the door, Kasavian's eyes never left him.

'Goodbye, Stuart.' Kasavian remained motionless, watching him leave. As Almon fled the building, he found it hard to shake the feeling that his fate was being sealed.

'We can't leave Almon in Kasavian's hands,' warned May as they reached the PCU. 'There's no telling what might happen to him.'

'Kasavian wouldn't be that stupid,' said Bryant. 'Besides, we can't stop him. All we have is a note from a dead woman who was undergoing psychiatric evaluation.'

'But if Giles gets his evidence of systematic poisoning—'

'It could take days to get that information from the labs without Kasavian himself signing off the budget. Jack just texted me; Kasavian's on the six p.m. Eurostar to Paris tomorrow evening. We have to arrest him before then.'

'Couldn't we trump up something? They arrested Al Capone for tax evasion. Hold him on some minor infraction until the lab report comes back?'

'And what if it's inconclusive? He has the backing of the Deputy Prime Minister. It'll take a lot more than that to stop him catching that train.'

As the detectives were shaking the sooty rain of King's Cross from their jackets in the hallway of the PCU, Maggie found them and launched into a fragmented, frantic plea neither of them could understand.

'I'm sorry,' said Longbright, trying to stop her, 'she just turned up here. I have no idea what she's on about.'

'It's OK, Janice,' said Bryant, 'I'll handle this. Slow down, you silly woman, I can't understand what you're saying.'

'Come with me, I'll show you.' The white witch seized Bryant's soggy arm and pulled him upstairs towards his office. Bryant shrugged back at his partner and allowed himself to be led.

Throwing open his volume of Greek mythology, she turned to the index and traced a finger down the list. 'You told me Mr Kasavian and his team own a company called Pegasus providing security intelligence to the scientific community. Pegasus and Theseus were half-brothers. You do remember Theseus, I take it?'

'Of course. Peter Jukes, the whistle-blower who died. That was where he worked.'

'Do you still have the file on him?'

'It's in the attic. Pass me my stick, the top stairs are a bugger.'

'Would somebody please tell me what's going on?' asked Raymond Land, hearing the commotion. 'What is that woman doing here? She's a bloody menace.'

'Yes, we know,' said Bryant. 'She's helping us on the case. We'll be in the attic.'

The PCU's top floor ran the full length of the building and was still partly filled with dust-sheeted crates left by previous tenants. 'Walk on the joists,' warned Bryant, 'or you'll fall through the ceiling. Dan rigged up a couple of lights for me. The switch should be somewhere on your right.'

'Madame Blavatsky,' Maggie whispered, pressing her

palm against the glass case housing a yellowed fortune-telling wax effigy of the Victorian medium. 'So she *is* here.'

Behind the dummy were stored the memories neither Bryant nor May had room for in their apartments, items they had salvaged from previous notable investigations: a giant prop lightning bolt from the Palace Theatre; a lamp that had hung in an underground canal tunnel; one of the seventy-seven brass clocks; a highwayman's hat; a snow globe from Devon; a pub sign; a pair of antlers; a Tube-station worker's yellow pith helmet; and the latest addition – a Mr Punch doll.

Bryant burrowed into the boxes and attempted to drag a box of files out. 'John, give me a hand with this,' he instructed. 'I kept hard copies of our reports because you were always tinkering with the unit's technology.'

It took the best part of an hour to find what they were looking for among the mildewed blue cardboard folders, tied off in bundles with bits of twine. 'Here we are,' said Bryant. 'We need more light.'

May pulled down one of the cabled bulbs and held it over the file. '"Dr Peter Jukes, Salisbury, Wiltshire",' he read, then précised: 'Body found by fishermen floating off Black Head on the Lizard Peninsula, Cornwall. He'd been missing for a week, but had only been in the water for a few hours. The local coroner gave a verdict of accidental death, although he acknowledged that there were unexplained injuries. Jukes's boat was found miles away, washed into a local harbour. There was a fair amount of dissent about the death. The coastguard concluded that he was unlikely to have been thrown from his boat, because local tides and currents would have taken them to shore together. I don't see what relevance—'

'Keep going,' said Bryant.

May held the light closer. 'Jukes told friends he was

going fishing with a colleague named . . .' He stopped and looked up. 'Oskar Kasavian.'

'Read on further.'

'Jukes had once belonged to some kind of Druid sect, and the press picked up on it. They tried to imply that he had fallen victim to Satanists. There was talk of a witch-hunt.'

'Now, there's an awful lot of misinformation about Satanists—' Maggie began.

'Don't start,' warned May, continuing hastily. 'Jukes's family admitted he had arcane hobbies, but were forced into denials to protect his reputation after it was suggested that there was some kind of connection between his injuries and his supposed interest in black magic. The press were curious because Jukes was a scientist involved with biological-defence experiments at the MOD's Porton Down laboratory. There had been a public scandal over part of the lab being outsourced to private companies. The police vindicated the coroner and agreed with the verdict of accidental death, promoting the idea that Jukes had become mentally unstable. He'd been suffering from clinical depression for a number of years, and had been recognized as a security risk.'

'The company he and Kasavian were outsourced to was called Theseus,' Bryant interpolated. 'It employed a number of epidemiologists studying the pathogenic spread of mutating viruses. The US National Science Advisory Board for Biosecurity wanted to shut the place down. They were worried that terrorists might be able to translate the scientific data into a weapon if it was published.'

'I remember that part – it was tangential to a case we were handling. But there's no real link between the drowning and Kasavian. He managed to prove that he was nowhere near Jukes's boat on the day.'

'Kasavian and his pals now run the sister company to Theseus,' said Bryant. 'One of the things they've been looking into is the protection of scientists developing genetic mutations of avian flu.'

'Arthur, there is still no direct connection. And you' – May pointed at Maggie – 'should not get involved in this kind of conspiracy theorizing. It only makes matters worse.'

'What if I told you that I'm holding the thread between the past and the present?' said Maggie.

'I don't care if you're holding the fireman's ball, if you're hiding any kind of information that could help us, then you're required by law—'

'I know where it can be found,' Maggie stated very loudly.

'I don't see how that's possible. There's nothing you could know about this case.'

'The girl who died in St Bride's' Church. Her name was Amy O'Connor, I believe.'

'That's right.'

'She was Peter Jukes's girlfriend,' said Maggie. 'She always believed he had been murdered. And she went to her death in London still looking for proof.'

41

THE BLOOD LINK

'How could you possibly have got hold of this information?' May demanded to know. 'We've been over everything time and time again and there was nothing . . .'

'You interviewed the churchwardens, then?' asked Maggie.

'Of course we did, all three of them.'

'But there are four. Jake Wallace was working in the basement that day. He and I have been friends for years. He used to attend my Mind and Spirit evenings when he was a penniless student, although to be honest I think he only came for the vol-au-vents. I mentioned I'd seen you, and he told me he'd seen Miss O'Connor on a number of occasions. She confided in him. People often do in churches, as you'd know if you weren't both such heathens.'

May looked flummoxed. 'I don't understand how we could have missed him.'

'I imagine nobody thought to tell you he'd switched to a different shift. Sometimes he took over the shop between one and two p.m., while the others were at lunch. Miss O'Connor said her boyfriend had told her to come to St

Bride's if anything happened to him. Soon after he died, she visited the church, but couldn't find out why he had sent her there. A couple of months ago she returned and started visiting more regularly. I imagine by then she assumed his request was more of a spiritual nature. Jake said it seemed as if Miss O'Connor found peace there, just knowing her partner had been in the same place. He got the feeling she wanted to talk to somebody but didn't know whom to trust. And she was running out of money. I think she was working in a bar and had had her hours reduced.'

'She told the warden all this?'

'There's something open and friendly about Jake that people in distress respond to. That's why he's a church-warden.'

'Why would Peter Jukes have sent her to St Bride's?' May asked.

'It's the journalists' church,' said Bryant. 'Perhaps he wanted her to discover something and spread the word. Had Jukes ever met your friend Wallace?'

'I don't know,' Maggie admitted.

'Dan checked every square inch of that place,' said May.

'What about the basement? St Bride's was badly damaged in the Blitz, but the bombs uncovered a sealed vault.'

'Nobody mentioned a vault.'

'No, it was closed up by the authorities in 1854, after a cholera outbreak. Dan only covered the ground floor. He was looking at a possible murder site, not studying archaeology. Is the basement open to the public?'

'I don't think so, no,' said Maggie.

'Then why is a warden posted down there?'

'Jake's helping an American radiography unit. There's a visiting professor analysing the bones and coffin plates.'

'An American scientist,' said May. 'Theseus had US connections. Perhaps Pegasus does, too. As much as I'm loath to allow you to go wandering off into church crypts, Arthur, I think you'd better get down there, if you're up to it.'

'Of course I'm up to it,' said Bryant, affronted. 'What are you going to do?'

'Someone has to keep an eye on Stuart Almon. I wasn't comfortable about leaving him with the Prince of Darkness.'

Samuel Simmons was a director of the Cincinnati Bioanthropology Research Unit, currently in charge of the Diagnostic Imaging Program being undertaken at St Bride's. Right now he was keen to analyse an abscessed jawbone belonging to a young girl who probably died of the pain alone. Instead, a rumpled old man in a sagging tweed hat was peering at him intently from behind a stack of coffin lids.

'Can I help you?' asked the bearlike Simmons, extending a paw.

'Arthur Bryant. Yes, you can.' He handed the professor his PCU card. 'I need that back, it's my only one.'

Simmons examined it and was clearly none the wiser. He returned it. 'You're a policeman?'

'As amazing as it may seem, yes. I understand you've been working down here for over two years?'

'On and off. It's a slow process.'

'Why, what exactly are you doing?'

'We're comparing the grave-marker plates found here with official death records to see if they accord. Then we X-ray the remains to see if the causes of death were accurate.'

'And are they?'

'Not very often. Between this and the charnel house crypt next door there must be the remains of around seven thousand bodies.'

Bryant leaned into one of the lead coffins as if choosing

something from the freezer. 'Come up with any surprises? Found anything you shouldn't have?'

'Like what?'

'I mean before you started excavating. We're looking for – well, I don't exactly know what we're looking for. Something a visitor could have left in the basement.'

'No members of the general public are allowed down here,' said Simmons. 'Many of these coffins once housed cholera victims. There's no risk of infection, but the by-laws require us to keep potential contaminants away from the public.'

'How about visitors from within the scientific community?'

'Yeah, we get a few of those. None lately.'

'Have you found anything at all that shouldn't be here? I'm thinking someone came by, used their company pass to gain entrance and left something to be collected.'

'You have no idea what this item might have been?'

'I'm afraid not.'

Simmons pulled off his gloves. 'Come with me. There's a box of stuff in the back. Everything on the site has to be annotated, and the items that remain unidentifiable get put in a junk box.' He pulled at a mud-stained cardboard carton and opened its flaps. 'It's mostly just debris, plus sweaters and books left behind by employees. But please, knock yourself out. There's a table lamp over there.'

Bryant picked his way through lost Tube passes, gloves, a pair of football boots, unallocated chunks of coping stone, loose change, paperbacks and folders of unfinished notes.

He was about to give up when he saw it, a small steel memory stick sealed in a clear plastic bag. There was no label. There didn't need to be. The bag had been tied with a strand of red wool.

He held it beneath the lamplight. 'Do you know where this came from?' he asked Simmons.

'No idea,' Simmons replied. 'People sometimes dropped in to see their partners. We had quite a number of interns helping us at the start.'

'Do you have a record of their names?'

'No need,' said Simmons. 'I remember them all. Try me.'

'Amy O'Connor.'

'The woman who died last week? Nope.'

Bryant passed over the dossier on Peter Jukes and showed him a photograph. 'Does this chap look familiar?'

Simmons shook his head. 'I'm just one of the guys here. I guess he could have visited while I was back in the States. The company name rings a vague bell.'

'His name was Peter Jukes.'

'The guy who drowned? I wasn't here at the time, but I heard he came up from the MOD in Wiltshire to see what we were doing. Must have been soon after we started. Somebody read about his death and remembered the name.'

'Do you have any idea what he wanted?'

'Apparently we had a team project in common.'

'What was that?'

'Blood. Yeah, I know, weird, huh? In the early days of our research we thought we might find a blood link through the bodies interred here. It seemed we might locate a hereditary disease passed through bloodlines because there were so many fathers and sons, mothers and daughters buried together. It didn't take long for the Ministry of Defence to start sniffing around. A whole bunch of guys turned up and started asking questions. Some time later, Jukes followed them.'

'What do you think they were all looking for?'

'C'mon, Mr Bryant, you're the detective, I think you know the answer to that one.'

'They were interested in any biochemical discoveries you might make, particularly with regard to military applications.'

'Can't think of any other reason why they would be interested, can you?' Simmons gave a lopsided grin.

Bryant was amazed. O'Connor had come back, knowing that her lover had directed her to the church, but had not thought to check in the basement.

'This was intended for the woman who died,' said Bryant, indicating the wool-tied bag. 'Why didn't she come downstairs to collect it?'

'That's easy,' said Simmons. 'If you didn't know about the crypt you sure wouldn't come looking for it. You enter the church and look around, and the ground floor is all you can see. The vault door's kept shut.'

Given his fractious relationship with technology, Bryant didn't trust himself to run the contents of the flash drive on Simmons's computer equipment. Pocketing the bag, he thanked the professor and took his leave, heading back out into the rain.

42

THE ROOFTOP

'She's not going anywhere tonight,' said Colin, looking up at the windows.

'How do you work that out?' Meera asked.

'Stands to reason, doesn't it? There was a Lovefilm DVD in her mailbox and she's just gone in with a bottle of plonk. Bet you there's a pizza delivery within the next half-hour.'

The pair were still camped outside Edona Lescowitz's apartment. 'How do you know it'll be a pizza?'

'She's a skinny European bird. They can really pack away the nosh without ever putting on weight. They eat green salads in restaurants and shovel down pasta at home, usually followed by a tub of ice-cream.'

'I've always been amazed by your sensitive understanding of women, Colin.'

'Thank you.'

'I mean you don't have any. We're an alien race to you, aren't we? A complete and total mystery. You're probably aware that we share the same number of limbs, if not appendages, and that's about it. So you and your mates down the pub can make up whatever you like about us and congratulate each other on being able to understand us. Incredible.'

'You'd be surprised, Meera. I understand more than you think. Especially about you.'

Meera folded her arms and leaned back against the dustbins. She waited while a drum-and-bass-deafened teen in a pimped-up van thudded past. 'Go on, then,' she challenged. 'Give me the benefit of your amazing male insight.'

'No, 'cause you'll get angry with me.'

'No, I won't.'

'Promise?'

'Yeah, just this once.'

'Say it. Full sentence.'

Meera hissed angrily through her teeth. 'I promise not to get annoyed with Colin Marlin Bimsley when he tells me what he thinks of me, all right? Is Marlin really your middle name?'

'It was my grandad's. All right. I'll tell you where your anger comes from. Your mum and dad favoured your sister. In their eyes she could never do anything wrong. Even though she always screwed up and let them down, especially when it came to fellas, they've always pretended not to notice or have quickly forgiven her, which winds you up, so they see you as the angry one. They were against you joining the force because they wanted you to get married to a nice Indian boy and give them grandchildren. When your sister said she wanted to open a restaurant they paid for it, even though they couldn't afford to, and you didn't speak to them for the best part of a year. Every time you try to put things right, they take it the wrong way. You try to control your anger, but you know it won't go away until your sister gets married and takes the pressure off you, and there's not much chance of that happening because she always picks the wrong blokes.'

'That is a complete load of the most – total . . .' Meera grasped for the words, fighting to control her temper.

'It's accurate, Meera. You know it and I know it.' He pointed at her accusingly. 'Every time you go home, you come back in a foul mood. Last time you even kicked Crippen, and it turns out she's a lady cat so that's not nice. And there's only one thing that ever calms you down.'

Meera's eyes narrowed. 'And what is that?'

'Being with me. I can lower your blood pressure in a matter of minutes. You know the first thing you always do when you come back from seeing your folks? You come into the common room to find me. You hang around scowling for a while, shooting the breeze, then you go to your room. And you don't even realize you do it. It's me, Meera.' He slapped his chest. 'Ever since you joined the PCU, I've been the one constant loyalty in your life.'

'Colin, that is so—'

'Sssh.' He held a forefinger gently to her lips. 'Don't say anything that will make you break your word.' He looked up. 'There you go, pizza-delivery man, ten minutes ahead of schedule.'

A black Triumph had pulled up at the kerb. Its rider dismounted and opened his red pillion panier. He removed a pizza box and went to the front door.

'At least she's getting something to eat,' said Colin. 'I'm bloody starving.'

Meera looked across the road in puzzlement. 'Why hasn't he turned his engine off?'

'I guess he's only going to be a minute.'

'No, nobody does that. I should know. And it's not the kind of bike pizza places use, it's way too high-powered.'

Colin threw down his coffee and ran across the road. The rider had already been admitted, but the door buzzer was still keeping the main entrance open. The empty pizza box had been dropped just inside the doorway. Colin took the stairs three at a time.

She had already opened the door to him. Edona was in her dressing gown, money in one hand, but he had his arm across her throat. Colin's rugby tackle was spectacularly foolhardy even by his standards. He brought them both down, which couldn't be helped, but the rider was up and swinging something sharp in his fist. Colin was wearing his father's old police-issue Doc Martens with steel toecaps, and punched one leg out hard, connecting with the rider's groin.

The blow must have been off though, because his opponent was up again, vaulting over him, heading for the staircase above.

Each of the terraced buildings had a different kind of roof access. This one had a short flight of steps to a steel fire door, which was already being opened.

Colin found himself faced with his worst nightmare. The roofscape before him was a darkened obstacle course of steep tarred slopes and brick gaps. He tried not to look down as he negotiated the slates, tried not to think of the spaces between objects that would recede or elongate, tricking his senses.

Diminished spatial awareness was the inherited inability to judge heights and widths, a form of Ménière's disease caused by the interaction of the eye, brain and inner ear. The problem had initially resulted in his rejection from the Met, but he had largely learned to control it – except at night, when everything appeared to flatten out.

His quarry had jumped across an alleyway, landing on the next angled roof, and Colin gamely followed, but could feel a familiar giddiness starting to kick in. Screwing his eyes shut he made the jump, but barrelled into a chimney stack. Ahead, the rider climbed a slated peak with ease and jumped down the other side. Colin followed, but his left leg was burning and felt as if he had gashed it. He climbed, trying to gain purchase on the dirt-crusted slates, pushing himself up on the rubberized soles of his boots.

On a flattened section of the next roof a group of teenagers were lounging in deckchairs, drinking beer. As Colin skittered down the rider dropped into the surprised circle and landed among them, scattering cans. Grabbing one of them, a young girl, he swung her over the edge in a single movement, as if she was weightless. He extended his arm and she screamed, trying to reach his neck.

Colin held the others back. The motorcyclist appeared to be the same one that he and Meera had chased, although that was impossible. He wore the same riding leathers, helmet and boots, even moved with the same rolling gait. PCU members were unarmed; all Colin could do was force the others to stay back beyond the rider's reach.

Once he saw that they were not going to try and rush him, the rider lowered his hostage to the roof and pulled her with him towards the adjoining building.

Colin followed as closely as he dared, but the game had changed. His main purpose now was to make sure that no harm came to the girl. She was being dragged backwards to the roof doorway, but was not making it easy for her abductor, yelling and kicking as she went.

As soon as he saw a chance to get closer Colin ran forward, throwing himself at the pair with the same un-directed energy that had got him banned from boxing matches. Crashing them all into a tangle of limbs, he tore at the rider's helmet, trying to reveal his features, and was struck on the side of the head for his efforts.

By the time he had recovered his senses, the rider was off again, a leather-clad rhinoceros thundering across the rooftops of Walthamstow, smashing through a washing line, then another, then – incredibly – straight through a barbed-wire fence that separated two roofs.

Tearing the wires away, he headed doggedly on, jumping to the terrace of an apartment building where each flat was divided from the next by a wooden partition. They came down like drawbridges beneath his boot, one after the other. Gnomes were scattered. A pair of plastic herons were poleaxed and a *faux* wishing well went for a burton.

What's it going to take to stop him? Colin wondered, trying to stay close. He would have kept the pace, too, had not an angry householder burst from his home to berate the detective constable over a flattened begonia bed.

As the old man blocked his path and threatened to call the cops, he could only watch helplessly as the rider dropped from the end of the terrace and vanished from view.

By the time he got free and managed to make his way downstairs, there was no sign of the pizza bike.

He found Meera in the stairwell. 'I wasn't close enough behind him,' she said. 'My bike was parked too far in the opposite direction.'

'Is she OK?'

'Fine – a bit shaken. Maybe we should take her back with us.'

'Take her on your bike,' said Colin. 'I'll follow behind. I nearly had him.'

'I know you did, Colin.' Meera smiled at him. 'Go on, I'll see you back at the PCU.'

43

LOW CASTES

'A whistle-blower,' John May exclaimed, 'that's what this is all about.' He had examined the material on the flash drive left in the crypt by Peter Jukes and was debriefing the team members in the chaotic common room. 'Jukes was worried that what he knew would place him in a dangerous position, so he copied the information and left it at St Bride's. The only person he was sure he could trust was his girlfriend. He told her part of what he knew but didn't give her proof. I guess he knew that doing so would have shifted the burden of knowledge to an innocent outsider. Instead, he left it in the church, thinking that he could send her there to collect it if things got bad.'

'But she didn't get the message?' said Land, confused.

'Perhaps she misunderstood, or he failed to make himself clear. He'd have been under a lot of pressure by then. Arthur believes she went to the church thinking he meant it was a place where they could connect on a spiritual level, whereas Jukes had something more practical in mind.'

'So what's on the drive?' Longbright asked, 'or are we still being kept on a need-to-know basis?'

'Theseus was developing a bioweapon for Porton Down that appears to have been banned under international law. It was codenamed "Scarlet Thread".'

'They chose the name of a biblical myth,' said Bryant.

'A mutation of avian flu that could be air-transmitted to affect everyone with a particular characteristic. Not in the blood, though – everyone's blood is the same. Variations in race occur because of melanin and tiny differences in DNA. The idea was to develop a weapon that could be used against specific ethnic groups. For example, you could infect a town of insurgents and not harm your own people.'

'Something similar was once practised here in London, albeit in a very different form,' Bryant added. 'People from specific Eastern European countries, most notably Romania, were employed to shift the coffins of plague victims because they were genetically immune to a particular strand of the disease found here.'

'Now here's the really nasty part,' said May. 'Jukes wouldn't have known about the specifics of the bio-experimentation if it hadn't been for the deaths of a number of low-level workers at the plant. He started his career as a journalist, and couldn't help noticing two connected elements. Those who died were all from the Indian sub-continent, and all suffered mental problems followed by death from drowning. The more he uncovered, the more worried he became. He realized they had died as the result of testing the Scarlet Thread.'

'Let me get this right.' Land ticked off his fingers. 'Oskar Kasavian worked for Theseus, so you're saying he knew about this. Then he found out that one of his employees was about to turn whistle-blower, so he ordered Jukes's death and made it look like an accident.'

'It certainly seems that way,' said May. 'Jukes died because he was planning to go public on the MOD breach. He was killed, then smeared by the tabloids for having "satanic" connections. Sabira found out that her husband had sanctioned murder. Either she went through his belongings at home or read something at his office. She told the photographer, Waters, because he worked for a press agency and she had no one else to confide in. When Sabira led us to the painting I simply thought of madness, but she was smarter than that. She was pointing to the Rakes' Club and to the Scarlet Thread experiment, a bioweapon that could make its targets lose their reason and commit suicide. Those who had been experimented on drowned themselves; according to Giles here, it would have been the least painful way to alleviate toxic symptoms. They shared the information online with each other. When Kasavian discovered that his wife knew about his involvement, he realized that he needed to discredit her. Which is exactly what he did.'

'Is Kasavian explicitly named in Jukes's documents?' asked Land.

'Jukes was more concerned with laying out the legal breaches in the case than with blaming any one man. His documentation is dry reading, but it states a clear case. However, it doesn't help us make an indictment.'

'You're telling me you have proof of murder—'

'Effectively, yes.'

'—and there's no way we can use it to put Kasavian in court?'

'That's about the gist of it.'

'You do realize that he heads to Paris tomorrow? If we attempt to indict him after he's appointed head of this international initiative, we'll cause an international scandal. One of the purposes of the Paris meeting is to monitor the movements of terrorists and prevent their access to biochemical products.'

'I'm aware of that,' said Bryant. 'I did warn you. We could arrest him on suspicion, let him know that we have Jukes's evidence and hope that he indicts himself.'

'No, there has to be something else,' said Land. 'Some other way of reeling him in.'

'Stuart Almon is our only remaining informant,' said May. 'He was prepared to name his boss. We need to try him again.'

'Jack's keeping an eye on his movements right now,' said Longbright. 'Maybe there's a reason to bring him in.'

Everyone looked to Bryant for approval. The PCU staff were under no illusions about who actually ran the unit. Bryant did not look happy. To be precise, he had a face like a codfish with a liver complaint.

'What's wrong?' asked May.

'All the way through this investigation, we've been several steps behind the Home Office,' Bryant said. 'That was my fault, sending us off on a wild-goose chase, but we haven't been moving fast enough. Kasavian had us thrown out of the Rakes' Club and we simply bowed to his superiority and walked away!' Bryant shook his head angrily. 'He knew that his director was about to tip us off, which means he either has to get rid of Almon – which he can't do – or enlist his aid in clearing things up. There must be something that tangibly links him to the Scarlet Thread deaths. I'm betting he'll use Almon to get rid of it. Kasavian has to be completely clean before the Paris meeting begins. If there's any more dirty work to be done, it has to be done tonight.'

'How do you know he won't just get rid of Almon as well?' asked Longbright.

'Because Almon is a player, and the ones with power are never at risk. Look at the victims: a bunch of low-paid Indian workers, a researcher, a backroom biochemist and his girlfriend, a freelance photographer and the immigrant wife of a civil servant. These aren't people who hold influential positions within the class system. They're outsiders, lower castes, which means that nobody will be questioning their loss in the House of Commons. But someone has to bring the whole ugly business to light, and it has to be us.'

44

CONFLAGRATION

Stuart Almon had never been to Whitechapel before, and was dismayed by what he saw there.

He lived in one of London's most expensive streets, situated at the top of Campden Hill in verdant Holland Park. Now he found himself walking past shuttered shops that sold Asian DVDs, plastic washing baskets and money orders. It was almost midnight and the street was busy. Groups of Indian kids were hanging around on corners, looking at him, he thought, with menace and malice in their eyes. The accountant slipped his hand over the smartphone in his pocket as he walked past them. The boys, however, were concerned with their own lives and barely noticed him. To them he was just another awkward-looking white guy heading home from one of the fancy restaurants that had begun to border the area.

Almon wanted to check the streetfinder on his phone, but did not dare to withdraw it from his jacket in case somebody spotted the light and mugged him. In all honesty, the only time he ever saw this many black faces was when he had to pass through a railway terminal.

What the hell was Kasavian doing leaving his 'Pandora's box' in such a poor neighbourhood anyway? Any number of private banks and financial institutions could have offered him twenty-four-hour access to a safety-deposit box. Of course they would have recorded the event, and there Almon had the answer. Around here half the street lamps were out and the CCTVs would fail to pick up identifiable images.

He looked for the warehouse and finally found it in a narrow backstreet that, over 120 years earlier, had been inhabited by the desperate women who had fallen foul of Jack the Ripper. Stepping over the sodden trash in the gutter, he searched for a way in. Someone had left a foil tray of curried rice on the steel code box, and it had leaked over the keypad. What was wrong with these people? Almon punched in the number Kasavian had given him and waited while the steel shutter rolled up.

He couldn't locate a switch inside, but his mobile had a torch app. Turning its beam around the bay before him, he found himself in the storeroom for the Spitalfields Art Fair. The smell of curry was pervasive.

A series of immense papier-mâché props peered out at him: Perseus, Heracles, Hippolyta and the Minotaur stared down from gold-sprayed pedestals. Behind them stood a pair of rearing stallions, a silver chariot, a scale model of the Parthenon, plus assorted braziers, columns and pediments.

This cloak-and-dagger stuff is absurd, thought Almon. He had not expected that his clumsy attempt to betray Kasavian would dump him in a rundown Whitechapel warehouse at midnight.

The safe stood in the rear corner of the bay, an absurdly theatrical affair made of green-painted iron, with a huge old-fashioned steel dial on the front. Using the code he had been given, he matched up the numerals and hauled on the safe's handle. It seemed to be stuck, but then it gave way with an agonized groan. Inside was a red wooden despatch box with a brass lock, into which he inserted the key.

At first he thought the box was filled with taxi receipts. There were lots of small rectangles of paper. When he saw the accompanying letterheaded notes from Porton Down, it crossed his mind that if he took the contents he might be able to gain the upper hand over Kasavian. But even as he considered the idea, he knew he would never be able to pull it off. He did not possess the Machiavellian gene. Something was bound to go wrong. If this was one incriminating box Kasavian had missed, there had to be others he'd managed to destroy, and even more he'd salvaged for his own purposes. It was better to follow the instructions he'd been given: burn the contents and have done with it.

Doing this was not as easy as he thought it would be. Almon was not a smoker, and had no lighter or matches on him. Searching around, he tried to find something on the shelves that would burn the paperwork.

Outside he heard footsteps and low laughter. Waiting in the shadows until a group of boys had passed, he tipped out one crate after another, searching for a light. Finally, in one of the artists' craft boxes, he found what he was looking for: a can of petrol and a box of matches.

He sprayed most of the can into the despatch box and struck a light, tossing it in. He figured he'd let it burn through, then kick the box lid shut, but the fiery explosion caught him by surprise. A moment later the flames had leaped up to the oil-burnished papier-mâché statue of the Minotaur. The fire swept over its great bull-thighs, jumping across to Perseus and Hippolyta, and within seconds the entire back wall of statuary and model buildings was alight.

Before his eyes, an ancient civilization was collapsing.

Almon jumped back, dropping his mobile and the paper with the code for the bay door. The fire was spitting oily droplets that spattered when they landed, spreading gobbets of molten lava. Acrid black smoke roiled across the low roof in silken folds. The statues were weakest at their legs and had already started to collapse, spraying more burning debris as they tipped and fell.

In the cramped rooms above the warehouse, Mandhatri Sahonta and his wife Jakari were asleep with their four children.

The couple had abandoned their village in Karnataka and moved to London to run a catering company. They worked long hours and sent most of their money home every month, with the result that they had not been able to move from the apartment they had lately come to despise. Jakari was the first to smell the varnish blistering on the flat's outer doors. Shaking her husband awake, she told him of her fears and they set about rousing the children.

Outside, Renfield saw the flames flare behind the warehouse windows. A light went on in the flat above, but was extinguished with a pop; the fire had already reached the building's electrics.

He needed to deal with Stuart Almon, but first there were families to warn. Renfield had spent some time behind the sergeants' desk at Bethnal Green Police Station, and knew how crowded many of these old houses were. Few had fire escapes, and most had only one narrow staircase in or out.

He reported the fire as he ran into the burning bay, seizing Almon just as his jacket caught alight. Swiftly cuffing him, he dragged the dazed, smouldering civil servant outside and left him against a lamp-post on the opposite side of the road while he went back in to find a way of reaching the building's imprisoned residents.

Stuart Almon watched in horror as the flames flared

to extraordinary heights, splintering windows and filling the night with the cries of the trapped. He had no interest in the lives he had placed at risk. All he could see was his career going up in flames.

45

DEAD IN THE WATER

Early on Thursday morning, Edgar Digby, a lizard-eyed lawyer with hair as shiny as a mackerel and a suit that cost more than the average annual wages of a fisherman, turned up at the unit to give Stuart Almon some outrageously expensive advice, to whit: *Don't say anything that incriminates you.*

'Christ, Digby, I think I could have figured out that part for myself,' muttered Almon as the pair sat before Raymond Land, Renfield and May. 'You're here to get me out.' He needed to talk to Kasavian urgently and update him on what had happened.

'I don't think that's going to be quite as easy as you think,' warned Digby. 'You were observed entering the warehouse and torching it.'

'I didn't "torch" it, I was . . .' But Almon could not say what he was doing. He decided to follow his lawyer's advice and shut up.

'It's thanks to the sergeant who was following you that you're not here on a manslaughter charge,' said John May. 'If he hadn't managed to evacuate the building—'

'Well he did,' Almon snapped impatiently. 'So what happens now?'

'Before we get to that,' said May, 'perhaps you'd care to tell us what you were doing setting fire to a Whitechapel warehouse in the middle of the night?'

'I was – it got out of hand. I dropped a match.'

'You don't have to say anything at this juncture,' reminded Edgar.

'Well, I'm not going to sit here and let myself be incriminated, am I?'

'Perhaps you would excuse us while I brief my client?' the lawyer asked Land.

'Forget it, that's not going to happen,' Renfield replied, turning to Almon. 'You nearly killed six people, matey. Four kids, two of them little girls aged six and four years old – they almost died because of you.'

'That wasn't my fault,' Almon complained unconvincingly. 'You can't keep me here without charging me. I know my rights.'

'We can hold you for thirty-six hours on the authority of a police superintendent, which we have,' said Land, 'but we can push arson under the Terrorism Act, which gives us the right to hold you for fourteen days. And we have all the evidence we need to keep you here without bail. So you can start by admitting the truth, or we'll make damned sure you're charged with attempted manslaughter.'

'There's no such thing as attempted manslaughter,' Edgar pointed out. 'You can't attempt to accidentally kill someone. In the event, nobody was hurt.'

'You were happy enough to name names yesterday,' said May. 'You struck a deal with Kasavian, didn't you? Clear up his mess and get back in his good books, something like that?'

'I had no choice,' said Almon pathetically. 'You don't understand how the Civil Service works. It's all about reciprocity. You get caught up in the favours you owe, and I'm in deeper than I ever intended to be.'

'I would *really* stop there,' warned Edgar.

The accountant would not be interrupted. 'Kasavian buried the story, but it resurfaced.'

'I'm afraid we're ahead of you, Mr Almon,' said May coldly. 'We're going to give you a chance to do the right thing. We need evidence that your boss was directly involved. Did he send you there?'

'I never said that,' said Almon, trying to buy enough time to think.

'We're going to find proof that he is a murderer. The documents you burned can be reconstituted.' May had no idea if they could be or not, but it was worth a try.

Almon was shocked by the bluntness of May's words. The language of the Civil Service was tapestried with euphemism and allusion. 'Oskar never gets his hands dirty. Nothing sticks to him. He commissions the services of others. I can't imagine for one moment that he'd leave a trail that could be followed back to him.'

'Well, he's having to act spontaneously now. He can't have thought of everything.'

'You still don't understand who you're dealing with,' said Almon. 'He has all the resources he needs to cover his traces, and they're not accessible to you except through official government channels. Your unit is answerable to him. He's made sure that it's impossible to bring him down. Why else would he have come to you in the first place? He knew you were incompetent.'

Renfield had never hit a civilian before but came close to it now. 'You were never going to give us what we needed,' he said. 'You were just trying to keep your own career from tanking. Well, it's over now. After this you won't be able to get a job cleaning toilets.'

'You still have nothing on Kasavian,' said Almon simply. 'And you won't find it, because there's nothing to find. He's wiped his prints from everything. You talk about my career being over? You're screwed, all of you. You've been played. You're floating corpses. Your unit is dead in the water, just how he planned it would be from the very beginning.'

That was when Renfield jumped at him.

46

SHADOW IMAGE

After Stuart Almon signed a statement negotiated to the satisfaction of all parties, he was charged with arson and reluctantly released on bail.

A sickly grey and yellow dawn broke over King's Cross. The clouds looked as if they had fallen down a flight of stairs and badly bruised themselves. The news reports promised heavy rain as the capital's traditional summer weather – squalls of disappointment with intermittent outbursts of gloom – returned.

The PCU team had worked through the night, but the atmosphere was one of defeat.

They knew they had nothing and would find nothing. As each lead was followed and came to a dead-end, the detectives saw just how carefully the web of their downfall had been constructed. Kasavian had clearly been testing them to see what could be uncovered, secure in the knowledge that even if his original crime was known, there was no way of connecting it back to him.

Colin and Meera had returned to tell of the night's events. 'They're ex-military lads, these bikers,' Colin told them. 'Freelancers, up for anything. I can tell the type.

Guys like that used to come to our boxing club. They were lousy at playing by the rules but they were tough as nails, the kind of men who trained out in the snow in shorts and vests. Kasavian probably found them through his old MOD connections. They're taught to keep their mouths shut no matter what happens. They're as solid as railway sleepers. I checked to see if they had a shared base here, somewhere they might meet or train together, but they're real loners.'

'I'm getting a warrant to turn the Rakes' Club inside out,' said Banbury, 'but the chances of finding anything there now are unlikely.'

'What time is Kasavian heading to the station?' asked Land.

'He's got a car coming at four thirty p.m.,' Longbright told him. 'I'm sending out for breakfasts.' She pressed the heels of her hands against her eyes and rose from her desk, where a tundra of reports had spread in the last dark hours.

'Arthur, can I get you something?' she offered, looking in on Bryant's office.

'Just a cup of tea. I'm not hungry.'

'Perhaps you should try and get some rest.'

'If I fall asleep now I may not wake up again.' Bryant peered blearily over a stack of books and rubbed at his wrinkles. 'Why is it that we always run to the fifty-ninth minute of the eleventh hour? Just once, I'd like to close an investigation a few days earlier than expected.'

'You still think we're going to wrap it up?'

'Yesterday I felt sure we'd arrest Kasavian before his departure. But there's something wrong here. I keep asking myself: Why is there no evidence?'

'You know the answer: everything was pre-planned.'

'No, Janice. He didn't know that Jukes had left something for his girlfriend to find. Nobody did. Fancy leaving it in a bloody crypt!'

May passed Longbright as she was leaving. 'I hope she just convinced you to eat.'

'Food makes me sleepy. I've got a quarter of pineapple cubes here.' Arthur rattled a paper bag. 'The sugar will keep me going. Tell me, John, you're absolutely sure there's nothing on that flash drive that can convict Kasavian?'

'No. There are some classified reports on Scarlet Thread and the inquiry findings on the research scientists' deaths. Jukes was more concerned with damning the science behind the project. There's nothing to indict Kasavian because Jukes didn't know he was going to die and leave us with no bloody proof.'

'We still have a few hours left to find something. But we won't, and I'm beginning to think I know why.'

May seated himself and waited patiently, but could finally bear the suspense no longer. 'Do you want to tell me?'

Bryant dug out a grubby hanky and blew his nose. 'No, because you'll really hate the answer.'

'You always say that. It's an incredibly annoying habit.'

'I know, isn't it? Let me at least try to elucidate. Come and sit beside me.'

May pulled a chair up next to his partner. 'What's that funny smell?'

'I got kebab juice down my vest last night, so I sprayed it with air freshener from the toilet.' He turned back to an immense sheet of paper covered in scribbled names. 'You have a rough idea of how my brain works.'

'Sort of. Yes. But not always,' May admitted.

'You know how much I trust your instincts and working methods.'

'Of course.'

'Well, we differ on one major point. You believe that from the outset of every investigation the most obvious facts point to the right solution. Occam's razor. I don't. Fair?'

'Fair enough,' agreed May.

'In this case, what did our instincts tell us?'

'That Kasavian wasn't to be trusted.'

'Exactly. Whether we were conscious of it or not, that was the agenda we were pursuing. And we got the result we wanted. We proved his guilt to ourselves. We've followed the line all the way from the sanctioned death of Peter Jukes, right through to poor, dim Stuart Almon setting fire to the evidence.'

'Except that we still can't make an arrest.'

'Indeed. I've been over the timeline from beginning to end and it's solid – and yet there's a shadow image behind it.'

'What do you mean?'

'Another theme, as it were. An undertone that contradicts everything we know.' His words hung ominously in the still air of the office.

'But the one thing we know is that Kasavian is guilty.'

'Yes, nothing can change that fact. He's implicated in a murder dating back to his time at Porton Down.' Bryant rubbed his forehead wearily. 'But suppose there was something we missed right from the start? Not a direct fact – something foggier and more obscuring.'

'You're losing me.'

'I know. What I'm trying to say is that there's another agenda at work here. I think it may have something to do with class. Perhaps this whole thing is really only about class.'

'Arthur, I've seen you reach this point many times before, and I still don't quite understand the journey you take. And I don't know what you're trying to say.'

'You know that at heart I'm an academic, not a criminologist. I'm out of my depth when it comes to the construction of empirical data. That's your speciality. But when I look at the victims and the suspects, you know what I see? Two entirely separate classes. Anna

Marquand, living in a run-down council house with her mother. Amy O'Connor, working in an East End bar. Jeff Waters, a barrow boy turned photographer. Sabira Kasavian, a disadvantaged Albanian kid whose parents worked in a smelting plant. Then there are the attacks on Edona Lescowitz and even on you and me. The ruling elite consider everyone here to be several classes below them, and therefore thrashable. But they're wrong. They've misjudged us.'

May was anxious to return the conversation to more solid ground. 'Do you think you can get something concrete on Kasavian before his delegation heads off?'

'I honestly don't know. We're trying to indict our own supervisor for murder, John. If it doesn't stick, it won't be just you and I who'll be thrown to the wolves.'

'But if we get him, we'll be exonerated once and for all. There will have to be a new regime.'

'I wish I had your faith, but I know nothing will change. My parents always obeyed the instructions of the authorities, from town-hall officials to railway clerks. It was a working-class habit I was determined not to take into my life.'

'Kasavian doesn't intimidate you, does he?'

Bryant blew a raspberry of defiance. 'No, of course not. At my age the only thing that still commands respect is death. But Kasavian makes me fearful for others with more to lose. I have no right to risk their careers.'

'They already gave you their vote. If we let him off the hook now, we'll be bowing to authority once more.'

'All right. There's something else that's been troubling me. Kasavian was involved in an illegal programme of research that resulted in sickness and suicide. But he couldn't have been alone in this; I imagine the whole thing was government sanctioned. His wife saw some papers that proved she was married to a man who was, at best, morally deficient. Why should he have cared?

I mean, really? Nobody was going to listen to her. She could tell a couple of friends, and nobody would listen to them, either. She had no solid proof. So why would he still go to the trouble of killing her?'

'Arthur, you cannot be this full of doubt at this late stage.'

'I'm afraid I am.'

'Well, I'm going to stop Kasavian from leaving the country, whether you give me reason to or not. So you'd better get back into those books and find whatever it is you're looking for, before it's too late. Find me your assassin's shadow image, or whatever you call it. And you'd better get a bloody move on because I'm leaving soon.'

Bryant watched his partner blast out of the office and felt suddenly alone. May was right to force his hand, but he had no idea how to give his partner the evidence he needed.

47

MR MERRY

John May stood on the corner of Euston Road with Banbury and Longbright, trying to shield his watch from a light drizzle of sooty rain. He felt as if he could hear the seconds ticking by in his heart. *Of course you're anxious,* he told himself. *Who wouldn't be? You're heading off to commit career suicide.*

'We can't leave it any longer,' he said finally. 'Let's go and do it discreetly, without any fuss.'

'Are you certain we've got cause for arrest, John?' Longbright was still uncomfortable with their line of inquiry. Hell, it hung on a web of slender threads including a bizarrely encrypted note from an unstable wife and some research carried out by a supposed suicide who believed in witchcraft. It wasn't anywhere near enough.

'No, I'm not,' May admitted.

Giles Kershaw was still waiting for the St Pancras Biomedical Centre's verdict on the contents of Sabira Kasavian's stomach, and Dan Banbury had found nothing of a chemically hazardous nature in her room at the Cedar Tree Centre. Without evidence of poisoning it would prove impossible to link Kasavian to his wife's demise.

No further evidence had come from May's wrecked BMW. Nothing more had come to light from Waters's flat. The delivery man's mobile phone had been found by the kids on the Walthamstow rooftop, but apart from that there was nothing. A sense of demoralization flooded over the melancholy group.

'We have enough to make Kasavian miss his train, but that's about all,' May replied. 'If we pick him up before he boards, we undermine his career and derail the initiative, he calls his lawyer and the burden of proof returns to us. I've never done anything like this before. It doesn't look like it will end well for anyone.'

'If we don't do anything and he gets his promotion, I imagine he'll become pretty much untouchable under European jurisdiction,' said Banbury. 'Did Mr Bryant say what he was going to do or how long he'd be?'

'He just said he was going out.' May checked his watch again. 'Why does he always have to cut it so fine? He was exactly the same in his early twenties, running for trains just as they were pulling out of stations.'

'I don't know where he's gone,' said Longbright. 'He just said he was going to follow up an idea. He wasn't in his office when I left, and his phone is going straight to voicemail.'

'Right.' May wiped rain from his face and headed towards the only spare staff car, a battered blue Fiat that Land kept for his exclusive use. Longbright had filched the keys. 'I guess we just have to pray he comes through with something in time. Dan and I will handle the actual arrest. You stay in the car. We make it as discreet as possible, call him down to the foyer and request that he accompanies us. He's going to go crazy but we have to hold our nerve.'

'Let's do it,' said Longbright, getting behind the wheel of the Fiat. 'I hope Arthur's really concentrating on getting us out of this.'

'Of course, one of the most common themes in early Christian writings was female subservience to men,' said Maggie Armitage, riding the escalator to the first floor. 'You'd expect it from the patriarchy of the Church. If any group chose not to conform to Christian teachings, they were immediately attacked. You know I'm doing this under protest, don't you?'

'I appreciate that,' said Bryant, stepping off the escalator and taking Maggie's hand as they made their way around the raised concrete circle. On either side, stone sections of the original London Wall thrust up between glass office buildings, preserved to remind the City's inheritors of their debt to the past. 'Go on.'

'There's nothing shared between the genders in traditional Christianity. Religious and financial power always returns to the hands of males, even now. You haven't met Mr Merry before, have you?'

'No,' said Bryant, 'I've only heard you speak of him.'

'He is my nemesis. It could be argued that we are of equal power, and therefore cancel each other out, but I believe he thinks he is stronger, which I can only pray is mere arrogance. He's certainly very bright. He teaches at the museum. Many years ago we trained together, but we took different paths. My studies took me to the light and his led him towards darkness.'

Bryant studied his old friend with great affection. Maggie had donned legwarmers of different lengths and colours, one pinned with an ankh, the other with a Star of David. It was easy to get distracted by her wayward wardrobe choices. 'Are you saying he's a Satanist?'

'That's such a slippery word. Mr Merry believes he is Ipsissimus, an equal of the gods. He uses his abilities for personal profit and the cruellest of pleasures, whereas I cannot take a penny from clients, and I never venture towards that borderland of pernicious and sinister influence wherein he operates. He believes in something called Paradox Philosophy, a psychological system that involves freeing yourself from so-called "old impediments": right and wrong, true and false. I wouldn't be taking you to see him if I thought there was any other way, believe me. There'll be a price to pay.'

In his desperation to break the deadlock of his stymied thinking, Bryant had called his old friend to ask for advice. After some considerable soul-searching, Maggie agreed to lead the detective to Mr Merry.

'We'll be safe here,' she said, pulling Bryant ahead, 'he won't be able to hurt us, not in a public place. Listen to me, Arthur, I must give you some rules to follow. Under no circumstances should you shake his hand. At no time must you come into contact with his person. If he reaches out to touch you, step out of his way. If he tries to get you to take anything, you must politely refuse. If he drops anything, do not pick it up. If he looks you in the eye, break contact. If he asks you a question, reveal nothing of yourself. It would be better if you didn't speak and let me do the talking. At least I know how to handle him.'

'OK, I'll keep my distance. That sounds easy enough.'

'It sounds easy but it won't be.'

'Why not?'

'Because you have to tell him everything about the case, just as you told it to me. You must do so honestly, or he'll know you're lying, and you can leave nothing out because he'll sense that, too. As you talk together, he'll start to lower his voice until it seems barely audible to you, and you'll find yourself moving in closer, straining to hear. You must not do this, because he'll be trying to plant subconscious commands in your mind. He is extremely manipulative.'

'You're making him sound like some kind of monster,' said Bryant as they reached the entrance to the Museum of

London. 'As a rationalist, I can't afford to start believing that such people have supernatural powers.'

'He certainly has a magnetic effect on people,' said Maggie. 'The majority of believers in Satanism are also avid readers of the Bible. Insecure people are drawn in by such readings as they look for something to believe in, and Mr Merry knows exactly how to exploit them. It's free admission: go around the ticket counter to the right and follow the stairs to the lower ground floor.'

The Museum of London does not merely hide its light under a bushel. It extinguishes the light, and then buries the bushel inside a series of unrelentingly grim walkways, making the building entirely invisible from its exterior. Even those who know of its whereabouts venture there by following other lost visitors.

'What makes you so sure Mr Merry can help us?' asked Bryant.

'I'm not,' Maggie replied. 'But he has a strange way of getting to the root of things.'

They reached the bottom of the staircase and pushed open glass doors into a dimly lit exhibition space. Taking up the entire wall facing them, millions of plague rats scampered in a moving carpet down a flight of stone steps. The video projections were designed to disgust, and the sight was met with appropriate noises of horror from a party of schoolchildren.

'The Great Plague of 1665 was caused by the fleas that infested the Dutch cotton bales, then travelled on rats and jumped on to humans,' intoned a deep, mellifluous voice. 'They bit into the flesh and spread the disease by sucking in and spitting out blood. The fleas lived on the rats, and the rats lived on the ships, and the ships arrived at the London Docks, in the poorest part of the city.'

Bryant looked around but could not see anyone speaking.

'They hopped and jumped across filthy floors and

dirty beds, into babies' cots and on to sleeping mothers, burrowing into unwashed hair and wriggling into sweaty clothes, and they bit and drank and spread their poison. The houses of the poor were close together, so the infection spread. The authorities ordered all the cats and dogs of London to be killed, and by doing that they destroyed the only creatures that could catch the rats. So then they told everyone to smoke, and to burn pepper and hops and frankincense, to kill the evil humours in the air. By now one-fifth of all the people in London had died. Then something happened that ended the plague – can you tell me what it was?'

'Fire!' shouted a few of the children.

'That's right, fire,' said the voice. 'The sparks leaped across the narrow lanes and the inferno roared through the city, gutting the grandest churches and the lowest slums, destroying ninety per cent of the houses it reached. And so one great evil cancelled out the other. So perhaps we could say that this second evil was a good one.'

From the centre of the schoolchildren rose an extraordinary figure, dressed as a pirate. Mr Merry was as round as a pudding. His barrel chest was covered by a gold-braided coat with turned-back sleeves. His great black bushy beard was sewn with coloured beads. He had smiling kohl-lined eyes and thick black eyebrows, possibly dyed. His large head was topped with a black felt tricorn hat, from which protruded a thick beaded ponytail. In his right fist he held a rat. The children screamed in delighted horror, but as many reached out to touch the stuffed rodent as fell back.

Before Maggie Armitage could stop him, Bryant had stepped forward to the edge of the children's circle. 'The Great Fire of London didn't end the plague,' he said cheerfully. 'The disease was already dying out before the fire started.'

Mr Merry slowly turned to look at him. The eyes missed

nothing. He smiled faintly, and the smile grew, and then he laughed, patting children on the head and shooing them away. 'To your drawing pads, you homunculi,' he boomed. 'I want to see works of genius, or I'll send all your souls to the Devil.'

As the children dispersed he turned his attention to the small wrinkled old man wrapped in an olive-green scarf who stood blinking at him, waiting for an answer.

'Who's to say if the flames really burned away the germs?' he replied. 'We were none of us there to witness the events of that terrible year. You are Arthur Bryant, I take it?' He held out a welcoming hand.

Bryant ignored the proffered palm. 'I don't think we've ever met.'

'No, but your reputation precedes you.' Mr Merry dropped his hand and smiled again. 'I imagine you've come to talk to me about your problematic case.'

As Mr Merry took a step forward towards him, Bryant took one back. 'I don't believe I've discussed it with anyone.'

'Of course you haven't. Perhaps Oskar mentioned it to me. We're very old friends, after all. But you mustn't let that deter you. I value an open mind above all else. Tell me the facts of the case, and I'll see if I can offer you some advice.' He looked around at the children. 'Oh, don't mind them. I've blocked their hearing. Your words will sound in their ears as meaningless gibberish.'

Despite Maggie's warnings, Bryant felt himself becoming unsettled. Whether or not Mr Merry actually possessed paranormal powers was beside the point. It was clear that, if nothing else, he was a devious psychologist.

Bryant glanced at Maggie, wondering how to begin. He could feel Mr Merry's black eyes fixed upon him, and sensed the importance of keeping the warlock within his peripheral vision, as one would an unshackled crocodile.

'Come, sit next to me.' Mr Merry dropped himself on to a leather bench and patted the space beside him.

Bryant felt the warning in Maggie's glance. 'I'll stand, if it's all the same to you,' he said. 'But I'll tell you what I know.'

Mr Merry crossed pale beringed fingers with black painted nails over his tight waistcoat while, behind him, his acolytes crawled on the floor with pens and paper. 'Please,' he said expansively, 'enlighten me.'

48

FINAL CALL

The foyer of the Home Office was as quiet as a fish tank. A cleaner was slowly wiping a rubber plant. The receptionist sat entranced by her desk monitor. She looked up at John May. 'It couldn't have been more than ten minutes ago,' she said. 'His car was early.'

'Who did he go with?'

The receptionist checked her screen again. She had very little neck and needed new glasses, so that craning forward to read it was an effort. 'Mr Almon, Mr Hereward, Mr Lang and a team of lawyers. He's not due back in the office until the middle of next week.'

'I've tried every route across London that exists,' said Longbright as they returned to the car, 'including one where you have to reverse a quarter of a mile down a one-way street, and I bet I can get there before an Addison Lee hired car.'

Slapping a siren on the roof of the Fiat that they were not technically entitled to use, they set off towards the north. Longbright drove like Ayrton Senna needing to find a bathroom. She overtook trucks on the inside, slid up on to the emergency lanes and shot through traffic lights with an abandon that made May screw his eyes shut tight.

'It's going to be a nightmare trying to find him in the station,' he warned.

'I don't think so,' said Dan Banbury. 'I borrowed his iPhone to check his contacts some while back. I reset it so that I could pinpoint him through my phone's SatNav. It's accurate to under a metre.'

'Why did you do that?'

'It was for his protection, so that if anything happened we could find him quickly. And I just wanted to see if it worked. I tried to do it to yours too, but of course that didn't work.' He turned his own phone on and checked Kasavian's progress. 'Looks like he's a mile ahead of us on the same road.'

The traffic was heavy, and fresh squalls of rain made the going slow. The only thing more depressing than driving to St Pancras International in a grey slush of drizzle, sprayed by the cars in front, was driving away from it in the knowledge that you weren't leaving the country.

Longbright parked on the pavement in Euston Road as a policeman ran up and warned her not to leave the vehicle there. 'You park it, pal,' she said, throwing him the keys and noting the ID code on his epaulette. 'I'll find you.'

'He's with a team of corporate lawyers,' said May. 'It's not the ideal situation for an arrest.'

The Paris counter clerk told them that Kasavian had already checked in his luggage and had gone through passport control to the business lounge. May took the others through the side office at immigration control and they made their way to the lounge, where the receptionist confirmed that Kasavian had just passed inside.

May hesitated before the glass doors. He could see the security head surrounded by lawyers from here. 'I'll go in and try to do this quietly,' he told Longbright and Banbury. 'I don't want it to look heavy-handed.'

'You're arresting him for conspiracy to murder, John, I don't see how it can be anything else,' said Longbright. 'The way everyone tiptoes around this guy is incredible. This is how people like him stay in power.'

'Yes, I know.' May pushed open the door. 'Try Arthur again and find out if he's got anything. Kasavian will be only too well aware of his rights. He'll soon see we haven't enough to hold him.'

The others watched as May crossed the floor and approached Kasavian. The ensuing confrontation unfolded in mime.

Kasavian passed through stages of puzzlement, incredulity and fury, and then summoned his lawyers. The momentary advantage May had gained through the power of surprise was swiftly lost.

'I don't know what Mr Bryant is doing, but he had better come back with something Kasavian isn't expecting,' said Banbury doubtfully.

Under Maggie's watchful eye, Bryant finished outlining the case, carefully keeping to impersonal facts. From time to time Mr Merry would lean forward on the bench, appearing to listen more intently to a particular detail, but it became clear that he was looking for something else as Bryant talked, a seam in the detective's armour. Finally Bryant reached the point where his partner was heading off to carry out the arrest, and Mr Merry sat back, considering what he had heard.

'You realize your mistake by now, I take it?' he asked. His hooded eyes made him appear half-asleep.

'I'm hoping you'll enlighten me,' said Bryant.

'You should have trusted the children more at the outset – the young ones always see what is happening far more clearly than adults. We overcomplicate matters. These dullard fathers walking around the museum have no understanding of their children. They hold the power to shape youthful souls and alter the course of destiny, and they waste the opportunity.' His dark eyes glittered with the thought of twisting young minds. He was Captain Hook and the crocodile rolled into one.

'The children,' Maggie prompted. 'What about them?'

Mr Merry's eyes refocused on the problem at hand. 'The game the pair of them were playing, *Witch Hunter*, it's based on the old precepts of medieval witch-finding in England. You understand how and why the creation of such witch-hunts came about?'

'We were discussing the subject earlier,' said Maggie.

Mr Merry was greedy for details. 'Where was this? How did the subject arise?'

'I don't understand why the principles of witchcraft keep recurring in this case,' said Bryant hastily.

'Come here and I'll tell you,' offered the black magician, beckoning. 'I shall whisper in your ear.'

A sharp warning glance from Maggie put paid to the idea. 'Tell me,' said Bryant.

'Very well, but it's nothing you don't already know, for your mind was set upon its subconscious course when you first heard about the death of the O'Connor woman. Do you remember what the children told you of how they picked her, and why? It was there that you got your first inkling of the truth.'

They spoke quietly together for some minutes while Maggie watched, carefully pushing Bryant back from Mr Merry's range whenever she felt they were drawing too close together.

'And now you see it, and must finish your work,' said Mr Merry, rising and looking down to where Bryant's scarf had fallen on the floor. Sweeping it up in his ringed hand, he handed it back. 'Yours, I think?'

'Why don't you keep it as a souvenir?' said Bryant.

'Very well, but if you succeed in closing the case now, I shall exact my fee,' said Mr Merry. 'Don't worry, I won't expect cash. There are other ways for you to pay.' Patting his tricorn back on to his head and pocketing the scarf, he clapped his hands and roused his children from their torpid studies.

'The thought of him working with children every day makes my blood run cold,' said Bryant. 'Why is he allowed to do so?'

'The staff say he's wonderful with them, but I'm not sure about that. I keep a watchful eye on him from a distance, and he knows it. I'm surprised he didn't affect you. You know how susceptible you are.'

'I turned down my hearing aid slightly,' Bryant admitted. 'I missed everything he said in a lower register.'

Bryant and the white witch exited the museum. 'Can I let you see yourself home?'

'Why do all my evenings with you end the same way?' Maggie sighed. 'Go on, be off with you. And a word of advice, keep Mr Merry's image out of your head tonight or he'll worm his way into your dreams. Where are you going now?'

'Claridge's,' said Bryant. 'I'm told they serve the most wonderful macaroons for afternoon tea.' Waving her off at the Tube station, he hailed a black cab and rang Longbright's number as one pulled over before him.

'Has John done it yet?' he asked, climbing aboard.

'He's in with Kasavian right now. It doesn't look as if it's going well. The lawyers are arguing and they've just announced the final call for passengers on the Paris train.'

'Janice, I need you to go in there and tell John something for me,' he said, settling back in his seat. 'He'll know what to do with the information.'

'What do you want me to say?'

'Tell him to let Kasavian go.'

49

WITCHCRAFT

There were few raised eyebrows as Arthur Bryant made his way across the thick grey carpet of Claridge's dining room. Shambling elderly men were part of the furniture in the esteemed hotel. Many of them were waiters. The pianist was tiptoeing through the tulips and planning a segue into 'Roses of Picardy'.

They were seated at their usual table in the corner, just where the maître d' had said he would find them. There were three of them: Anastasia Lang, Cathy Almon and Emma Hereward, halfway through a teatime spread of tiny fairy cakes decorated with lurid arabesques of intestine-pink and acid-green icing sugar, accompanied by diaphanous leaves of brown bread, compotes, salads, savouries and glasses of thick yellow Chardonnay. They barely bothered to look up when he arrived at their table. Bryant was just another member of staff, indistinguishable from waiters, cleaners and concierges, made a necessary evil by his usefulness.

'Mr – Brighton, isn't it?' said Ana Lang after a brief moment of thought. 'What a strange coincidence. Who are you with?'

'I'm not dining here, Mrs Lang. My wages wouldn't cover a sausage on a stick in a joint like this.' He took a chair from another table and dragged it over to theirs, which was something one might do in a public house but never in Claridge's. 'I hope you don't mind me joining you for a minute. I'm knackered. My knees are on their last knockings.' He peered at Mrs Lang's side plate of cheeses with interest. 'What are those leaves?'

'It's a rocket garnish,' she said through perfect clenched teeth.

'You know, that stuff was the first thing to grow back over bomb-sites after the war. My mum used to bundle it up by the bushel. God knows what she did with it; no greens ever found their way on to our plates. She mainly did cruel things to suet. And now I bet they charge a tenner for that.'

'I don't suppose you came here to discuss the cuisine.'

'Indeed not. I wanted to let you know that the funeral of Sabira Kasavian can finally be planned, and should be able to take place next week, depending on her husband's availability.'

'You mean you've concluded the post-mortem,' said Mrs Lang, insufficiently hiding her surprise.

'The verdict of the inquest will be made public later today,' said Bryant. 'I don't suppose there's any chance of a cup of tea? Nothing fancy, builders' will be fine.'

'Well, what was the conclusion?' asked Cathy Almon, withholding her pastry-fork.

'Oh, exactly what we thought it would be.' Bryant waved at a waiter and failed to catch his rheumy eye.

'What killed her?' Emma Hereward enunciated impatiently.

'Now, that's rather interesting. I can't remember these things, so Giles wrote it down for me.'

Bryant patted his pockets and located a rumpled piece of paper. Donning smeary reading glasses, he squinted at the page. 'Basically a mix of SSRIs. Or selective seratonin reuptake inhibitors, as they're known. She was stuffed to the gills with anti-depressant drugs. The problem with high-dose combinations of Prozac, Paxil, Zoloft, Luvox and the rest is that they can cause akathisia. It's – let me quote here – "a state of physical and mental agitation that can spark off fits of violent, self-destructive behaviour". Many of the terrible killing sprees that occur in America are carried out by people misusing prescribed drugs.'

'So I imagine you'll be looking to indict her doctor for over-prescription,' said Mrs Hereward.

'No, because her doctor didn't prescribe them.'

'Don't tell me poor Oskar is under arrest,' said Mrs Almon. 'I knew it.'

'No, not at all. I believe he's on his way to Paris at this moment.'

'Well, I don't understand,' said Mrs Lang. 'Why exactly are you here?'

'To be honest, I've always been a bit of a theatre buff, and sometimes we're offered free tickets. I was given a matinée seat for the new RSC production of *Macbeth* today, but I couldn't use it. People are always fascinated by the character of Lady Macbeth, but for me it was always about the witches.'

Miraculously, a waiter had heard his plea for tea and had crept over with a pot. Bryant poured and took a sip, but it was too hot to drink. 'Historically speaking, the number three has always had magical qualities. It appears several times through the play: three witches; three prophecies; three apparitions; and the "weird sisters" repeat their incantations three times.'

To everyone's horror he poured his tea into his saucer and slurped it through his false teeth. 'Of course, the word "witches" is rather ambiguous. The Folio text refers to them in stage directions and speech prefixes as witches, but really they represent the three Fates of ancient mythology, weaving the threads of human destiny, foretelling the future and altering the paths of men's lives.'

Anastasia Lang was visibly losing patience. 'If you have something to tell us, perhaps you'd be so good as to do so?'

'Sorry, of course. It would help if I explained my thinking a little. There's a governing rule of investigation: *lex parsimoniae*, the law of succinctness. It means that the simplest and most likely explanation, the one which feels organically right, is usually most likely to be correct. In this case we had an answer suggesting itself from the outset. Given our past dealings with Oskar Kasavian and knowing how Machiavellian he could be, it seemed highly likely to me that he poisoned his own wife. Even though he hired us to look into the case, and seemed genuinely distraught when he heard of her death, everything always pointed back to him.' The room's background banter and tinkling teacups receded into silence as he explained. It seemed that everyone was listening.

'But, you see, there was another, deeper level of *lex parsimoniae* at work, perhaps less rational, and it was something I began to realize I had sensed from the outset. A certain, shall we say, distaff element to the case. The subject of witches kept arising.

'Amy O'Connor was reading *Rosemary's Baby*, and had folded down the corner of page 145, in which Rosemary opens a book called *All Of Them Witches*. Lucy and Tom, the children who hunted her in the courtyard of St Bride's Church, were playing a game called *Witch Hunter*, loosely based on the medieval instructions for searching out witches. Sabira Kasavian told me it was witchcraft, that she had been placed under some kind of spell, but she was at a loss to explain how it worked. And whenever she talked about the cause of all her problems, she said "they", never "he". Which was odd, considering she knew her husband had covered up a government-sanctioned murder. Even Oskar told me that his wife was the subject of a witch-hunt.'

Now it appeared that even the pianist had ceased playing and was watching them with interest.

'Then, of course,' Bryant continued, 'the three of you told my detective sergeant you always protect your men. You described in some detail how you controlled your weak husbands from the sidelines. And yet there we were, ignoring this deeper truth for the more obvious idea that Oskar Kasavian was the culprit. After all, he had been responsible for covering up a whole series of deaths, although technically they were suicides. And perhaps Peter Jukes drowned himself too. For all I know, he may have been an unwitting part of the test group.

'Sabira Kasavian discovered the truth about her husband, as wives are wont to do. She named him in the code she sent us, but I should have realized that she wasn't referring to the architect of her own troubles. She was talking about his work. He had murdered in the course of duty. How did she find out about him, I wonder? Did he toss guiltily in his sleep, speaking of the terrible burden that still haunted him? That seems rather unlikely. It's obvious Oskar sleeps pretty easily at night. It was more likely something prosaic. Sabira often went to the Pegasus offices to wait for her husband, and it's likely that she read something she shouldn't. Little notes that looked just like taxi receipts. She was a bright girl. She quickly realized what he was hiding. But who could she tell? Not any of you, all of whom she hated, because you were a class above – and she knew you were searching for opportunities to advance your husbands.

'I don't know which of you first discovered that Mr Jukes's girlfriend was in London, but I imagine her arrival was enough to stir you into action once more. Your husbands were all on the board and in the club; any exposure would taint them. You had taken steps to hide the past before, hiring a former member of the Russian militia to remove my biographer. You didn't even need to get your hands dirty. Such men live invisibly, work cheaply and are untraceable, so why not do it again, and close the circle by getting rid of Miss O'Connor? You weren't to know that she knew nothing. You just knew that she was Jukes's girlfriend, and had started visiting his old haunts.

'That should have been the end of it. Except there was Sabira again, making accusations, talking to strangers, throwing tantrums, being *common*, and you couldn't just have her whacked. You hired someone to watch her and report back. No wonder she felt persecuted! As you took turns to visit her you poisoned her mind against her husband, and then you poisoned her body with your helpful ministrations. "Try taking two of these every morning, Sabira, they've always worked for me." "Take one of these before bedtime." As for Oskar, well, I imagine that when he heard about O'Connor's death he assumed the government cover-up was continuing without him, never realizing that you were taking care of the business, destroying his wife and undermining his career.'

The women stared and stared at him, frozen to their chairs, all thoughts of food forgotten.

'Apart from switching the folder of evidence Sabira found with one full of taxi receipts, all you had to do was make the odd phone call to an untraceable number and draw out some cash. But Sabira had a mouth on her. She talked to the photographer, who traced O'Connor to the church. She talked to her girlfriend, and things just kept getting more complicated. Best not to think of it as murder, you told yourselves, more like an act of self-preservation. But here's the funny thing. If you hadn't interfered in the first place, you could have let events unfold naturally and most of your problems would have been taken care of.

'One thing puzzled me. If Sabira suspected the three of you, why on earth did she take your pills and your poisoned advice? And then I realized what I should have known from the start; that it was a class issue. Even though Sabira was afraid of you, she obeyed you because you were posh. Oh, she complained about you to me, but whenever you arrived full of apologies and turned on the charm and *deferred* to her, she thought that she might finally be gaining acceptance. But you weren't accepting her. You were killing her. The children were right. There really are witches.'

A waiter dropped a tray, making everybody jump.

'What children?' said Ana Lang, confused.

For once, the women were dumbfounded. They looked even more shocked when a pair of constables from Savile Row nick appeared at the end of the table ready to take them into custody, but Bryant suspected it was more to do with the embarrassment of being arrested in Claridge's than any real resentment at discovery.

'To save time and energy,' said Bryant, 'I'd rather we didn't have to go through the tedium of denials. You covered your tracks, but of course the Russians like to know who they're dealing with and did some checking up on you. They recorded your calls. Guess whose mobile just got handed in?'

The wives rose with the little dignity they could muster. 'John, put this on my bill, would you?' Mrs Lang told the maître d' with an impressive level of imperiousness.

'Do you need a taxi, madam?' asked the maître d'.

'No, we'll probably walk if the rain has stopped.' Ana Lang leaned into Bryant as she passed. 'I'll tell you what will happen now, you nasty little old man. First, the lawyer. Then, your head.' She brought her hand up swiftly and would have slapped his face had not one of the constables been quick enough to stop her.

'On second thoughts,' said Bryant, 'you'd better hand-cuff the three of them together. They're clearly dangerous.'

So it was that the county wives of the Home Office were removed from the dining room of Claridge's locked to one another like common criminals, as the clientele watched in open-mouthed amazement.

50

THE OUTSIDERS

The detectives took everyone, including Crippen, to the Nun and Broken Compass that night. Oskar Kasavian was in Paris representing the views of the British government, and Raymond Land had agreed to stick the Home Office with the drinks bill.

Jack Renfield unloaded the beer tray and squeezed in beside Longbright as they raised their glasses. It was the British version of a midsummer's evening: rain fell against the windows and there was a fire in the grate. Through the window they could see umbrellas turning inside out.

'What do you think will happen now?' he asked Bryant, tearing open a packet of crisps.

'Oskar will get the new position, the wives will be indicted and the department will be swept clean,' said Bryant, sipping his porter. 'It's a perfect opportunity for HMG and GCHQ to be seen to be putting their houses in order while burying the past. Nothing will actually change.'

'Except that Kasavian will have to follow through

on his promise to grant us full status under the City of London,' said May.

'In that case I'd like to propose a toast,' said Maggie Armitage, who had wedged herself next to Raymond Land. 'May the purple candle of friendship neutralize the effects of karmic retribution.'

As toasts went it didn't strike a very upbeat note, but everyone raised their glasses, and much beer was spilled. Did they realize, as they sat huddled together in the corner of the snug, that they were all outsiders in one way or another? Marked apart by the fierceness of their curiosity, they moved among the docile majority unacknowledged, mistrusted and unloved to the point where they only found solace in one another's company.

'Where did you suddenly disappear to this afternoon?' asked May.

Bryant glanced across at Maggie. 'I went to see someone who confirmed my theory. He told me to re-examine everything through the eyes of the children. They were hunting witches. And so were we. As soon as I changed perspectives, everything made sense.'

May's mobile suddenly rang. He checked the text and frowned. 'Arthur, it seems that somebody wants you,' he said, holding up the screen. The message read: 'Send Bryant outside'.

Just at that moment, something crackled and glowed beyond the pub window. Everyone rose and headed for the door.

On the rain-spattered pavement before them was a trail of fire. As it began to die down, they could read the words it had formed:

TIME TO PAY MY FEE – MR MERRY

'Do you have any idea what that means?' asked May.

Bryant caught Maggie's eye and silenced her. He turned

to his partner, his wide blue eyes swimming with the innocence of one whom London has made truly devious. 'No idea at all,' he said. 'My round, I think.'

Back inside the pub, Crippen gave birth to nine kittens.

Christopher Fowler
The Bryant & May Mysteries

'Witty, sinuous and darkly comedic storytelling
from a Machiavellian jokester'
Guardian

FULL DARK HOUSE
In the Peculiar Crimes Unit's first great case, the
detectives Bryant and May are caught up in a bizarre
gothic mystery, which begins when a beautiful
dancer is found without her feet . . .

THE WATER ROOM
An elderly woman's body is found. Her demise
seems to have been peaceful but for the fact that
her throat is full of river water . . .

SEVENTY-SEVEN CLOCKS
Strikes and blackouts ravage the country and members
of an aristocratic family are being disposed of in various
grotesque ways . . . but what have seventy-seven
ticking clocks got to do with it?

TEN-SECOND STAIRCASE
A controversial artist is found dead, displayed as part
of her own outrageous installations. No suspects, no
motive and no evidence – just a witness who swears the
killer was a masked highwayman on a black horse . . .

WHITE CORRIDOR
A blizzard sweeps the country, trapping Bryant and May on Dartmoor while back at the Peculiar Crimes Unit HQ in London, one of the team has been found dead. As the snow thickens a deranged killer is on the prowl . . .

THE VICTORIA VANISHES
On the trail of a killer who targets women at London pubs, Bryant and May find themselves on the pub crawl of a lifetime – and come face to face with their own mortality . . .

BRYANT & MAY ON THE LOOSE
A decapitated body is found in a shop freezer. Then a second corpse is found, again minus its head, and Bryant and May are called in to find the missing body parts – and the killer.

BRYANT & MAY OFF THE RAILS
Bryant and May are on the trail of an enigma: a young man with a false identity. All they know is that somehow he escaped from a locked room and murdered one of their best and brightest.

BRYANT & MAY AND THE MEMORY OF BLOOD
The defenestration of an impresario's young son was definitely not the best way to end the play's first night party. And the crime scene itself was most unusual. A locked bedroom. No sign of forced entry. No prints or traces of blood. Just a sinister, life-size puppet of Mr Punch lying on the floor . . .

PAPERBOY
By Christopher Fowler

Superman, Dracula, Treasure Island, The Avengers . . . when you're ten years old you can fall in love with any story, so long as it's a good one. But what do you do if you're growing up in a home without books? Christopher Fowler's childhood memoir captures life in suburban London through the eyes of a lonely boy who spends his days between the library and the cinema devouring novels, comics, cereal boxes – *anything* that might reveal a story. But it is 1960, and after fifteen years of post-war belt-tightening, his family's not quite ready to indulge a child cursed with too much imagination . . .

Caught between an ever-sensible, exhausted mother and a DIY-obsessed father fighting his own demons, Christopher takes refuge in words. His parents try to understand their son's peculiar obsession but they fast lose patience with him – and each other. As the war of nerves escalates to include every member of the Fowler family, something has to give, but do the tough lessons of real life mean a boy must always let go of his dreams?

The memoir of a childhood at once eccentric and endearingly ordinary, this does for storytelling what Nigel Slater's *Toast* did for food.

'*Paperboy* is fabulous, and I hope it sells forever'
Joanne Harris